Advanced Topics in Information Technology Standards and Standardization Research
Volume 1

Kai Jakobs
Aachen University, Germany

OVERNORS STATE UNIVERSITY
NIVERSITY PARK
. 60466

UBLISHING
rne • Singapore

Acquisitions Editor: Michelle Potter
Development Editor: Kristin Roth
Senior Managing Editor: Amanda Appicello
Managing Editor: Jennifer Neidig
Copy Editor: Shanelle Ramelb
Typesetter: Diane Huskinson
Cover Design: Lisa Tosheff
Printed at: Integrated Book Technology

Published in the United States of America by
        Idea Group Publishing (an imprint of Idea Group Inc.)
        701 E. Chocolate Avenue, Suite 200
        Hershey PA 17033
        Tel: 717-533-8845
        Fax: 717-533-8661
        E-mail: cust@idea-group.com
        Web site: http://www.idea-group.com

and in the United Kingdom by
        Idea Group Publishing (an imprint of Idea Group Inc.)
        3 Henrietta Street
        Covent Garden
        London WC2E 8LU
        Tel: 44 20 7240 0856
        Fax: 44 20 7379 3313
        Web site: http://www.eurospan.co.uk

Library of Congress Cataloging-in-Publication Data

Advanced topics in information technology standards and standardization
    research / Kai Jakobs, editor.
        p. cm.
    Includes bibliographical references and index.
    Summary: "A collection of articles addressing a variety of aspects related
to IT standards and the setting of standards"--Provided by publisher.
    ISBN 1-59140-938-1 (hardcover) -- ISBN 1-59140-939-X (softcover)
-- ISBN 1-59140-940-3 (ebook)
    1. Information technology--Standards. 2. Standardization.   I. Jakobs,
Kai, 1957-   .
    T58.5.A375 2005
    384.30218--dc22
                                    2005023871

British Cataloguing in Publication Data
A Cataloguing in Publication record for this book is available from the British Library.

Advanced Topics in Information Technology Standards and Standardization Research

Volume 1

Table of Contents

Section V: Applications

Section VI: The Economic Perspective

Section VII: After Standardization

Preface

More than five years have passed since Idea Group Publishing (IGP) published my first edited book on IT standards and standardisation. Back then, this was a fairly arcane topic, discussed in only a comparably small number of papers, in very few books, and at even fewer conferences.

Luckily, this situation is improving. More and more people realise the true importance of the role standards play in almost all aspects of our lives. I quite like to quote documents published by the European Commission at such occasions, and this one is no exception: "Voluntary standards, properly used, can help establish the compatibility of innovative concepts and products with related products and so can be a key enabler for innovation." It seems that (even) policy makers are realising the potential of properly done standards, and the impact standards have on innovation. It remains to be seen whether we will also eventually see policy making being informed by standardisation.

Of course, this holds particularly for information and communication technologies. We are surrounded by ICT artefacts and services: Cameras are controlled by microprocessors, and so are cars and telephone handsets. The same also holds for an ever-increasing number of other gadgets that by now are integral parts of our daily lives. And as many of these devices need to interoperate in one way or another, it is safe to say that (interoperability) standards form one of the pillars of what in Europe is called the "Information Society." The Internet, which may be considered another such pillar, probably would not even exist without standards.

Moreover, standards setting is big business today. The costs of developing one ICT standard may easily run into seven- or eight-digit numbers, and winning or losing a standards war may well be a matter of life or death even for large companies. One should think this fact alone to be sufficient to trigger massive interest in IT standardisation.

Yet, knowledge about the issues surrounding IT standards and standardisation is still rather underdeveloped. Curricula are few and far between, and the number of researchers and scholars with an interest in the subject, albeit increasing, is still fairly small. Possibly even worse, relevant knowledge also seems to be scarce within industry.

I would love to claim that this book will change this situation. It will not, but it may well contribute to a slight improvement. The chapters compiled in this volume should be of interest to practitioners and academics alike; they cover a wide range of highly relevant topics, from the perhaps more theoretical question of what exactly establishes an open standard to aspects of extremely high practical relevance (and equally high financial implications), like IPR and antitrust problems in standards setting.

The book is subdivided into seven sections, each of which comprises one to three chapters.

In Section I, some background will not hurt. In fact, it is quite important here. Therefore, we will start with a bit of theory. The section comprises two chapters that discuss aspects relating to the classification of standards and the terminology used. Together, they should contribute to an improved communication between those involved in standards setting, research, or implementation.

Section II is titled "Coordination in Standards Setting." Obviously, a standards body's working group should be the place for cooperation between the different stakeholders. However, it may also be (mis)used as a platform for infighting. We do not have any chapters on that, though. Rather, the three chapters explore various forms of coordination, also including alternatives to the traditional committee-based approach.

The three chapters of Section III discuss the issue of speed. Speed, or rather the lack of it, has always been a major point of criticism of the formal standards-setting process. The three chapters look at this issue from different but complementing angles, from both the top and the trenches. They identify pros and cons of "slowness," as well as some reasons for it.

IPR problems are (among) the most pressing issues in IT standards setting, and are addressed by the two chapters of Section IV. The inherent conflict between keeping one's IPR and making it available through its incorporation into open standards is hard to resolve. The chapters analyse different aspects of the problem, and these analyses should be another important step on the way to truly useful IPR policies in standards setting.

Section V looks at the application of standards. After all, an IT standard is not a means in itself, but a means to an end: interoperable applications. The three chapters discuss three different application domains with very different characteristics. It appears that the particularities of the respective application domain need to be taken into account when developing standards.

Of course, standards also have an economic dimension. Accordingly, the economic perspective is discussed in the three chapters of Section VI, with network externalities being the common ground here. They are playing a crucial role, specifically in the case of compatibility standards.

Work does not stop once a standard has been approved. Rather, quite a bit is left to do after standardisation. This is addressed by the chapter that forms Section VII. Specifically, it looks at problems and issues surrounding conformance testing, and presents a working solution.

Kai Jakobs
Aachen, 2005

Section I

A Bit of Theory

Chapter I

IT Standards Typology

Henk J. de Vries, Erasmus University, The Netherlands

ABSTRACT

A scientific community needs well-defined and agreed-upon concepts with related terms as a basis for a sound academic discourse. The field of IT standards and standardization research is still developing in this direction. This Chapter aims to contribute to this process by proposing a definition for the area of research as well as classifications of IT standards. It appears that classifications of standards can be based on the subject matter concerned, on the process of developing the standards, or on the intended use of the standards. IT standards are IT standards because of the subject matter addressed in the standards. The Chapter concludes by discussing how the scientific as well as the professional IT standardization community might use the findings.

NEED FOR A STANDARDS TYPOLOGY

A scientific community needs well-defined and agreed-upon concepts with related terms as a basis for a sound academic discourse. Unfortunately, the area of IT standards and standardization research, or more generally, standards and standardization research, has not yet achieved this. There is variety in concepts, more or less well defined, and there is even more variety in terms. In this chapter we will try to bring some structure in this field of research by developing a typology of IT standards. To illustrate the existing variety of concepts in (IT) standardization literature, we will first discuss some examples of standards classifications. We have to base our typology on a sound definition of standard, but here also there is a lack of consensus on a common definition. We will choose one and develop a typology by combining this definition with classifications found in literature.

First, we will give some examples of the existing variety. Tassey (2000) classifies standards according to the basic function they fulfill in standards for quality and reliability, information, compatibility and interoperability, and variety reduction. Moreover, he distinguishes design-based and performance-based standards, and product-element and nonproduct standards, the latter consisting of basic, laboratory, transfer, and industry standards.

Jakobs (2000) distinguishes de facto and de jure standards. The first ones are established through the creation of large market shares for the implementation of a technology. He also mentions voluntary vs. statutory standards, public vs. industry vs. proprietary standards, proactive vs. reactive standards, and base vs. functional standards (the latter being valid for certain IT standards only). Krechmer (2000) mentions four categories that relate to four historic periods of technical development: unit and reference standards, similarity standards, compatibility standards, and "etiquettes." Sherif (2001) distinguishes standards for units (reference), standards for variety reduction (similarity), standards for interactions (compatibility), standards for evolution (flexibility), and standards for performance and quality. With respect to the product or service life cycle, the standards can be anticipatory, participatory, or responsive.

From these examples, it can be concluded that standards can be studied from different perspectives. These need not exclude each other. However, the overall picture is not clear. The distinction between de jure and de facto, for instance, mixes apples and oranges, the apples being the organization that develops the standard (whether or not operating a voluntary process) and the acceptance of the standard in the market: Consensus in a committee where all parties are represented does not guarantee that the resulting standard will be accepted in the market. Tassey (2000, p. 593) defines basic standards as "the most accurate statements of the fundamental laws of physics." They "qualify as pure public goods and hence are provided entirely by government" (Tassey, 2000, p. 593). However, Einstein's discovery $E = mc^2$ is not a standard ('t Hooft et al., 2005), and many standards on quantities and units have been issued by nongovernmental bodies (Teichmann, 2001). So the overall preliminary conclusion may be that concepts and related terms used to describe IT standards are not only diverse but also confusing. Therefore, this chapter attempts to offer some structure to this area of research by offering an IT standards typology.

DEFINING STANDARD

A typology of standards must start with the definition of standard. Jakobs (2000, p. 17) develops this definition by discussing six other definitions. Jakobs defines a standard as:

a publicly available definitive specification of procedures, rules and requirements, issued by a legitimated and recognized authority through voluntary consensus building observing due process, that establishes the baseline of a common understanding of what a given system or service should offer.

Some elements of this definition may be disputed:
- Certain standards for data security, for example, specifying an encryption technique, should preferably not be publicly available.
- IT standards may concern more than just procedures, rules, and requirements, for example, the ASCII character set or the OSI model.
- Should standards be "issued by a legitimated and recognized authority"? This would exclude industrial consortia. In general, specifications set by such organizations are called standards as well.
- Is "voluntary consensus building observing due process" needed? In consortia, these characteristics are not self-evident, and in the case of company standards like the software specifications set by Microsoft, these characteristics are absent. Still, such specifications may be called standards.
- Does each IT standard relate to "what a given system or service should offer"? Standards on quantities and units, and ergonomic standards that specify how to behave in order to avoid RSI (repetitive strain injuries) seem to be (rare) exceptions.

Tassey (2000, p. 588) defines an industry standard as "a set of specifications to which elements of products, processes, formats, or procedures under its jurisdiction must conform." This definition may be also questioned.
- Some categories of standards seem to be excluded, for example, management-system standards, standards for quantities and units, and terminology standards.
- The term must suggests that standards are obligatory by nature.
- Laws that set specifications in the categories mentioned seem to be included.

The International Organization for Standardization (ISO) and the International Electrotechnical Commission (IEC, 2004, p. 8) have developed an official definition: A standard is a:

document, established by consensus and approved by a recognized body, that provides, for common and repeated use, rules, guidelines or characteristics for activities or their results, aimed at the achievement of the optimum degree of order in a given context. Note — Standards should be based on the consolidated results of science, technology and experience, and aimed at the promotion of optimum community benefits.

It is not clear whether this definition concerns standards issued by formal standardization organizations only or also covers documents issued by other parties. Moreover, the definition might also apply to laws. De Vries (1997) has developed a definition of standardization from which the following definition of standard can be derived. It is an:

approved specification of a limited set of solutions to actual or potential matching problems, prepared for the benefits of the party or parties involved, balancing their needs, and intended and expected to be used repeatedly or continuously, during a certain period, by a substantial number of the parties for whom they are meant. (De Vries, 1999, p. 15)

Because the process of standards development and the reasons for having a standard need not be visible in the standard itself, the phrase "prepared for the benefits of the party or parties involved, balancing their needs" might be omitted, but then the economic qualification of standardization would not be clear. This qualification makes the difference with laws.

Some differences with the other definitions mentioned are the following.
- no restriction of issuing authorities
- the characterization of the kind of problems for which solutions are chosen: matching problems. This encompasses more than procedures, rules, and requirements (Jakobs, 2000), or specifications of "elements of products, processes, formats, or procedures" (Tassey, 2000)
- balancing the needs of parties involved instead of consensus and community benefits
- no restriction to public availability
- no exclusion of voluntary standards
- addition of "intended and expected" mass use.

The chosen definition of standard can form the basis for classifications of standards. It consists of elements related to the subject matter (solutions, matching problems), to people and their activities and needs (preparation, benefits, needs, the parties involved), and to the wider circle of interested parties and their activities (parties for whom the solutions are meant or by whom they are used). Classifications of standards may relate to either the first, the second, or the third group of elements: subject-matter-related classifications, standards related to standards development, and classifications related to standards use.

SUBJECT-MATTER-RELATED CLASSIFICATIONS

Standards concern solutions for matching problems. A matching problem is a problem of determining one or more features of different interrelated entities in a way that they harmonize with one another, or of determining one or more features of an entity because of its relation(s) with one or more other entities (De Vries, 1997). An entity is any concrete or abstract thing that exists, did exist, or might exist, including associations among these things, for example, a person, object, event, idea, process, or so forth (ISO & IEC, 1996a, p. 8). Thus, standards can be classified according to these entities. An entity may be the following:

- a person or group of persons
- a thing such as an object, an event, an idea, or a process. Note that both the physical and the virtual domain are included. Things include plants and animals. Gaillard (1933) provides a rather complete list of possible entities.
- a combination of the first two kinds of entities (for instance, a PC [personal computer] with a keyboard, a screen, and a printer).

It can be concluded that matching problems, because they concern interrelated entities or relations between entities, can concern the following:

- matching a thing to a thing (for instance, plugs and sockets),
- matching man to a thing (for instance, safety or ergonomic requirements), or
- matching man to man (for instance, procedures and management systems).

Classifications Related to Differences in Entities

In the case of IT standards, at least one of the entities concerned is an IT system or a part thereof. Matching a thing to a thing is a technical IT standard. Standards that concern the matching of man and IT systems include:

- terminology standards (language use related to IT systems),
- standards that help to describe IT systems (e.g., symbols for flow charts),
- standards that provide help to design IT systems (e.g., reference models such as the Open Systems Interconnection Model [ISO & IEC, 1994], architectures such as in the framework and taxonomy of international standardized profiles [ISO & IEC, 1998a, 1998b, 1998c; Morris & Ferguson, 2000], and programming languages),
- standards for user interfaces and ergonomics,
- personal identification and biometric standards,

- process standards such as procedures and instructions, and
- management-system standards (e.g., software quality standards and standards for information security management).

International Classification of Standards

Formal standardization organizations use the *International Classification of Standards (ICS)* (ISO, 2001) to structure their standards catalogues. ICS is based on a classification of entities. However, by definition, standards concern the matching between two entities. This causes difficulties in the use of ICS. This problem has been solved by listing most standards under two headings. A further complication is that ICS confuses the entities concerned with fields of activity; consequently, ICS mixes entities with the human use of them. The subgroups of ICS Group 35, Information Technology and Office Machines, can be seen at http://www.iso.org/iso/en/CatalogueListPage.CatalogueList?ICS1 =35&ICS2=&ICS3.

Horizontal and Vertical Standards

Standards bodies often distinguish between horizontal and vertical standards. Horizontal standards set general requirements for a collection of different entities, for instance, requirements on electromagnetic compatibility (EMC) on electrotechnical and electronic equipment. Vertical standards set several requirements for one kind of entity, for instance, a PC monitor or an EDI message. In the IT field, functional standards are an example of vertical standards. Functional standards (or profiles) provide a selection out of the options offered in more general standards (Kampmann, 1993).

Classifications Related to the Requirements

Basic, Requiring, and Measurement Standards

A distinction can be made between basic standards, requiring standards, and measurement standards.

- **Basic standards:** Provide structured descriptions of (aspects of) interrelated entities to facilitate human communication about these entities, and/or to facilitate use in other standards. Examples are
  - terminology standards;
  - standards providing quantities and units, for example, SI (Système Internationale d'Unités or International System of Units);
  - standards providing classifications and/or codes, for example, ISO 7372: Trade Data Interchange—Trade Data Elements Directory (ISO, 1993);
  - standards providing systematic data, for example, ergonomic standards describing man's characteristics and abilities, such as dimen-

sions of the human body (these data are used in other standards; Schultetus, 1997); and
- standards providing reference models.
- **Requiring standards:** Set requirements for entities, or relations between entities. These can include specifications of the extent to which deviations from the basic requirements are allowed. There are two subcategories: performance-based standards and design-based standards (Tassey, 2000). In terms of our definition, these can be described as the following.
  - **Performance-based standards:** Set performance criteria for the solution of matching problems. They do not prescribe solutions. Performance-based standards can include specifications of the extent to which deviations from the basic requirements are permissible.
  - **Design-based standards:** Describe solutions for matching problems.[1]
- **Measurement standards:** Provide methods to be used to check whether requiring-standards criteria have been met.

Interference, Compatibility, and Quality Standards

Simons (1994) distinguishes between the following types of standards:
- interference standards, which set requirements concerning the influence of an entity on other entities. Examples are safety, health, environmental, and EMC standards.
- compatibility standards, which concern the fitting of interrelated entities to one other in order to enable them to function together, for example, standards for local-area networks and GSM mobile-phone specifications.
- quality standards, which set requirements for entity properties to assure a certain level of quality. ISO 9000 quality-management standards, measurement standards, and company procedures are examples of quality standards.

Basic standards are missing in Simons' (1994) classification. His interference, compatibility, and quality standards are particularizations of requiring standards. Compatibility standards are always descriptive; interference and quality standards can be performance-based standards as well as design-based standards. Measurement standards are included in Simons' quality standards. In fact, they are a particular kind of requiring standard, namely, standards that describe a solution for measuring.

Compatibility Standards

In IT, compatibility standards are of the utmost importance. Therefore, several authors propose refinements of this category. Wiese (1998) distinguishes between horizontal and vertical compatibility.

- Horizontal compatibility concerns the fit between functionally equivalent objects. Examples are two Lego bricks or two telephones.
- Vertical compatibility concerns the fit between functionally different things. Examples are hardware and software, or tracks and trains.

Indirect horizontal compatibility results from the common fit of functionally equivalent objects to functionally different objects (Wiese, 1998), for example, the connection between Telephone A, the telephone system, and Telephone B.

Unit and Reference Standards, Similarity Standards, Compatibility Standards, and Etiquettes

Based on the above, we can discuss Krechmer's (2000) proposal to distinguish between unit and reference standards, similarity standards, compatibility standards, and etiquettes.

- Unit and reference standards started with units of weight and measure, and developed into, for instance, the metric system. These are a subset of the basic standards mentioned before.
- While unit standards (e.g., gallon) may define the units to measure the carrying capacity of a barrel, similarity standards define how similar in construction one barrel is to the next. Similarity standards describe similar realizations with common properties (Krechmer, 2000). Similarity standards are standards for minimal admissible attributes (David as cited in Krechmer) so they are design-based standards: a common solution for different entities. Horizontal compatibility standards fit within this category. Performance-based standards are not similarity standards because performance criteria may be met using dissimilar solutions.
- Compatibility standards describe a relationship between two or more dissimilar entities (Krechmer, 2000). They are design-based standards as well, including vertical compatibility as well as interference standards.
- Krechmer's (2000) last category concerns an etiquette: a protocol of protocols. It enables separate systems to find compatible modes of operation. In a later paper (Krechmer, 2001), he calls these adaptability standards as they are standards that specify a negotiation process between systems that include two or more compatibility standards or variations, and are used to establish communications. Krechmer adds some IT examples. In fact, an etiquette or adaptability standard is an example of a procedure. Again, this is a design-based standard.

Comparing Krechmer's (2000) classification with the other classifications related to the requirements of a standard, it can be concluded that his classification does not cover all standards: not all categories of basic standards (e.g., terminology standards), and not the performance-based standards.

CLASSIFICATIONS RELATED TO STANDARDS DEVELOPMENT

For standards development, two groups of actors are relevant: the inner circle of those involved in standardization, and the wider group of parties interested in the development of the standard. The first is a subset of the second.

Classifications related to standards development may base distinctions on:

- actors that are interested or involved,
- organizations that set the standards, or
- characteristics of the process of developing standards.

Classification Related to Actors that are Interested or Involved

Parties having a stake in standards development are usually producers and customers, and some other parties such as governmental agencies, pressure groups, consultancy firms, scientists, and organizations involved in testing and certification. The set of potentially interested actors can be determined by identifying which (groups of) people have a direct or indirect relation to the entities involved during the entities' life cycle. In a system of interrelated entities, this life cycle may differ per kind of entity. De Vries, Verheul, and Willemse (2003) provide a method for the systematic identification of the relevant stakeholders. Additionally, they classify stakeholders on the combination of the characteristics of power, legitimacy, and urgency (Mitchell, Agle, & Wood, 1997). Such classifications of stakeholders interested or involved in the process of standards setting do not directly relate to a classification of the resulting standards.

Company, National, Regional, or International Standards

The geographic spread of the actors may lead to a classification related to the level of standardization. Verman (1973) distinguishes the individual, company, association, national, and international level. Nowadays, the regional level is often added, for example, NAFTA (North American Free Trade Agreement) countries, the European Union and European Free Trade Association, and ASEAN (Association of Southeast Asian Nations) countries. In formal standardization, this classification corresponds to the geographic spread of the parties that are able to get involved. A first reason why this may cause confusion is that organizations like the U.S.-based organizations IEEE SA (Standards Association of the Institute of Electrical and Electronics Engineers) and ASTM (American Society for Testing and Materials), as well as national standards bodies like DIN (Deutsches Institut für Normung) have opened their membership to people and organizations from other countries. A second reason for confusion is that the standards can be used by actors in other geographic areas too, whereas, conversely, actors within the geographic area will not always use the standards. The A and B series of paper sizes, for example, have been laid down in international standards, but in the North American region, different sizes are used. In de facto standardization, it is more difficult to use a geographic classification: The actors involved in preparing the standard may even be limited to one company. The Windows versions, for example, can be regarded as company standards of Microsoft. Such a company may be based in one city, but others have offices in many countries. Thus, it appears that the level of standardization is not very convenient to be a common classification criterion.

Classification Related to Organizations that Set the Standard

Formal, De Facto, and Governmental Standards
    As stated above, the distinction between de jure and de facto standards is not clear. According to Webster's dictionary (Gove, 1993), de facto means in fact, whereas de jure means by right or by a lawful title. For instance, a de jure war has been officially declared by government whereas a de facto war just happens. Applied to standardization, this would mean that a de jure standard has an official seal of approval by a recognized authority, whereas a de facto standard just happens to be there. However, almost all standards have a form of approval, for instance, from a standards-development organization, a consortium, or a single company. Then the question remains whether or not this organization has been recognized in the right way. Until recently, several of the official standards-development organizations had no governmental recognition at all: They originated from private initiatives and remained fully private organizations. Now the national governments have appointed many of them as WTO (World Trade Organization) enquiry points.[2] On the other side, certain consortia may have governmental recognition. A more practical term might be formal standards. A formal standard is a standard issued by (De Vries, 1999):
- ISO or IEC,
- their national members (being national standardization organizations like ANSI [American National Standards Institute], BSI [British Standards Institution], and AFNOR [Association Française de Normalisation]),[3]
- regional standards-development organizations related to these national members,
- ETSI (European Telecommunication Standards Institute), or
- ITU (International Telecommunication Union).

    Unfortunately, the term formal may also cause confusion. Does it mean being recognized, for instance, by governments or by the international standardization organizations ISO, IEC, or ITU? Or does it refer to a more or less democratic, consensus-based decision-making process and openness to all interested parties? The latter applies to many consortia as well. The literature provides no clear answers to this question. According to Stuurman (1995), being formal refers to the organization's recognition. This should be seen not as governmental recognition, but as recognition, in the case of national standards bodies, by ISO or IEC, and in reverse, recognition by national standards bodies (which applies for ISO, IEC, CEN, and CENELEC). In our definition of formal, we exclude standards-setting organizations accredited by a national standardization organization. This applies in particular to the American situation where ANSI has accredited many sector-specific standardization organizations.

Other standards categories related to the organization that issues these standards include the following (De Vries, 1999):

- **Consortium standards:** Set by an alliance of companies and/or other organizations (see also Egyedi, 2003)
- **Sectoral standards:** Set by an organization that unites parties in a certain branch of business
- **Governmental standards:** Set by a governmental agency other than a formal standardization organization
- **Company standards:** Set by a company. A company standard may have the form of (a) a reference to one or more external standards officially adopted by the company, (b) a modification of an external standard, (c) a subset of an external standard (for instance, the company's choice of competing possibilities offered in an external standard), (d) a standard reproduced from (parts of) other external documents, for example, suppliers' documents, or (e) a self-written standard.

Classification Related to the Process of Developing Standards

In standards development, a design process and a decision-making process are combined. The processes of standards making may differ in the issues of:

- when the standard is made,
- whether or not a new design is made or an existing one is chosen,
- how decision making is done,
- who is allowed to participate, and
- whether or not the process can be observed by nonparticipants.

Moreover, several stages in this process can be distinguished.

Anticipatory, Concurrent, or Retrospective Standardization

Standardization freezes solutions for matching problems. Three typical situations can apply (Stuurman, 1995).

1. **Anticipatory (or prospective) standardization:** In anticipation of an expected future matching problem, a standard is developed so that the matching problem can be solved from the outset (Bonino & Spring, 1991)
2. **Concurrent standardization:** Matching problems are solved as soon as they occur.
3. **Retrospective standardization:** Standardization solves present matching problems.

This corresponds to the above-mentioned distinction by Sherif (2001) in anticipatory, participatory, or responsive standardization.

Designing or Selecting Standardization

In general, standardization combines a design and an approval process. However, in many cases, the design is available already and the only thing the standards-setting organization has to do is to decide to adopt a document developed already by a company or another standardization organization. So, we can distinguish between the following:

1. **Designing standardization:** Standardization directed at creating a limited set of new solutions to solve matching problems
2. **Selecting standardization:** Standardization directed at establishing a set of preferred solutions out of already available solutions to the matching problems.

In anticipatory standardization, solutions have often not yet been developed. In that case, standardization includes designing solutions. In retrospective standardization, there are often several solutions, and standardization entails the selection of one or some of these. The development of new solutions is also an option in retrospective standardization. The combination of existing solutions and the modification of existing solutions are in-between options.

Consensus or Nonconsensus Standards

Many standardization organizations use the principle of consensus during meetings. Consensus is:

a general agreement, characterized by the absence of sustained opposition to substantial issues by any important part of the concerned interests and by a process that involves seeking to take into account the views of all parties concerned and to reconcile any conflicting arguments. Note: consensus need not imply unanimity. (ISO & IEC, 2004, p. 6)

In summary, a consensus is an agreement to not disagree any longer.

Open or Closed Standards

In the case of open standards, all interested parties have been welcome to participate, whereas participation has been restricted in the case of closed standards (Stuurman, 1995). Krechmer (2006), however, mentions nine other possible meanings of open standards related to standards development or related to standards use; apparently, this term may cause confusion.

Transparent or Nontransparent Process

Transparency is the possibility for nonparticipants to influence standardization, for example, by providing an opportunity for making comments on draft standards. Walsh (1997) lists four ways for achieving transparency in the case of Standards Australia.

Visible or Invisible Process

Another issue concerns the existence of a public gallery, real or virtual. Are files and/or draft documents available? Are journalists allowed to join meetings and report on these (Baskin, 2001)?

Stages in the Process

Several standardization organizations distinguish several stages in the process of standards development, and provide the documents resulting from each stage with tags that indicate their degree of maturity. ISO (2004), for instance, distinguishes between NP (New Work Item Proposal: proposal stage), WD (Working Draft: preparatory stage), CD (Committee Draft: committee stage), DIS (Draft International Standard: committee stage and enquiry stage), FDIS (Formal Draft International Standard: approval stage), and IS (International Standard: publication stage).

CLASSIFICATIONS RELATED TO STANDARDS USE

Now we will discuss several ways to relate standards to the intended users based on:
- the function(s) of the standard concerned,
- the business sector where the standard is to be applied,
- business models,
- the extent of the availability of the standard, and
- the degree of the obligation of the standard.

Functional Classifications of Standards

Kienzle (1943; as cited in Hesser & Inklaar, 1997, pp. 39-45) provides a functional classification of standards. According to him, a standards function is "the inevitable link between a standard as independent variable and the consequences that depend upon its content" (ibid, p. 39). So, the standard itself and its functions are at the centre. In another functional approach developed by Susanto (1988, p. 36), it is not the standard but its use that is the focal point: "Standardization functions are taken to mean the relationship between the actual state before standardization (input variable) and the results of standardization (output variable) of a set of circumstances (system)."

We will speak about intrinsic functions of a standard when we use Kienzle's (1943) definition. The functions of standards according to Susanto's (1988) definition will be called standards' extrinsic functions. A third category is subjective functions, which indicate actor-specific interests related to a standard.

Intrinsic Functions

Though their definitions of functions are clear, Kienzle (1943) and Susanto (1988) confuse these three different functions. Combining Kienzle, Susanto, and Bouma (1989), intrinsic standards' functions can be concluded to be:

- describing a set of agreed solutions to a matching problem,
- recording these,
- freezing them during a certain period, and
- providing elucidation to them.

The first three apply to all standards.

Extrinsic Functions

Extrinsic functions differ per standard and can include:

- enabling interchangeability,
- enabling interoperability,
- creating an installed base,
- matching the life cycle of different entities[4],
- controlling assortment,
- providing transparency (by laying down unambiguous descriptions),
- facilitating information exchange between people and/or institutions,
- storing know-how and keeping it accessible,
- enabling repetition of the solution laid down in the standard,
- enabling dissemination of the solution laid down in the standard,
- enabling economies of scale,
- serving as a benchmark (for instance, in process management to be able to decide between approval and disapproval), and

- assuring performance (by setting, for instance, certain quality or safety characteristics).

Subjective Functions

Apart from the above functions, there are subjective functions related to the interests of specific actors: external stakeholders, the organization itself, and/or departments and people within that organization. Subjective functions of IT standards include:

- enabling interworking and portability of IT systems (making use of interconnection and interoperability),
- facilitating technological innovations (by using a standards architecture that permits changes in parts of the entity systems without affecting other parts of these systems, or by describing good R & D [research and development] practice),
- improving the maintainability of IT systems,
- contributing to quality management,
- reducing costs,
- facilitating processes (for instance, by using standard procedures), and
- contributing to knowledge management.

In the case in which external actors are included, for example, customers, suppliers, or regulators, the subjective functions of IT standards include:

- enabling interworking and the portability of IT systems (making use of interconnection and interoperability), and in this way, improving cooperation with suppliers, customers, and other stakeholders,
- enabling a company to continue selling its IT products,
- creating barriers to new entrants and/or competitors,
- stimulating price competition between suppliers,
- enabling customization at acceptable costs (by assembling standardized modules in IT products, methods, and marketing tools),
- eliminating barriers to trade (for instance, by harmonizing national requirements),
- creating barriers to trade (for instance, by creating a regional standard different from an existing international standard),
- providing transparency in the supply of products or services (by means of standardized descriptions of them),
- enlarging consumer safety,
- avoiding extra legal safety requirements,
- facilitating the meeting of legal requirements,
- providing reliable testing,
- enabling reuse, and
- improving the maintainability of products or systems.

Although these lists of subjective functions are longer than other existing lists, they are not complete. This is virtually impossible since certain actors may have particular interests. In both lists, the first item mentioned is by far the most important one. These functional classifications include much more than the four categories mentioned by Tassey (2000).

Standards Related to Business Sectors

As many of the standards are mainly used in a particular business sector and/or professional discipline, many standards classifications refer to such stakeholder groups. For instance, a standard may be in one of the three dimensions in the often-cited standards classification of Verman (1973). Verman mentions engineering, transport, housing and building, agriculture, forestry, textiles, chemicals, commerce, science, and education. IT might be added to this list and refinements are possible, for example, by adding sector classifications developed by statistical institutes such as the Standard Industrial Classification (SIC).

In practice, classifications that relate standards to business sectors are not unambiguous because:

- a professional discipline may be developed around aspects, such as environmental aspects, from which a separate business sector may subsequently emerge,
- as most standards relate to two or more stakeholder groups, most of them fit into two or more categories unless all the stakeholder groups share the same general category, and
- standards may be used by groups for which they have not been developed.[5]

It appears that such standards classifications are not really fundamental, though they may be practical for bibliographic reasons as well as for standards bodies that use market segmentation.

Classifications Related to Business Models

A French classic publication on standardization (AFNOR, 1967) includes a classification of objectives of standardization based on a business typology. The dimensions of this typology are (a) activity rhythm (seasonal fluctuations in production), (b) product complexity, (c) characterization of the added value, (d) production techniques, (e) production speed, and (f) market and customer characteristics. The typology is obsolete, but it illustrates the possibility of relating a standards classification to a company model. For companies in service sectors, such a model has been developed by De Vries (1999).

Cargill (1990) offers two standards classifications related to company practice. His second classification distinguishes between the following:

- **Regulatory standards:** Standards having some form of statutory enforcement behind them
- **Business or marketing standards:** Standards to gain a business or marketing advantage, or to avoid a business or marketing disadvantage
- **Operational standards:** Standards to structure day-to-day operations of an organization.

In the case of regulatory standards and operational standards, the company can obtain the standards needed from outside and/or make company standards. The only reason to get other parties involved is that they might be facing the same problems so they could cooperate in finding solutions. In the case of business or marketing standards, a strategy is needed to handle the situation of different parties having different interests. Consequently, three different situations are possible.
1. the party making its own (company) standard
2. the party cooperating with other parties having the same interests
3. the party trying to find its way in an arena with different parties having different interests.

Although it provides some insight into actors' interests, Cargill's (1990) classification is not unambiguous and mixes apples and oranges.

One may try to relate standards to accepted business models, such as Porter's (1985) value chain. Without explicitly talking about business models, *Enjeux* ("Les Functions de l'Entreprise," 1992) groups standards into areas of business activities. Feier (1995) links standards to hierarchical levels within a company. However, a standard that sets requirements for a software module sold by Company A, for example, may also be applied by Company B that integrates this module in its IT system. For A, the entity is a product; for B, it is a part of a system used for supporting its business processes. This demonstrates that the same standard may be placed into different classes in business-model-related standards classifications depending on the party using them. Therefore, we have not tried to develop such a classification. In describing the interests of different stakeholders, however, it would be useful to relate a standard to their business processes.

Classification by Extent of Availability

Public or Nonpublic Standards
Public standards are accessible to all third parties, whereas nonpublic standards are accessible only to parties involved in drafting them (Stuurman, 1995). Unfortunately, the term public might also relate to the organization that

has developed the standard (Jakobs, 2000), or to the sector that applies the standard, for example, "standards for public life" (http://www.public-life.gov.uk).

Another question is whether or not one has to pay for the standard. Many standards-setting organizations offer their standards for free on the Web, but in other cases the user has to buy the standard.

Licensed and Nonlicensed Standards

Aggarwal and Walden (2005) conceptualize IT standards as bundles of patents. Though this is exaggerated, it is a fact that many standards lay down requirements for which patents apply (Clarke, 2004). A licensed standard is created when a company (or group of companies or agencies) establishes a new design, gains patent or copyright protection for it, and explicitly sets out to persuade other companies to use the same one (Crawford, 1991). Standards and patents both describe a mostly technical solution. A standard, however, is intended to be used by all parties for which it is meant, whereas a patent is only used by the patent-holder and, via licenses, by third parties chosen by him or her who usually have to pay for this use. Standards and patents have in common the fact that they provide information to prevent reinventing the wheel. A standards classification can specify whether or not such intellectual property rights apply.

Classification by Degree of Obligation

Many authors classify standards by the degree of obligation. Obligation can be due to market forces, contracts, or law. In practice, the difference between legally enforced and voluntary standards is not strict. In the European New Approach, for instance, voluntary standards are related to European directives, which makes them almost obligatory in practice. A standard can be voluntary for one actor and obligatory for another party. For example, Company A may use the Capability Maturity Model (CMM) on software quality assurance as a benchmark in its quality-management policy (voluntary use), whereas Company B may be forced by its customers to meet CMM Level 2. Therefore, classification by rate of obligation is not fundamental, although it may be of help to describe the interests of different stakeholders.

CONCLUSION

The scientific community in the area of (IT) standards and standardization research uses a variety of concepts for standards. Partly, this is inevitable since there are many possible scientific approaches, and the classification needed depends on the focus of the research. However, to a certain extent, the research community is also locked in on the use of concepts or terms that in some cases appear to be inconsistent, such as the distinction between de facto and de jure standards.

A classification of IT standards starts with a definition. We have chosen the definition of De Vries (1999) and propose that a standard is an IT standard in the case that it provides solutions for matching problems between entities where at least one of these entities is an IT system or part thereof. This is decisive for the difference between IT standards and other standards. IT standards are not IT standards because of the stakeholders involved in preparing the standard, or a business sector using the standard.

Literature shows several attempts to develop a standards typology, whether or not restricted to IT standards. Some classifications appear to be useful. Other ones are irrelevant or inconsistent, such as classifications based on:
- the level of standardization (company, national, regional, international),
- business sectors,
- business models,
- the degree of the obligation of the standards, and
- the distinction between de facto and de jure standards.

We found and developed distinctions that seem more relevant grouped into entity-related and actor-related standards classifications. The latter can be split into classifications related to standards development and classifications related to standards use. We have tried to add terms to the concepts, but this is tricky because many terms with different meanings are used.

Subject-Matter-Related Classifications

In additional to a classification of entities themselves, the major subject-matter-related classification of standards is related to the requirements in the standards and includes the following:
1. basic standards
2. requiring standards
   2.1 performance-based standards
   - interference standards
   - quality standards
   2.2 design-based standards
   - interference standards
     - compatibility standards
       - horizontal compatibility
       - vertical compatibility
   - quality standards
3. measurement standards.

Sets of requiring standards can also be divided into horizontal vs. vertical standards.

Classification Related to Standards Development

Classifications related to standards development can depend on the following:

- actors that are interested or involved
- organizations that set the standards: formal, de facto, and governmental standards. De facto standards can be distinguished in consortium, sectoral, and company standards
- characteristics of the process of developing standards, for example,
  - anticipatory, concurrent, or retrospective standardization
  - designing or selecting standardization
  - consensus or nonconsensus based
  - open or closed standards
  - transparent or nontransparent processes
  - visible or invisible processes.

Classifications Related to Standards Use

Standards fulfill different functions. Intrinsic, extrinsic, and subjective functions can be distinguished. Because a standard, in general, fulfils different functions for different stakeholders, classifications related to standards use are not feasible from a scientific point of view. For practical reasons some of them may be valuable to a certain extent.

DISCUSSION

We have found a standards typology in the sense of a classification based on a comparative study of types (Gove, 1993). This structure might be further discussed in the scientific community. Finding acceptable terms is even more of an effort because of the installed base of current jargon use.

We expect that our findings equally apply to IT and non-IT standards. In IT the main issue is compatibility and interoperability. That is different from many other areas, so especially at the point of subjective functions of standards, there are differences. For example, stock control and environmental care are not listed here whereas these are major issues in the case of, respectively, standards for the process industry (Simpkins, 2001) and standards that provide methods to measure pollution (Nielsen & Nielsen, 2003).

In fact, defining common concepts and terms and getting these accepted is a standardization activity, and the question is which form of standardization fits best. In the first place, there is ISO Technical Committee 37, Terminology and Other Language Resources. This committee has developed *ISO/IEC Guide 2: Standardization and related activities — General vocabulary* (ISO & IEC, 2004). This guide can be of help, but it is incomplete and in some cases incorrect.

It is important to use this formal standardization channel in order to stimulate the scientific and professional communities to use the same concepts and terms. Another way for the scientific community to increase commonality in concepts and terms could be along the line of reviews of scientific papers. The editors of the two scientific journals *International Journal for IT Standards and Standardization Research* and *EURAS Yearbook of Standardization* can especially play a role in this. The same applies to chairs of scientific conferences, but there the people may differ per conference. A third channel may be the organization of standardization researchers EURAS (European Academy for Standardization).

In this chapter, some new classifications have been introduced and relevant ones from the literature have been sorted out. Researchers can use them as a help to describe their area of research so that at least within a scientific paper it is clear what it is about. Too often, scientists claim conclusions about standardization in general or IT standardization in particular whereas they just studied one segment of the research area and seem to be unaware of that. For instance, Shapiro and Varian (1999) provide an excellent analysis of the "standards wars" but talk about just one category of them, namely between competing products for which compatibility issues apply in markets where network externalities exist. There are many other standards wars, for example, between competitors on hindering market access by means of design-based standards (Delaney, 2001), or between rich and poor countries concerning the product or process performance level to be laid down in standards (Commonwealth Secretariat & International Trade Centre, 2003).

Practitioners in standards development can use our findings as follows.

- to better define criteria for the inclusion of proposed work items in their work program
- to describe proposed and current standardization projects. Entity-related classifications can be used to describe a standard's scope, and actor-related classifications can be used to describe its field of application.[6]
- to create coherent sets of standards that correspond to relevant entity structures, matching problems therein, and actors and their interests

REFERENCES

Aggarwal, N., & Walden, E. (2005). Standards setting consortia: A transaction cost perspective. *Proceedings of the Thirty-Eighth Annual Hawaii International Conference on Systems Sciences* (p. 204).

Association Française de Normalisation (AFNOR). (1967). *La normalisation dans l'entreprise.* Paris.

Baskin, E. (2001). Reporting on standards committee meetings: A step toward open standards. *Proceedings from the 2ⁿᵈ IEEE Conference on Standardization and Innovation in Information Technology* (pp. 144-148).

Bonino, M. J., & Spring, M. (1991). Standards as change agents in the information technology market. *Computer Standards & Interfaces, 12*(2), 97-107.

Bouma, J. J. (1989). Standaardisatie, een vak apart? *Normalisatie-magazine, 65*(3), 8-10, 19.

Cargill, C. F. (1990). Justifying the need for a standards program. In R. B. Toth (Ed.), *Standards management: A handbook for profits* (pp. 1-10). New York: ANSI.

Clarke, M. (2004). *Standards and intellectual property rights: A practical guide for innovative business.* London: NSSF.

Commonwealth Secretariat & International Trade Centre. (2003). *Influencing and meeting international standards: Challenges for developing countries: Vol. 1. Background information, findings from case studies and technical assistance needs.* London/Geneva, Switzerland: Commonwealth Secretariat/International Trade Centre UNCTAD/WTO.

Crawford, W. (1991). *Technical standards: An introduction for librarians* (2nd ed.). Boston: G. K. Hall & Co.

David, P. A. (1987). Some new standards for the economics of standardization in the information age. In P. Dasgupta, & P. L. Stoneman (Eds.), *Economic policy and technological performance* (pp. 206-239). London: Cambridge University Press.

Delaney, H. (2001). Standardization and technical trade barriers: A case in Europe. In S. M. Spivak, & F. C. Brenner (Eds.), *Standardization essentials: Principles and practice* (pp. 161-166). New York: Marcel Dekker, Inc.

De Vries, H. J. (1997). Standardization, what's in a name? *Terminology: International Journal of Theoretical and Applied Issues in Specialized Communication, 4*(1), 55-83.

De Vries, H. J. (1999). *Standardization: A business approach to the role of national standardization organizations.* Dordrecht, The Netherlands: Kluwer Academic Publishers.

De Vries, H. J., Verheul, H., & Willemse, H. (2003). Stakeholder identification in IT standardization processes. *Proceedings of the Workshop on Standard Making: A Critical Research Frontier for Information Systems* (pp. 92-107).

Egyedi, T. M. (2003). Consortium problem redefined: Negotiating "democracy" in the actor network on standardization. *International Journal of IT Standards and Standardization Research, 1*(2), 22-38.

Enjeux les functions de l'entreprise. (1992). *123,* 28-59.

Feier, G. (1995). Einbindung der normung in die unternehmensstrategie. In *Normung wird unverzichtbar für erfolgreiche unternehmungsführung,* Normenpraxis (pp. 3-1–3-14). Berlin, Germany: ANP Ausschuß in DIN/ DIN Deutsches Institut für Normung e.V.

Gaillard, J. (1933). *A study of the fundamentals of industrial standardization and its practical application, especially in the mechanical field.* Delft, the Netherlands: NV W. D. Meinema.

General Agreement on Tariffs and Trade (GATT). (1980). *Agreement on technical barriers to trade.* Geneva, Switzerland.

Gove, P. B. (1993). *Webster's third new international dictionary of the English language unabridged.* Cologne, Germany: Könemann.

Hallett, J. (2004). From bone screws to pacemakers: The changing patterns of implants. *ISO Focus, 1*(5), 24-25.

Hesser, W., & Inklaar, A. (1997). Aims and functions of standardization. In W. Hesser, & A. Inklaar (Eds.), *An introduction to standards and standardization* (pp. 33-45). Berlin: Beuth Verlag.

International Organization for Standardization (ISO). (1993). *ISO 7372: Trade data interchange—Trade data elements directory* (2nd ed.). Geneva, Switzerland.

International Organization for Standardization (ISO). (2001). *International classification for standards (ICS)* (5th ed.). Geneva, Switzerland.

International Organization for Standardization (ISO). (2004). *ISO/IEC directives: Pt. 1. Procedures for the technical work* (5th ed.). Geneva, Switzerland.

International Organization for Standardization (ISO) & International Electrotechnical Commission (IEC). (1994). *ISO/IEC 7498-1: Information technology-Open systems interconnection-Basic reference model: The basic model.* Geneva, Switzerland.

International Organization for Standardization (ISO) & International Electrotechnical Commission (IEC). (1996). *ISO/IEC 2382-17: Information technology–Vocabulary: Pt. 17. Databases.* Geneva, Switzerland.

International Organization for Standardization (ISO) & International Electrotechnical Commission (IEC). (1998a). *ISO/IEC TR 10000-1: Information technology—Framework and taxonomy of international standardized profiles: Pt. 1. General principles and documentation framework* (5th ed.). Geneva, Switzerland.

International Organization for Standardization (ISO) & International Electrotechnical Commission (IEC). (1998b). *ISO/IEC TR 10000-2: Information technology–Framework and taxonomy of international standardized profiles: Pt. 2. Principles and taxonomy for OSI profiles* (5th ed.). Geneva, Switzerland.

International Organization for Standardization (ISO) & International Electrotechnical Commission (IEC). (1998c). *ISO/IEC TR 10000-3: Information technology–Framework and taxonomy of international standardized profiles: Pt. 3. Principles and taxonomy for open system environment profiles* (5th ed.). Geneva, Switzerland.

International Organization for Standardization (ISO) & International Electrotechnical Commission (IEC). (2004). *ISO/IEC guide 2: Standardization and related activities-General vocabulary* (8th ed.). Geneva, Switzerland.

Jakobs, K. (2000). *Standardization processes in IT: Impact, problems and benefits of user participation*. Braunschweig: Vieweg.

Kampmann, F. (1993). *Wettbewerbsanalyse der normung der telekommunikation in Europa* (Vol. 1360). Frankfurt am Main, Germany: Peter Lang.

Kienzle, O. (1943). Normenfunktionen. *STN, Blatt 6*. Berlin, Germany: Technische Hochschule Berlin.

Krechmer, K. (2000). The fundamental nature of standards: Technical perspective. *IEEE Communications Magazine, 38*(6), 70-80.

Krechmer, K. (2001). Standards, information and communications: A conceptual basis for a mathematical understanding of technical standards. *Conference Proceedings 2nd IEEE Conference on Standardization and Innovation in Information Technology*, 106-114.

Krechmer, K. (2006). Open standards requirements. In K. Jakobs (Ed.), *Advanced topics in information technology: Standards and standardization research, Vol. 1* (pp. 27-48). Hershey, PA: Idea Group Inc.

Mitchell, R. K., Agle, B. R., & Wood, D. J. (1997). Toward a theory of stakeholder identification and salience: Defining the principle of who and what really counts. *Academy of Management Review, 22*(4), 853-886.

Morris, C. R., & Ferguson, C. H. (2000). How architecture wins technology wars. *Harvard Business Review, 78*(2), 86-96.

Nielsen, G. L., & Nielsen, D. M. (2003). How safe is your drinking water supply? The role of ASTM standards in monitoring ground-water quality. *ASTM Standardization News, 31*(5), 20-23.

Porter, M. E. (1985). *Competitive advantage: Creating and sustaining superior performance*. New York: Free Press.

Schultetus, W. (1997). Standards for your comfort: Do we want to standardize man? *ISO Bulletin, 28*(7), 9-12.

Schwamm, H. (1997). Worldwide standards. *ISO Bulletin, 28*(9), 12-28.

Shapiro, C., & Varian, H. R. (1999). *Information rules: A strategic guide to the network economy*. Boston: Harvard Business School Press.

Sherif, M. H. (2001). A framework for standardization in telecommunications and information technology. *IEEE Communications Magazine, 39*(4), 94-100.

Simons, C. A. J. (1994). *Kiezen tussen verscheidenheid en uniformiteit.* Rotterdam, the Netherlands: Erasmus University Rotterdam.

Simpkins, C. R. (2001). Reengineering standards for the process industries: Process industry practices. In S. M. Spivak, & F. C. Brenner (Eds.), *Standardization essentials: Principles and practice* (pp. 191-204). New York: Marcel Dekker, Inc.

Standards Engineering Society (SES). (1995). *SES 1: Recommended practice for standards designation and organization-An American national standard.* Dayton, OH.

Stuurman, C. (1995). *Technische normen en het recht: Vol. 17. Reeks informatica en recht.* Deventer, the Netherlands: Kluwer.

Susanto, A. (1988). *Methodik zur entwicklung von normen.* Berlin, Germany/ Cologne, Germany: DIN Deutsches Institut für Normung e.V./Beuth Verlag GmbH.

Tassey, G. (2000). Standardization in technology-based markets. *Research Policy, 29*(4-5), 587-602.

Teichmann, H. (2001). *1901-2001—Celebrating the centenary of SI: Giovanni Giorgi's contribution and the role of the IEC.* Geneva, Switzerland: International Electrotechnical Commission.

't Hooft, G., Susskind, L., Witten, E., Fukugita, M., Randall, L. Smolin, L., et al. (2005). A theory of everything? *Nature, 433*, 257-259.

Verman, L. C. (1973). *Standardization: A new discipline.* Hamden, CT: Archon Books, The Shoe String Press Inc.

Walsh, P. (1997). Reengineering the standards preparation process. *ASTM Standardization News, 25*(10), 14-15.

Wiese, H. (1998). Compatibility, business strategy and market structure: A selective survey. In M. J. Holler, & E. Niskanen (Eds.), *EURAS yearbook of standardization* (Vol. 1). *Homo Oeconomicus, 14*(3), 283-308.

ENDNOTES

[1] The Agreement on Government Procurement (coming out of the Uruguay Round along with the World Trade Organization) advocates performance standards rather than standards that describe solutions (Schwamm, 1997). Companies and other stakeholders in standardization in general share this policy, but most developing countries prefer descriptive standards with a large number of technical details (Hesser & Inklaar, 1997). The percentage of performance-based standards is growing at the expense of design-based standards (e.g., Hallett, 2004).

2   Each member of the World Trade Organization should assure that an
        enquiry point exists that is able to answer all reasonable questions from
        other members and interested parties in other members, as well as to
        provide the relevant documents regarding technical regulations, standards,
        and conformity assessment (General Agreement on Tariffs and Trade,
        1980).

3   In most industrial countries, these national members are private organiza-
        tions; in many developing countries and former state economies, national
        standardization organizations are part of the governmental administration.
        Governmental national standardization organizations in industrial countries
        include the Japanese Industrial Standards Committee (JISC), the National
        Standards Authority of Ireland (NSAI), and the Standards Council of
        Canada (SCC). Though these are governmental bodies, their activities
        concern voluntary standardization.

4   According to Bouma (1989), standardization is directed at matching the life
        cycles of entities having different speeds of change: the infrastructure,
        which is rather stable in time, components, which are subject to rapid
        changes, and man in relation to these entities, who, in general, prefers a
        certain amount of stability.

5   EAN (International Article Numbering Association) bar codes, for in-
        stance, were initially developed for the retail sector to be placed on
        consumer products, but have found their way to business-to-business
        logistics, too.

6   These are ambiguously indicated in many current standards. Often the
        application field is missing. The Standards Engineering Society (SES, 1995)
        advises distinguishing between scope, purpose, and application. Application
        can be related to the above-mentioned intrinsic and extrinsic functions of
        standardization, and purpose to the subjective functions.

Chapter II

Open Standards Requirements[1]

Ken Krechmer,
International Center for Standards Research, USA

ABSTRACT

An open society, if it utilizes communications systems, requires open standards. The personal-computer revolution and the Internet have resulted in a vast new wave of Internet users. These new users have a material interest in the technical standards that proscribe their communications. These new users make new demands on the standardization processes, often with the rallying cry, "Open standards." As is often the case, a rallying cry means many different things to different people. This chapter explores the different requirements suggested by the term open standards. Perhaps when everyone agrees on what requirements open standards serve, it will be possible to achieve them and maintain the open society many crave.

INTRODUCTION

Open systems, open architecture, open standards, and open source, all sound appealing, but what do they mean?

The X/Open Company, Ltd. provided an early public usage of the term "open." X/Open Company, Ltd. was a consortium founded in 1984 to create a market for open systems. X/Open initially focused on creating an open standard operating system based on UNIX to allow the 10 founding computer manufacturers to better compete with the proprietary mainframe operating systems of IBM (Gable, 1987). Later, its direction evolved (and IBM joined) to combine existing and emerging standards to define a comprehensive, yet practical, common applications environment (CAE) (Critchley & Batty, 1993). X/Open managed the UNIX trademark from 1993 to 1996 when X/Open merged the Open Software Foundation (OSF) to form The Open Group.[2]

Perhaps the genesis of the confusion between open standards and open source developed with the similarly named Open Software Foundation (OSF) and the Free Software Foundation (FSF), two independent consortia based in Cambridge, Massachusetts. While the OSF addressed creating an open standard UNIX operating system, the FSF supported software that could be used, copied, studied, modified, and redistributed by creating the GNU (GNU's Not UNIX) licensing for software.[3] GNU licensed software is open source software.

The term "open architecture" also evolved from a related consortium. The Open Group is responsible for The Open Group Architecture Framework (TOGAF) which is a comprehensive foundation architecture and methods for conducting enterprise information architecture planning and implementation. TOGAF is free to organizations for their own internal, noncommercial purposes.[4] Considering this history, it is not surprising that there is some confusion between open systems, open architecture, open standards, and open source.

Some definitions are needed: Standards represent common agreements that enable information transfer, directly in the case of IT standards and indirectly in the case of all other standards. Open source describes an open process of software development. Often, open-source systems make use of open standards for operating systems (OSs), interfaces, or software-development tools, but the purpose of open source is to support continuous software improvement (Raymond, 2000) while the purpose of open standards is to support common agreements that enable an interchange available to all. Open architecture refers to a system whose internal and/or external interfaces are defined by open standards and/or available under open source license. Open systems embody each of these concepts to support an open systems environment.

Originally, the IEEE standard POSIX 1003.0 (now ISO/IEC TR 14252) defined an open system environment as "the comprehensive set of interfaces, services, and supporting formats, plus user aspects, for interoperability or for portability of applications, data, or people, as specified by information technology standards and profiles" (Critchley & Batty, 1993).

A few other definitions are needed. The term "standards-setting organization" (SSO) refers to any and all organizations that set, or attempt to set, what the market perceives as standards. The term "recognized SSO" refers to any

SSO recognized directly or indirectly by a government. Consortia is the term used for SSOs that are not recognized directly or indirectly by a government.

There are many requirements for an IT standard. The basic requirements—that the standard be consistent, logical, clear, and so forth—are more easily agreed upon. In an attempt to provide a definitive view of the more complex aspects of open standards, this chapter considers open standards from three vantage points on the open system environment.

Standardization consists of more than the process of standards creation; standardization includes implementations of the standard (by implementers) and use of the implementations of the standard (by users). As an example, it is common for a user organization to say, "We have standardized on Microsoft Word," meaning that they have agreed to use Microsoft Word software implementations throughout their organization. And Microsoft often refers to their implementation of Word as an open standard, meaning that they make their implementations of Word widely available to users (Gates, 1998). Certainly Microsoft does not plan to make the creation of Microsoft Word open in a manner similar to the OSF or the FSF. But standards must be implemented and utilized to exist; standards without implementations or open users are not usually considered a standard. Thus, the perspective of the implementers and users of an open standard is as necessary as the perspective of the creators of an open standard.

THE EMERGENCE OF IMPLEMENTERS

As recognized SSOs developed in the late 19th century, they focused, often with government approval, on supporting the open creation of standards and not on the open implementation or use of standards. This was quite reasonable as often the only stakeholders of standards at that time were the creators of the standards. As examples, the railroads, utilities, and car manufacturers were major creators, implementers and users of standards in this period. In the 19th and early 20th centuries, the significant standardization policy issue was the conversion from multiple company specifications to single SSO standards (Brady, 1929).

After the middle of the 20th century, large integrated organizations (companies that bring together research and development, production, and distribution of their products or services, e.g., IBM, AT&T, Digital Equipment Corp., British Telecom, France Telecom, NTT) had engineers who functioned, often on a full-time basis, as the integrated organization's standards creators. These standards creators supported specific recognized SSOs necessary for the broad aims of the integrated organization (Cargill, 1989, p. 114).

In the later 20th century, the increase in technology created a need for many more standards. Because of the growth of personal computing and the Internet, the number of implementers and users of standards increased dramatically. The stage was set for major changes in standardization activity and processes. By the middle of the 1980s, a new industrial movement emerged where larger integrated organizations began to devolve into segmented organizations where the overall organization exerts minimum unified management. Each segment of the overall organization focuses only on its own market(s) and therefore only supports the SSOs that appear necessary for their specific product-development requirements (Updegrove, 1995). This new industrial movement marked the rise of the implementers' activity (independent product-development group) in standardization and with it the rise in consortia standardization. At the same time, the overarching integrated organization's standardization organization was disbanded in many cases (e.g., AT&T, IBM, US PT&T's BellCore).

Since the 1980s, the technical communications standardization processes have been in transition from being driven by standards creators (standardization participants who are motivated to develop new standards) to being driven by standards implementers (standardization participants who are motivated to produce new products that embody a standard). In addition, the users of implementations standards (who usually do not participate in the IT standardization process) have a growing interest in seeing the concept of openness address their requirements. This view is confirmed in the 1994 report sponsored by the U.S. National Science Foundation which described an open data network as being open: "to users, to service providers, to network providers, and to change" (NRENAISSANCE Committee, 1994). Considering service providers and network providers as examples of implementers (perhaps also creators and users), this report identifies the three major perspectives on open standards: creators, implementers, and users.

Individual product-development groups in segmented organizations have no history or allegiance to a specific SSO and choose to support any SSO that best fits their specific product-development and marketing needs. Often, such a fit is made by sponsoring a new SSO to address the standardization requirements of a specific developer's product implementation (Updegrove, 2004). However, product implementers have very different interests than the standards creators they have replaced. What a product implementer considers as an open standard may be quite different from what a standards creator considers as an open standard. And both views are also different from what a user might consider as an open standard.

THREE VIEWS OF STANDARDIZATION

Each of the requirements of open standards relates to one or more of the stakeholders: creators, implementers, and users. To identify all requirements of an open standard, it is necessary to understand what standards creators, implementers, and users would consider the broadest reasonable requirements of an open standard. Each group of stakeholders is driven by specific economic desires:

- The creation of standards is driven by potential market development and control issues.
- The implementation of standards is driven by production- and distribution-cost efficiencies.
- The use of implementations of standards is driven by the potential efficiency improvement, due to the standard, on the user.

While there is some overlap among these economic drivers, e.g., market development and distribution cost efficiency, each stakeholder has a distinct economic motivation. Thus is it necessary to consider each stakeholder class separately. In the following, requirements of each stakeholder class is evaluated and each open standards requirement derived is given a name, usually shown in parentheses (e.g., open IPR).

THE CREATOR'S (SSO) VIEW OF OPEN STANDARDS

It is easiest to identify the view of recognized SSOs about open standards. Many SSOs' Web sites state what openness of standardization is to them:

The Institute of Electrical and Electronic Engineers (IEEE), "For over a century, the IEEE-SA has offered an established standards development program that features balance, openness, due process, and consensus."[5]

The European Telecommunications Standardization Institute (ETSI), "The European model for telecom standardization allows for the creation of open standards..."[6]

The American National Standards Institute's (ANSI, 2002) *National Standards Strategy for the United States*, "The process to create these voluntary standards is guided by the Institute's cardinal principles of consensus, due process and openness..."[7]

It is interesting to contrast these views with a view from the European Commission,[8] which, as a government, represents the user and implementer views:

The following are the minimal characteristics that a specification and its attendant documents must have in order to be considered an open standard:

- *The standard is adopted and will be maintained by a not-for-profit organization, and its ongoing development occurs on the basis of an open decision-making procedure available to all interested parties (consensus or majority decision, etc.).*
- *The standard has been published and the standard specification document is either available freely or at a nominal charge. It must be permissible for all to copy, distribute, and use it for no fee or at a nominal fee.*
- *The intellectual property — i.e., patents possibly present — of (parts of) the standard is made irrevocably available on a royalty-free basis.*
- *There are no constraints on the re-use of the standard.*

Most SSOs follow rules to ensure what they consider an open standards creation process by requiring open meetings, consensus and due process. These represent the standards creators' view of the requirements for open standards. Most SSOs do not suggest "the standard is made irrevocably available on a royalty-free basis" (the highest level of open IPR).

THE IMPLEMENTERS VIEW OF OPEN STANDARDS

Consider a commercial developer of software or a manufacturer of hardware as examples of implementers. They want the ability to compete on an equal basis with their competitors. This concept has often been termed a "level playing field." An implementer of a standard would call a standard open when it is without cost to them (open IPR, open documents), it serves the market they wish (open world), does not obsolete their prior implementations (open interface), does not preclude further innovation (open change), and does not favor a competitor. These five requirements ensure a level playing field.

The standards creators requirements of open meeting, due process, and consensus assist the standards implementer in meeting their last two requirements, but do not address the first three. Many recognized SSOs are national or regional while many implementers' markets are international. Most recognized SSOs allow intellectual property to be included in their standards. This is not in the interests of the implementers who do not have similar intellectual property to trade.

Consortia have, in many cases, done a better job of addressing the needs of standards implementers. Many consortia offer international standardization and allow IPR negotiation (which recognized SSOs do not). Since consortia address more of the implementers' requirements, this may be one reason for the increase in standardization by consortia.

THE USERS VIEW OF OPEN STANDARDS

Consider an office or a factory of an organization that uses implementations of the standard. The simple goal of the user is to achieve the maximum possible return on their investment in the implementation of the standard. While many aspects of the users' return on investment are not related to the standardization process (e.g., the implementation quality), four aspects are. A user of an implementation of a standard would call a standard open when:

- The implementation operates and meets local legal requirements in all locations needed (open world, open use, open documents).
- New implementations desired by the user are compatible with previously purchased implementations (open interface, open use).
- Multiple interworking implementations of the standard from different sources are available (open interface, open use).
- The implementation is supported over the user desired service life (ongoing support).

It is worth noting that users do not often participate in IT standardization. Perhaps that is because many of the requirements specific to users are not even considered in most SSOs.

UNDERSTANDING ALL THE REQUIREMENTS OF OPEN STANDARDS

The above analysis identified 10 separate requirements that are listed in Table 1. The names of these requirements are arbitrary and it is possible to imagine combining different groups of requirements into one requirement category, so there is no importance to number 10. What is important is that these requirements, by whatever name, are the requirements that standards creators, implementers, and users have rational reasons to desire.

Table 1. Creators, implementers, and users see openness differently

	Requirements/ Stakeholders	Creator	Implementer	User
1	Open Meetings	x		
2	Consensus	x		
3	Due Process	x		
4	One World	x	x	x
5	Open IPRs	x	x	x
6	Open Change	x	x	x
7	Open Documents		x	x
8	Open Interface		x	x
9	Open Use		x	x
10	Ongoing Support			x

Table 1 shows that the requirements of the major stakeholder groups are sometimes similar and sometimes divergent. Users have little interest in how a standardization process was conducted. The concept that open meetings, consensus, and due process support the development of multiple sources of implementations of a completed standard is recognized but rarely supported by users. Users are focused on being able to purchase compatible equipment from multiple sources. History does suggest that the Open Meetings, Consensus, and Due Process facilitate the creation of multiple sources. In the case of Requirement 5 (open IPRs), even though all the stakeholders have an interest, their interests are quite different. Creators appear satisfied to support reasonable and nondiscriminatory (RAND) SSO IPR policies. Commercial implementers require a means to identify and control their IPR cost, which RAND does not offer. This pushes implementers to form consortia. Users are often unwilling to pay high prices caused by multiple IPR claims. So users may be put off by the impact of RAND policies.

As Table 1 identifies, the first three requirements are oriented to the stakeholders focused on standards creation. The first four requirements are also at the heart of the World Trade Organization (WTO) *Agreement on Technical Barriers to Trade, Code of Good Practice*.[9] The fourth requirement, Open World, is supported by ANSI but not required. The ANSI open-standards concept requires the first three requirements for all ANSI-accredited standards organizations (ANSI, 1998). The fifth requirement, Open IPR, has been formally added to the U.S. standards-development process by ANSI and many SSOs.

Currently, the widest interest regarding open standards focuses on Open World and Open IPR. Open World addresses standards as barriers to trade or enablers of trade. Open IPR impacts the profitability of all communications-equipment companies today. The additional five requirements (6 through 10) represent open standards requirements that are emerging but are not yet supported by most SSOs. Table 1 identifies that the five additional requirements are more oriented to the implementation and use of standards.

In the following descriptions of the 10 requirements, an attempt is made to quantify each specific requirement. Certainly the details and quantifications proposed are open to further consideration, but some quantification of each requirement is useful to identify how well the different SSOs support the concepts of open standards. Using these quantifications, Table 3 summarizes the support of these 10 requirements for several SSOs.

1. Open Meetings

Open Meeting — all may participate in the standards development process. Currently, openness of meetings is deemed to be met (i.e., under many SSO requirements) if all current stakeholders may participate in the standards-creation process. But, as technology has become more complex, user participation in standards creation has declined significantly (Foray, 1995). When the largest number of stakeholders (users) no longer participate, such a definition of open meetings is no longer functional.

"All stakeholders can participate" is a mantra of many recognized SSOs. But this mantra does not address all the barriers to open meetings. Recent social-science research has identified 27 barriers to open meetings and grouped these into five categories: the stakeholders themselves, the rules of recognized standardization, the way the process was carried out, the role of the technical officers of the committee, and the culture of the committees (de Vries, Feilzer, & Verheul, 2004).

One major barrier to standardization participation is economic. Some recognized SSOs (e.g., the International Telecommunications Union, ITU) and many consortia (e.g., the World Wide Web Consortium, W3C) require membership before attendance. Paying to become a member is a significant economic barrier when a potential standardization participant is not sure they are interested in attending a single meeting. Participation expenses, unless quite low, are part of the real barriers to participation for students, many users, and even start-up companies in the field. Currently, only a few SSOs such as the Internet Engineering Task Force (IETF), the standardization organization for the Internet, and the IEEE offer low cost per meeting participation.

Economic access is a useful way to quantify Open Meetings. There are two broad classifications of the levels of economic access in Open Meetings.

1.   Any stakeholder can pay to become a member (current status of many SSOs).
2.   There is an acceptable cost to join on a per-meeting basis.

2. Consensus

Consensus — all interests are discussed and agreement found, no domination. Different SSOs define consensus differently. In general, consensus requires that no single stakeholder group constitute a majority of the membership of an SSO. Consensus may be identified by a vote of the standardization committee or may mean there is no active and informed opposition. Surprisingly, the IETF, which many find to be an example of a more open SSO, does not meet this criteria as the IETF area directors have a dictatorial level of control over the standardization decisions in their area (IETF, 1998).

Consensus is quantified only by the requirement (1) or lack of requirement (0) in each SSO.

3. Due Process

Due Process — balloting and an appeals process that may be used to find resolution. Different SSOs describe due process differently. In general, it requires that prompt consideration be given to the written views and objections of all participants. A readily available appeals mechanism for the impartial handling of procedural complaints regarding any action or inaction is part of the due-process requirement. As explained, the three requirements — Open Meetins, Consensus, and Due Process — are considered fundamental by recognized SSOs to the openness of their standardization process.

Due Process is quantified by the requirement (1) or lack of requirement (0) in each SSO.

4. Open World

Open World — the same standard for the same capability, worldwide. This requirement is supported by the WTO to prevent technical barriers to trade. The International Federation of Standards Users (IFAN, 2000) also supports uniform international standards. However, politically this can be a very contentious area. There are national standards for food processing that are based on religious beliefs (e.g., halal and kosher). There are standards for the environment, health, medical care, and social welfare that cause an imbalance in cost between countries that implement them (often richer) and countries that do not (often poorer). To avoid these contentious issues, most recognized SSOs currently support but do not require coordination of their standards work with worldwide standards. This allows but does not favor divergent regional or national standards.

In the richer countries, the rise of consortia, the decline of publicly funded research, and aggressive commercialism make it more difficult to achieve a single standard for a single function worldwide. The five different, incompatible wireless technologies of the 3G (third-generation) cellular standards (W-CDMA, cdma2000, UWC-136, TD-CDMA, FD-TDMA) are an example of these effects. Initially, these five 3G versions will operate in different geographic areas, but eventually users will demand worldwide compatibility. It appears likely that standardization organizations will continue to create incompatible standards for similar capabilities. This may be viewed as an indication of the failings of recognized standardization (Cargill & Bolin, 2004), or as an indication of the need to increase the support of open interfaces (see below).

An Open World is quantified by identifying the geographic operating area of each SSO. International scope is rated 1 and national/regional scope is rated 0.

The first three requirements of open standards have been addressed and in large measure resolved in most SSOs. The requirement for Open World is supported by the three recognized worldwide SSOs, ISO, IEC, and ITU, but many nations cling to the view that giving up their national standardization prerogatives would be giving up an aspect of their nation's sovereignty. Consortia standardization, which is unimpeded by political issues, usually creates worldwide standards.

5. Open IPR

Open IPR — how holders of IPR contained in the standard make their IPR available. Most recognized SSOs and many consortia consider that Open IPR means that holders of intellectual property rights (IPRs) must make available their IPR for implementation on reasonable and non-discriminatory (RAND) terms. Five levels of quantification of open IPRs currently are identified (0 through 4).

0. Commercial licensing may be the most prevalent way to use IPR legally. It is also the least open. In this case, the holder of an IPR and the potential implementer of the IPR agree privately on commercial terms and conditions for the implementer to use the holder's IPR.
    Band (1995) describes four additional levels of increasing openness relating to IPRs:

1. Microsoft believes that interface specifications should be proprietary but will permit openness by licensing the specifications to firms developing attaching (but not competing) products.

2. The Computer Systems Policy Project (CSPP) also believes that interface specifications can be proprietary but will permit openness by licensing the specifications on RAND terms for the development of products on either side of the interface.

3. The American Committee for Interoperable Systems (ACIS) believes that software interface specifications are not protectable under copyright, and that therefore reverse engineering (including disassembly) to discern those specifications does not infringe the author's copyright.

4. Sun Microsystems believes that critical National Information Infrastructure (NII) software and hardware interface specifications should receive neither copyright nor patent protection. This quantification is discussed further under "Open Change."

The range of possible refinements in this contentious area is probably limitless. Two further variations that might assist in resolving contentious IPR issues are:

•   Approach #2: the manner of operation of most recognized SSOs. This might be more acceptable to implementers if an IPR arbitration function existed when IPR is identified during the creation/modification of a standard (Shapiro, 2001).

•   Approach #4: this might be more acceptable to implementers if claims on basic interfaces were precluded but IPR on proprietary extensions were allowed. This could be technically feasible using the concepts of Open Interfaces.

Summary of the First Five Requirements

The fifth requirement, Open IPR, is the most divisive of the first five requirements as RAND (the current practice of many SSOs) is not sufficient to allow implementers to determine the impact of standards-based IPR on their costs. The semiconductor industry, where manufacturing cost drops rapidly with volume, exacerbates IPR problems for implementers. In the case of semiconductors, IPR costs may be based on fixed-unit charges. Such charges can be the largest single cost component of a semiconductor. Semiconductor implementers must control their IPR costs. It seems only fair that any implementer have a right to determine the exact cost of IPR to them before accepting its inclusion in a new standard. This issue has led many implementers to consortia as consortia often require joint licensing of related IPRs. This practice defines the cost of the IPR to the implementer. While commercial licensing ma seem the least open process, it may not be more costly than the RAND approach to specific implementers.

For emerging countries, RAND policies also appear to be causing an undesirable situation. The Chinese are rapidly developing into major suppliers of communications systems and equipment but do not have a portfolio of intellectual property rights that can be used to trade with commercial organizations in more developed countries who have IPR in communications standards. This may cause the Chinese to consider developing non-standard technologies for existing communication systems so that they do not have to pay for previous IPR standardization decisions that they did not participate in (Updegrove, 2005).

The Web site Cover Pages maintains a section on open standards and collects many different descriptions of open standards (including the 1998 version of this paper).[10] The view of open standards from the SSOs quoted on this site follows the first five requirements. The view of other organizations is more divergent and includes requirements that are discussed below.

6. Open Change

Open Change — all changes to existing standards are presented and agreed upon in a forum supporting the five requirements above. Controlling changes is a powerful tool to control interfaces when system updates are distributed over the Internet and stored in computer memory. Even with the most liberal of IPR policies identified (4 in Open IPR), Microsoft would still be able to control its Windows application programming interfaces (APIs) by distributing updates (changes) to users that updated both sides of the API interface. But without a similar distribution *at the same time*, competing vendors' products on one side of the same API could be rendered incompatible by such a Microsoft update.

The only way that interfaces can remain open is when all changes are presented, evaluated, and approved in a committee that supports the first five requirements. Considering today's environment of computers connected over the Internet, identifying and requiring open change is vital to the concept of open standards. Surprisingly, this is not widely understood. The original U.S. judicial order to break up the Microsoft PC- (personal computer) OS (operating system) and application software monopoly did not address this key issue (United States District Court, 1999). On March 24, 2004, the European Commission (EC) announced its decision to require Microsoft to provide its browser (Explorer) independently of the Windows operating system and make the related Windows APIs available to others.[11] This EC decision did not address the need for open change. The EC announced on June 6, 2005, the receipt of new proposals from Microsoft addressing Microsoft's support of interoperability.[12] Unfortunately, these proposals still do not directly address Open Change.

Open change is quantified as either supported (1) or not supported (0). Consortia that do not support the first three requirements of open standards are identified as not supporting open change.

7. Open Documents

Open Documents — committee documents and completed standards are readily available. Open Documents is the requirement for a stakeholder to be able to see any documents from an SSO. The openness of a standardization meeting to outsiders is closely related to the availability of the documents from the meeting. All standardization documentation falls into two classes: work-in-progress documents (e.g., individual technical proposals, meeting reports) and completed standard documents (e.g., standards, test procedures). Different stakeholders need to access these different classes of documents. Standards creators do not require open documents as they are involved in the creation of all the documents. Standards implementers need access to standards work-in-progress documents to understand specific technical decisions, as well as access to completed standards. Implementation testers (users and their surrogates) need access to completed standards.

The Internet Society (ISOC) supports a non-government-recognized standards-making organization, the IETF, that has pioneered new standards-development and -distribution procedures based on the Internet. While the IETF does not meet the criteria for Consensus and Due Process, the IETF is perhaps the most transparent standardization organization. Using the Internet, the IETF makes available on the Web both its standards, termed RFCs, and the drafts of such standards at no charge. Using the facilities of the Internet, IETF committee discussion and individual technical proposals related to the development of standards can be monitored by anyone, whose responses may be offered. This transparent development of IETF standards has been successful enough that some other SSOs are doing something similar. In July 1998, ETSI announced that its technical committee TIPHON (Telecommunications and Internet Protocol

Harmonization over Networks) would make available at no charge all committee documents and standards drafts.

Ultimately, as technology use expands, everyone becomes a stakeholder in technical standards. Using the Internet, access to committee documents and discussion may be opened to all. In this way, informed choices may be made about bringing new work to such a standards committee, and potential new standardization participants could evaluate their desires to attend meetings.

Three levels of transparency of open documents can be quantified:
1. Work-in-progress documents are only available to committee members (standards creators). Standards are for sale (the current state of most formal SSOs).
2. Work-in-progress documents are only available to committee members (standards creators). Standards are available for little or no cost (the current state of many consortia).
3. Work-in-progress documents and standards are available for reasonable or no cost (the current state of IETF).

8. Open Interface

Open Interface — supports proprietary advantage (implementation); each interface is not hidden or controlled (implementer interest); each interface of the implementation supports migration (user interest). Open Interface is an emerging technical concept applicable to compatibility standards used between programmable systems. The open-interface requirement supports compatibility to previous systems (backward compatibility) and to future systems (forward compatibility) that share the same interface. The idea that open standards should embody such a principle is new, but interest in open interfaces is increasing due to the considerable success of open interfaces in facsimile (T.30), telephone modems (V.8 and V.32 auto baud procedures), and digital-subscriber-line (DSL) transceivers (G.994.1 handshaking).

One way of achieving open interfaces is to implement a fairly new technique termed etiquette (Krechmer, 2000). Etiquettes are a mechanism to negotiate protocols. While a protocol terminates an X.200 (OSI) layer, an etiquette, which may negotiate multiple OSI-layer protocols, does not terminate (replace) any protocol layer function. An etiquette is used only for negotiating which protocol, options, or features to employ. The purpose of etiquettes is connectivity and expandability. Proper etiquettes provide the following:
•   Connectivity, negotiating between two devices in different spatial locations to determine compatible protocols.
•   Means to allow both proprietary and public enhancements to the interface that do not impact backward or forward compatibility.
•   Adaptability, so that a communications system can become compatible with a different communications system.
•   Easier system troubleshooting by identifying specific incompatibilities.

As long as the etiquette is common between the equipment at both ends, it is possible to receive the code identifying each protocol supported by the equipment at a remote site. Checking this code against a database of such codes on the Web or in a manual, the user can determine what change is necessary in his or her system, or the remote system to enable compatibility.

One of the earliest etiquettes is ITU Recommendation T.30, which is used in all Group 3 facsimile machines. Part of its function includes mechanisms to interoperate with previous Group 2 facsimile machines while allowing new features (public as well as proprietary) to be added to the system without the possibility of losing backward compatibility. Another etiquette is the ITU standard V.8, which is used to select among the V.34 and higher modem modulations. More recently ITU G.994.1 provides a similar function in DSL equipment.

As an example of the usefulness of open interfaces, consider Microsoft APIs. Assume that a standard based upon the Microsoft Windows API is created. Then any vendor could create an OS to work with Microsoft's applications or create applications to work with Microsoft's OS. If any vendor (including Microsoft) identified a new function such as short message service or videoconferencing that was not supported across the basic API, that vendor could then offer the new function, as an identified proprietary feature across the API, to users that purchase that vendor's OS and applications. Since an open interface supports proprietary extensions (Krechmer, 2000), each vendor controls the way the new function is accessed across the API but does not change the basic compatibility of the API. In this manner, a vendor is able to maintain control and add value based on the desirability of the new function.

Some aspects of the issue of open interfaces were explored in technical detail in 1995 (Clark, 1995). Since then, seven technical aspects of social interfaces have been identified (Krechmer, 2000). Currently, open interfaces have only been applied at the standard committee level and not addressed at the SSO level, so no detailed quantification is offered here.

9. Open Use

Open Use describes the assurance a user requires to use an implementation. Open Use identifies the importance to users of known reliable standardized implementations. Often, a user will trust an implementation based on previous performance, its "brand," or simply familiarity with the requirements. When such trust is not reliable, compliance, conformance, and/or certification mechanisms for implementation testing, user evaluation, and identification may be necessary. This more exact form of Open Use is termed conformity assessment. The Open Use requirement is supported by ANEC, a European organization that focuses on consumer issues associated with standardization (ANEC, n.d.). ISO/IEC 17000 defines conformity assessment as the "demonstration that specific requirements relating to a product, process, system, person, or body are fulfilled" (ISO/IEC 17000). Conformity assessment procedures, such as testing, inspection, and certification, offer assurance that products fulfill the requirements specified in the appropriate regulations or standards.

Open Use covers all possible parameters that may need to be identified as conforming to a standard for accurate, safe and/or proper use. Such parameters could include physical access (e.g., access by people with disabilities), safety (e.g., CE or UL mark, the European and US indications that equipment is designed safely) and correct weights and measures (e.g., certification of scales and gasoline pumps). To achieve Open Use may require testing by implementaters, regulators, users or their testing agencies as well as known and controlled identification marks (e.g., UL, CE) to indicate conformity to certain requirements.

Open Use may represent requirements on the standardization process as well as requirements on implementations that use the standard to identify and assure compliance and, if necessary, conformance. For a manufacturer of a scale to measure weight, a self certification process traceable to national standards may be required. For a communications equipment or communications software manufacturer, an interoperability event may be needed (often termed a plug-fest) to test that different implementations interoperate. For the user, a simpler mark of conformity is often desirable. As example, in the European Union (EU), the CE marking is the manufacturer's indication that the product meets the essential (mostly safety) requirements of all relevant EU Directives. This specific marking indicating compliance reduces the user's safety concerns. Many consortia support plug-fests and compliance testing as part of their members' desire to promote associated products. Two levels of Open Use are quantified:

1.   Open Use via plug-fests or over-the-Internet testing (implementer)
2.   Open Use via conformance marking (user). This may include the first level of use.

10. Ongoing Support

Ongoing Support — standards should be supported until user interest ceases rather than when implementer interest declines. Ongoing Support of standards is of specific interest to standards users as it may increase the life of their capital investment in equipment or software with standard interfaces. The user's desire for implementer-independent ongoing support is noted by Perens (1999) as one of the desirable aspects of open-source software. The support of an existing standard consists of four distinct phases after the standard is created (Table 2).

This list may be used to quantify the ongoing support that a specific SSO provides by identifying which steps of the ongoing-support process are widely

Table 2. SSOs' phases of support during a standard's lifetime

Phase	Activity	Description	Major Interest Group
0.	Creation of standard	The initial task of SSOs	Creators
1.	Fixes (changes)	Rectify problems identified in initial implementations	Implementers
2	Maintenance (changes)	Add new features and keep the standard up to date with related standards work	Users
3.	Availability (no changes)	Continue to publish without continuing maintenance	Users
4.	Rescission	Removal of the published standard from distribution	Users

announced by a specific SSO. This is a difficult requirement to quantify as different SSOs have different procedures for making this process public, and many older SSOs may not make good use of the Internet to distribute such information to users.

It is difficult to interest users in the first phase of standards development (creation) shown in Table 2 (Naemura, 1995). Even the second phase, Fixes, may be of more interest to the developers and implementers than the users. The next three phases, however, are where users have an interest in maintaining their investment. Possibly greater user involvement in the ongoing support of standards would be practical by taking advantage of the Internet to distribute standards and allow users to keep abreast of the work in standards meetings. Increasing the users' involvement with these aspects of the standardization process may also present new economic opportunities for SSOs. The ITU-T Telecommunications Standardization Bureau Director's Ad Hoc IPR Group Report, released in May 2005, includes "Ongoing support — maintained and supported over a long period of time" as one element of its Open Standards definition.[13]

COMPARING THESE 10 REQUIREMENTS WITH OTHER DEFINITIONS OF OPEN STANDARDS

Standards are a multi-disciplinary field. An academic view of the requirements of open standards should address each of the related disciplines — economics, law, engineering, social science, and political science. From a legal perspective, each of these 10 requirements may be a legal right of a specific group. As West (2004) notes, each of these requirements has an economic cost and a benefit to specific stakeholders. From an engineering perspective, two of these requirements (6, 8) directly impact communications equipment compatibility and design. From a social science perspective, the dynamics of different stakeholders may be examined in terms of each requirement. From a political science perspective, the first three requirements are basic to any fair political process, including standardization.

West (2004, p. 7) defines "'open' for a standard as meaning rights to the standard are made available to economic actors other than the sponsor." This definition offers a succinct economic view of open standards. But economic rights cannot be maintained without supporting political rights such as balance, consensus, and due process. In order for the economic rights associated with compatibility standards to be available, some technical process (change control) and technical functionality (open interfaces) are also required. In order for specific economic rights associated with Intellectual Property Rights (IPR) to be available, specific SSO procedures must be defined.

Perens (n.d.) offers a software engineering perspective of Open Standards. He presents six requirements and related practices. The requirements proposed are: (1) availability, (2) maximize end-user choice, (3) no royalty, (4) no discrimination, (5) extension or subset, and (6) no predatory practices. The 10 requirements proposed can be compared to the six principles proposed by Perens:

- Availability is addressed by Open Documents.
- Maximum end-user choice is addressed by Open Use.
- No royalty is addressed under Open IPR.
- No discrimination is addressed by Open Meeting, Consensus and Due Process.
- Means to create extension or subset is addressed by Open Interface.
- Means to prevent predatory practices is addressed by Open Change.

The six principles proposed by Perens map fully onto 8 of the 10 requirements of Open Standards proposed. In the six principles, Perens does not directly address the desires for or against Open World or the end user desire for Ongoing Support.

HOW OPEN ARE DIFFERENT SSOs?

Table 3 offers the author's quantification based on a review of the SSOs' documentation (as of September 2004) of the specific requirements these SSOs support. By quantifying 9 of the 10 requirements of Open Standards, it is possible

Table 3. Rating "openness" at different SSOs

Reqmts.	Consortium (Note 1)	ITU (Note 2)	IEEE (Note 2)	ATIS T1 (Note 2)	ETSI (Note 2)	W3C	IETF
1 OM	1	1	2	1	1	1	2
2 Con	0	1	1	1	1	1	1
3 DP	0	1	1	1	1	0	0
4 OW	1	1	0	0	0	1	1
5 OIPR	0	2	2	2	2	4	3 (Note 3)
6 OC	0	1	1	1	1	1	1
7 OD	2	1	2	3	3	1	3
8 OI							
9 OU	2	0	0	0	1	1	1
10 OS	1	2	2	2	4	4	4
Score	7	10	11	11	14	14	16

Note 1: This hypothetical consortium is modeled on the description found at ConsortiumInfo.org (http://www.consortiuminfo.org/what, September 2004).
Note 2: The ITU, ETSI, IEEE, and ATIS are recognized SSOs.
Note 3: The IETF IPR policy desires a royalty-free model, but is flexible.

to examine any SSO to determine what requirements are supported and what requirements are not, and how this impacts standards creators, implementers, and users. Then the political, social, economic, technical, and practical implications of the standardization process and the machinations of all the stakeholders may be more rigorously analyzed and understood.

CONCLUSION

Microsoft is an example of a commercial organization that uses the term open standards when referring to its products (Gates, 1998). In fact, Microsoft's products are considered by many (including the European Commission) to be examples of non-openness. It seems clear that defining all the requirements of openness is necessary to avoid such misuse and the confusion it causes. Table 3 identifies that no SSOs discussed meets all of the 10 requirements described, and each SSO differs significantly in which requirements they do meet. So it should not be surprising that implementers and users consider all SSOs with a jaundiced eye. This attempt to rate how well SSOs support open standards suggests this implementer and user view is a wise one.

The 10 basic requirements presented here are the broadest possible view of the meaning of Open Standards. Are fewer requirements sufficient? That question can only be answered when each stakeholder understands the consequences of what they may be giving up. The comprehension of the requirements that are supported by each SSO is usually buried in the fine print of the procedures of each SSO. Until each SSO clearly indicates which requirements of Open Standards they support and at what level, open standards will be just another marketing slogan.

ACKNOWLEDGMENTS

The author would like to thank Joel West and Henk de Vries for their continuing efforts to improve this chapter.

REFERENCES

American National Standards Institute (ANSI). (1998). *Procedures for the development and coordination of American national standards.* Washington, DC: American National Standards Institute.

ANEC. (n.d.). *The European consumer voice in standardization.* Retrieved from http://www.anec.org

Band, J. (1995). Competing definitions of "openness" on the NII. In B. Kahin & J. Abbate (Eds.), *Standards policy for information infrastructure* (pp. 351-367). Cambridge, MA: The MIT Press.

Brady, R. A. (1929). *Industrial standardization.* New York: National Industrial Conference Board, Inc.

Cargill, C. (1989). *Information technology standardization.* Bedford, MA: Digital Press.

Cargill, C., & Bolin, S. (2004). *Standardization: A failing paradigm.* Paper presented at the Standards and Public Policy Conference, Chicago.

Clark, D. C. (1995). *Interoperation, open interfaces, and protocol architecture.* Retrieved from http://www.csd.uch.gr/~hyp490-05/lectures/Clark_interoperation.htm

Critchley, T. A., & Batty, K. C. (1993). *Open systems: The reality.* Hertfordshire, UK: Prentice Hall.

De Vries, H., Feilzer, A., & Verheul, H. (2004). Removing barriers for participation in formal standardization. In F. Bousquet, Y. Buntzly, H. Coenen, & K. Jakobs (Eds.), *EURAS Proceedings 2004* (pp. 171-176). Aachen, Germany: Aachener Beitrage zur Informatik.

Foray, D. (1995). Coalitions and committees: How users get involved in information technology (IT) standardization. In R. Hawkins, R. Mansell, & J. Skea (Eds.), *Standards, innovation and competitiveness* (pp. 192-212). UK: Edward Elgar Publishing Limited.

Gabel, H. L. (1987). Open standards in the European computer industry: The case of X/Open. In H. Landis Gabel (Ed.), *Product standardization and competitive strategy.* Amsterdam: North-Holland.

Gates, B. (1998, June 13). Compete, don't delete. *The Economist,* 19.

International Federation of Standards Users (IFAN). (2000). *IFAN strategies and policies for 2000-2005.* Retrieved from http://www.ifan-online.org/

Internet Engineering Task Force (IETF). (1998). *IETF working group guidelines and procedures: RFC 2418.* Retrieved from http://www.ietf.org/rfc/rfc2418.txt

ISO/IEC 17000: 2004. *Conformity assessment: Vocabulary and general principles.* Geneva: International Organization for Standardization.

Krechmer, K. (1998). The principles of open standards. *Standards Engineering, 50*(6).

Krechmer, K. (2000). The fundamental nature of standards: Technical perspective. *IEEE Communications Magazine, 38*(6), 70.

Naemura, K. (1995). User involvement in the life cycles of information technology and telecommunications standards. In R. Hawkins, R. Mansell, & J. Skea (Eds.), *Standards, innovation and competitiveness.* UK: Edward Elgar Publishing Limited.

NRENAISSANCE Committee (1994). Computer and Telecommunications Board, National Research Council. *Realizing the information future.* Washington, DC: National Academy Press.

Perens, B. (n.d.). *Open standards principles and practice.* Retrieved from http://perens.com/OpenStandards/Definition.html

Perens, B. (1999). The open source definition. In C. DiBona, S. Ockman, & M. Stone (Eds.), *Open sources voices from the open source revolution* (pp. 171-189). Sebastopol, USA: O'Reilly & Associates.

Raymond, E. S. (2000). *Homesteading the Noosphere: Section 2.* Retrieved August 25, 2000, from http://www.csaszar.org/index.php/csaszar/interesting/the_open_ source_reader

Shapiro, C. (2001). Setting compatibility standards cooperation or collusion. In R. C. Dreyfuss, D. L. Zimmerman, & H. First (Eds.), *Expanding the boundaries of intellectual property.* Oxford, UK: Oxford University Press.

United States District Court (1999). *United States District Court: District of Columbia Civil Action No. 98-1232 (TPJ).* Retrieved from http://www.usdoj.gov/atr/cases/f3800/msjudgex.htm

Updegrove, A. (1995). Consortia and the role of the government in standards setting. In B. Kahin, & J. Abbate (Eds.), *Standards policy for the information infrastructure* (pp. 321-348). Cambridge, MA: The MIT Press.

Updegrove, A. (2004). Best practices and standard setting (How the "pros" do it). In S. Bolin (Ed.), *The standards edge, dynamic tension* (pp. 209-214). Ann Harbor, MI: Sheridan Books.

Updegrove, A. (2005). China, the United States, and standards. *Consortium Standards Bulletin, IV*(4). Retrieved from http://www.consortiuminfo.org/bulletins/apr04.php#editorsnote

West, J. (2004). *What are open standards? Implications for adoption, competition and policy.* Paper presented at the Standards and Public Policy Conference, Chicago.

ENDNOTES

[1] This chapter is a significant revision of the paper published in the *Proceedings of the 38th Annual Hawaii International Conference on System Sciences (HICSS)*, January, 2005. In turn, the HICSS paper is a major revision of Krechmer (1998).

[2] Wikipedia http://en.wikipedia.org/wiki/X/Open

[3] Wikipedia http://en.wikipedia.org/wiki/Free_software

[4] Wikipedia http://en.wikipedia.org/wiki/The_Open_Group

[5] IEEE Web site http://standards.ieee.org/sa/sa-view.html

[6] ETSI Web site http://www.etsi.org/%40lis/background.htm

[7] ANSI Web site http://www.ansi.org/standards_activities/overview/overview.aspx?menuid=3

[8] The principle of "open standards" is formulated in European Interoperability Framework section 1.3 "Underlying principles," derived from the eEurope Action Plan 2005 as well as the Decisions of the European Parliament, November, 2004, http://xml.coverpages.org/IDA-EIF-Final10.pdf

[9] WTO Web site http://www.wto.org/english/tratop_e/tbt_e/tbtagr_e.htm#Annex%203

[10] http://xml.coverpages.org/openStandards.html, May 2005.

[11] European Union, http://www.eurunion.org/news/press/2004/20040045.htm

[12] Eupopean Union, http://www.eurunion.org/index.htm, see Microsoft EU Tests Proposals.

[13] http://www.itu.int/ITU-T/othergroups/ipr-adhoc/openstandards.html

Section II

Coordination in Standards Setting

Chapter III

Block Alliances and the Formation of Standards in the ITC Industry

Alfred G. Warner, Pennsylvania State University – Erie, USA

ABSTRACT

This chapter extends the examination of block alliances in standard setting from market-driven to formal or committee-based processes in the information and communications industry. Formal-process block alliances are argued to emerge in anticipation of institutional failure, that is, from the prospect that formal standardization will not yield a timely or correct solution. These block alliances organize around particular standards or more general technology streams and have distinctive characteristics. These include a clear articulation and separation of marketing and technical specification roles. Finally, block alliances in formal standard setting exhibit a governance form corporate in nature and distinct from the star or clique forms exhibited in market-based alliances. Some potential causes of this are examined.

INTRODUCTION

Recent research has begun to examine the role of block alliances in the generation of product standards. In the conflict between Sony and JVC/Matsushita, the ability of the latter to assemble a broad group of producers is credited with being key in the eventual victory of the VHS format (Cusumano, Mylonadis, & Rosenbloom, 1992). The contest over RISC chips between Mips and Sun has been explored both in terms of the structure of the alliances (Gomes-Casseres, 1994) and the characteristics of the firms within the alliances (Vanhaverbeke & Noorderhaven, 2001). The emergence of digital-video-disc (DVD) technology featured a contest between producers Sony and Philips vs. the product and content alliance of Toshiba, Matsushita, Time-Warner, MGM-UA, and MCA, among others (De Laat, 1999). Ironically, the Sony-Philips alliance had already successfully sponsored compact-disc technology against single entries from Telefunken and JVC (Hill, 1997). All of these studies have focused on conflicts begun and often settled in direct competition between alliance blocks and the subsequent creation of market standards. Yet firms are increasingly turning toward formal standard-setting and standards-development organizations as a means of circumventing market battles. In these arenas, block alliances also play an important, though little explored, role.

In this chapter, I establish motivations for the emergence of block alliances in market-based standards battles. I then contrast this process with block alliances in formal standard setting, which are argued to have two objectives: (a) to create and support a market and (b) to generate a technical standard. In market-based battles, the standard follows success in the market by definition. The standard technology is recognized ex post as the survivor. However, in formal environments, standards are the product of negotiation and are often pre-competitive and anticipatory. The actual standardization process may have little to do with the support and generation of a market. In fact, block alliances in this environment may play either or both roles: In terms of developing standards, the alliances have important coordination and development functions but may also serve as a signal of incipient market power to competing technology groups and prospective adopters. I also examine how alliance organizational structure differs from the star or centralized alliances, or clique or network structures found in market battles. A preliminary model will be developed of block alliances in formal environments using the rich structure of the information and communications technology industry as illustration.

Since a discussion of the information and communications technology industry unavoidably calls on technologies and institutions with complex names, acronyms are used for convenience. Table 1 lists the acronyms used in this chapter and their referents.

BACKGROUND

Standards are dominant designs that ostensibly represent a sole solution in the market. As earlier noted, they are generally classified as market derived or committee derived (Farrell & Saloner, 1988). The studies of VHS and Betamax,

Table 1. Organizations, technologies, and their acronyms

ATM	Asynchronous Transfer Mode
DOCSIS	Data Over Cable Service Interface Specification
DSL	Digital Subscriber Line
ETSI	European Telecommunications Standards Institute
IEEE	Institute of Electrical and Electronics Engineers
IETF	Internet Engineering Task Force
IMTC	International Multimedia Teleconferencing Consortium
ITU	International Telecommunication Union
MCNS	Multimedia Cable Networking System
OIF	Optical Internetworking Forum

DVD, and RISC chips are examples of battles in the market, the winner of which becomes the standard. The compelling feature of such battles is that unlike other industries where a dominant design may be that technology with more than 50% market share (Anderson & Tushman, 1990), these conflicts are winner take all. This is a function of increasing returns to adoption.

Increasing returns to adoption arise through the presence of network externalities, from learning by doing, and from coordination by technical groups (Cowan, 1991; Islas, 1997; Mookherjee & Ray, 1991). Network externalities mean that each buyer of a technology receives greater benefits as the user network increases in size. Examples of this include telephone service and fax machines. Similarly, support services for a product (such as the availability of mechanics for certain types of automobiles) or complementary goods (such as software) can constitute an externality. Experience and learning are also examples: This knowledge is more portable and available to prospective new users the more widely the technology is adopted (Farrell & Saloner, 1985; Katz & Shapiro, 1986, 1992).

Increasing-returns markets differ significantly from more conventional decreasing- or constant-returns markets. In particular, they are "tippy"; network markets are not shared. In other words, incompatible contenders in such markets will not find a stable interior solution where both coexist in the market: One will fail. Also, historical events are not averaged away but matter in the ultimate outcome. As Arthur (1989) notes, increasing-returns markets magnify chance events and are unpredictable. This tippiness implies that these markets are disposed toward the emergence of market-based technology standards unless negotiated standards are established first (Katz & Shapiro, 1994).

This effect has several interesting implications for product contests. If users of a technology are forward looking, their purchase decision will be affected by their beliefs about the ultimate nature of the market. There are two effects to be considered. First, buyers may be reluctant to adopt a new technology in a network market since the value of their choice depends on the choices others make later. Rational-expectations models sometimes generate the result that no buyer moves to the new technology, no matter how attractive it may be, because buyers will purchase a new technology only if private benefits exceed the cost of purchase. Benefits are derived from the stand-alone value of the product and its network value (Kristiansen, 1998). If there are no other users in the network, it is irrational to make the first purchase if the stand-alone value does not exceed purchase price. A related cause is the fear potential users have of adopting the wrong technology and being subsequently stranded. This problem is termed excess inertia (Katz & Shapiro, 1994).

The opposite effect is excess momentum in adoption. The degree to which markets can tip has generally been discussed in dichotomous terms: Markets will either tip or they will not. In fact, increasing-returns markets may tip at different rates as a function of the natural size of the network, expectations about that size, or the usefulness (Besen & Farrell, 1994) or availability of complementary goods (Schilling, 2002). Thus, adopters may flock to a new technology without regard for the costs imposed on users of an existing technology if the new technology offers advantages over the old, even absent a fully developed network (Farrell & Saloner, 1986). Also, if the technologies are regarded as substitutes rather than complements, such small network adoption can occur (Shy, 1996). Finally, potential adopters may use the decisions of others as low-cost information rather than collecting it on their own. This can generate a sequence of adoptions that cascade through a market, creating a fad or frenzy (Bikhchandani, Hirshleifer, & Welch, 1998).

Market-Derived Standards and Block Alliances

This tension between inertia and momentum under the prospect of a winner-take-all outcome tends to make prestandard competition in network markets fierce, with tactics such as penetration pricing and product preannouncements used to sway or hold up user selection (Besen & Farrell, 1994; Farrell & Saloner, 1992; Katz & Shapiro, 1994). The race here is to build the installed base of users or to generate a widespread secondary market in complementary goods. Put differently, if the market can be built, the standard will follow. Unsuccessful firms in these races incur, at the least, the losses of sunk costs in R & D (research and development) and learning as the firm strives to adopt the winning technology. At worst, the firm may be locked out of future competition and fail (Schilling, 1998, 1999).

If firms can gain competitive advantage through deploying resources that are valuable, rare, and difficult to imitate or substitute (Barney, 1991; Peteraf, 1993), then the first step is to secure the value of the assets. No matter how intrinsically interesting or effective a technology might be, it is not valuable if it loses in a standards war. Thus, though Betamax technology was accounted by some to be a better solution than VHS, it had little value to Sony or early adopters after VHS emerged as dominant. The issue for firms is to make certain that the technology is at least valuable (i.e., successful) so competition can be carried out on other levels such as production, marketing, or support. If entry to generate success in an increasing-returns market is risky, firms can manage the vulnerability through alliance if the payoff is sufficiently high (Eisenhardt & Schoonhoven, 1996). Block alliances emerge as a way for firms to hedge the risk of battling alone in an increasing-returns market.

Vanhaverbeke and Noorderhaven (2001) describe block alliances as groups of firms linked together for a common purpose in contrast to the more common (or commonly explored) dyadic alliance. Here, competition occurs between groups rather than between particular firms. Innovators in an increasing-returns market may seek to reduce risk by assembling a group of cooperative firms to acquire and exercise market power and subsequently generate a standard architecture. In other words, they seek to establish a bandwagon effect both in terms of partners and adopters (Wade, 1995).

One way innovators can do this is to disseminate the technology at a relatively low cost to other producers in the industry. The use of licensing as such a means has increased due to higher costs of R & D, decreased product life cycles and time to market, and increasing competition from current and new competitors (Carayannis & Alexander, 1999). Licensing can increase the flexibility of firms in responding to innovation: While licensees will develop competing products, they are within the architecture, not in competition with it. This can help solidify the market base.

To illustrate, Ethernet was developed in the labs of Xerox but was effectively brought to market in the early 1980s through the combined efforts of Digital Equipment, Intel, and Xerox (DIX). This DIX consortium fought the prevailing environment of proprietary local-area network (LAN) technologies by licensing the technology to producers for only $1,000 (Hiltzik, 1999), and Ethernet became one of the three standard LANs. Similarly, JVC/Matsushita drove the spread of VHS through active licensing to partners (Cusumano et al., 1992; Hill, 1997). The licensees all made competing versions of VHS players and recorders, but they did not produce Betamax or other video technologies.

Firms can also actively seek alliance with providers of complementary assets. Schilling (1998) observes that a lack of complementary goods when they are important can lead to market failure since these affect buyers' perceptions of the technologies. This is illustrated by the decline of the Macintosh personal computer as a function of limited software based on a closely held operating system. Similarly, as noted earlier, the Toshiba DVD alliance included not only hardware producers, but the owners of the content as well. By lining up so many of the major content houses, this alliance presented the threat of complementary-good failure to the Sony-Phillips alliance (De Laat, 1999). Collectively, these measures can serve as a strong signal of market power to both competitors and prospective customers. A sufficiently large alliance of producers and complementary-good suppliers can make competition a fait accompli, insofar as competitive technologies are concerned, as smaller firms or groups may find themselves locked out of appropriate resources. This may compel these competitors to compromise or exit the market. Similarly, block alliances can signal consumers that the technology will succeed, thus overcoming incipient problems of inertia. Thus, by implication or fact, the objective of block alliances is to generate market power sufficient to establish a winning standard.

Threats to Markets and Recourse to Formal Standards

Still, if there are many prospective solutions, or contenders can assemble equivalent alliances, markets can fail to generate a standard. The example of stereo AM (amplitude modulation) in the United States is a widely used example of how multiple competing and exclusive technologies can each fail to attain sufficient market mass to compel a standard (Besen & Farrell, 1994; Farrell & Saloner, 1988; Katz & Shapiro, 1994). More recently, the conflict between US Robotics (USR) and Rockwell over 56K modem technologies featured alliances of users and producers on both sides. US Robotics lined up computer manufacturers that would install modems in new systems, and Internet-service providers that pledged support for the USR technology. Rockwell led a group of chip manufacturers including Intel and Motorola in the Open 56K Forum with the offer of an open specification to all interested users (Josifovska, 1997). Some observers argued that the battle deterred growth in modems (i.e., since no one knew which to buy, few were sold, and the market was not being determined) and opened the door to other technologies such as digital subscriber lines (Korzeniowski, 1998; Rafter, 1997). Ultimately, a compromise standard incorporating aspects of both technologies was developed by the ITU after an appeal by the alliances and other interested parties.

The prospect of market failure, or losing a standards race, can drive firms to seek institutional intervention from governments or standard-developing organizations (SDOs) (Farrell & Saloner, 1988; Hawkins, 1999). The institution can act to select a single winner in the market or to reconcile or harmonize competing approaches as in the 56K modem example. Recently, it has been clear that the cost of entry into an increasing-returns market with the concomitant possibility of loss or failure is sufficient for firms to seek standardization prior to going to market to reduce risk. These are so-called anticipatory standards (David & Greenstein, 1990; David & Shurmer, 1996). Block alliances still play a significant role in all of these cases, though their function is distinct from alliances competing in the market. In the next section, I describe how alliances emerge from the prospect of institutional failure and assume one or both of the marketing and specification roles depending on the circumstances around the standardization process.

Formal Standards and Institutional Failure

If alliances in market battles seek to generate momentum through adoption or implication of adoption, alliances in formal standard-setting environments have a somewhat different role. The twin tasks of market generation and technical specification are made explicit here. As noted earlier, in market battles, a standard emerges as a survivor and is encoded in proprietary designs or in the terms of licenses held by producers of similar and complementary goods. Formal standards are public documents, explicit directions on how to produce goods that are compliant with the specification.

SDOs are generally organized around principles designed to facilitate open exchange, which includes an emphasis on due process, ordered and transparent debate and participation, consensus among stakeholders with regard to decisions, and voluntarism (David & Shurmer, 1996). The SDOs are structured around committees and work groups composed of engineers who develop and submit the proposed standards. As David and Shurmer (1996, p. 795) note, the underlying ethos in this approach is a "strong belief that engineers could find a unique solution which…represented the best of the technical options (and) would be agreed upon by all parties." This is not to suggest that standards organizations are apolitical since voting members are often also employees of competing firms. Farrell, Monroe, and Saloner (1994) argue that this is one reason standards organizations are considered slow to resolve problems. More pointedly, observers have occasionally argued that corporate loyalty can supersede the objectivity of the standards process (Didio, 1993; Gold, 1993).

Standardization in advance of general market introduction is used to prevent problems with excess inertia, but this can generate problems of its own. First, the consensus rule means that standardizing new or radical technologies can be time consuming due to uncertainty over functionality. It may not be clear what the best technical solution actually is. Second, the deregulation of the ITC industry and the concomitant growth in the number and diversity of SDO participants means that consensus under any circumstance is more difficult to achieve (Balto, 2001; David & Shurmer, 1996). Some participants care much more about outcomes than do others, and it is the latter, according to Genschel (1997), that have the power. Because they are needed to generate sufficient majority, they can hold out for more favorable terms or play bargainers against each other. This can generate obstructionism and opportunism. Finally, participants with strong preferences reflecting specific firm assets or resources can mean that voters are less willing to compromise. New standards that eliminate particular approaches can put firms at a competitive disadvantage relative to those whose resources are supported by the standard (Genschel, 1997). As a general rule, when the perceived mutual gain from a universal standard is low, the speed of standardization decreases (David & Shurmer, 1996).

Put differently, if participants in standards setting believe that the SDO process will fail to deliver a standard in the time frame needed, or that compromise and consensus will deliver the wrong standard, they are anticipating a failure of the institution, similar to the failure of markets. When a process that has been described as unstructured and as such makes it "nearly impossible to set tight schedules and to adhere to them" (McCarron, 1998, p. 17) is combined with shrinking product life cycles (Carayannis & Alexander, 1999), this failure becomes more probable. When time is important, Farrell and Saloner (1988, p. 240) show that a hybrid model is the best performer: "[W]hile committees are better than the pure bandwagon system, they are even better if they can be subverted by preemptive action outside the committee." Although they present no examples of such subversive action, they do note a potential solution in explicit interfirm cooperation to facilitate the development and adoption of a product system standard either through a separate standards organization or, conceivably, as a stand-alone effort (Farrell & Saloner, 1988).

BLOCK ALLIANCES IN FORMAL STANDARD SETTING

Block alliances are a way for firms to circumvent prospective institutional failure. The form and specific objectives vary with where in the standardization process they focus their attention. In general, I argue that block alliances emerge to support a specific standard (supplemental alliances) or to more generally support a technology (developmental alliances). Supplemental alliances are intimately involved in the workings of an SDO, while developmental alliances generate front-end specifications for a variety of SDOs through coordinating and winnowing research within a general technology stream. Finally, a third form, breakaway alliances, occasionally emerges as an alternative to the SDO process (Figure 1). All have marketing and technical-specification roles.

The marketing role that most alliances have is twofold: to generate awareness of the technology vis-à-vis competitive technologies, and to coordinate interoperability testing and certification. Schilling (1999) has argued that in introducing a new technology, large firms have an advantage over small and existing firms in educating and convincing consumers about features and benefits. She notes that the disadvantages of size and age can be overcome through alliance with incumbents and linking with those firms' credibility. Block alliances serve to promote the interacting technologies of members in a consistent and comprehensive fashion through trade shows, advertising, and white papers.

Figure 1. Block alliances and the SDO process

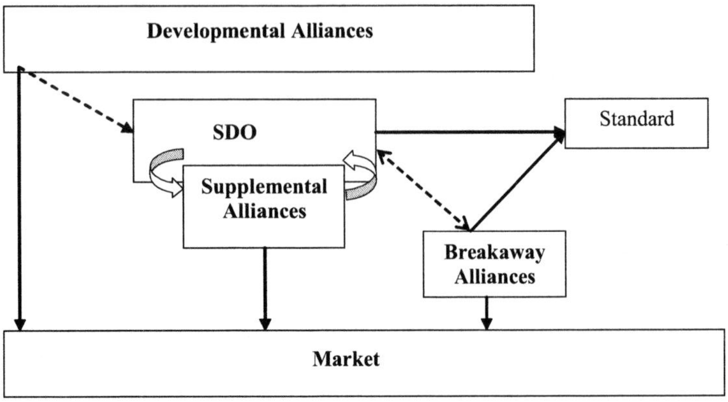

As important as raising consumer awareness about a technology is, establishing reliable interoperability may be the more important in generating credibility for the technology. As one industry executive put it, "interoperability on paper is vastly different than interoperability in the field...the only true test (is) linking equipment from different vendors" (Sansom & Juliano, 1993, p. 23). If standardization leads to economic benefits for buyers through reduced costs, it is essential that they have confidence that mix-and-match purchasing will result in a workable system (Balto, 2001; Besen & Farrell, 1994). The difficulty in interoperability is compounded by the complexity of the technologies in question.

The market-based battles mentioned earlier can be characterized almost universally as alliances organizing around a focal technology and the firm that controls that technology. These are centralized (Vanhaverbeke & Noorderhaven, 2001) or engineered (Doz, Olk, & Ring, 2000) alliances. In contrast, network technology may include several key technologies such as silicon from a set of IC makers, routers or hubs from another set of hardware makers, and media and related products from a third set. While the specification or standard may articulate how they are to work collectively, interfacing fundamentally proprietary solutions can be difficult. Therefore, a key component of almost all block alliances in the formal environment is a process for demonstrating and assuring interoperability. For instance, many alliances turn to the University of New Hampshire Interoperability Laboratory for independent testing and certification. Other alliances have dedicated testing facilities, such as the Multimedia Cable Network Systems' CableLabs, for assessing compliance with the cable modem DOCSIS standard. An emerging view is that broad interoperability among multiple vendors ought to occur before a standard is ratified (Hochmuth, 2001b).

Writing technical specifications is another way in which formal-standards block alliances differ significantly from market-based alliances. As noted earlier, in market battles, the standard is the ex post result of competition. In the SDO process, the standard is a public document and, in the case of anticipatory standards, precompetitive. That block alliances should have any role in writing standards seems at odds with the process described above, but in fact, they do have such a role. The engagement can vary from publicly available specifications submitted for consideration by developmental alliances to the fundamental role some supplemental alliances play in the process.

Finally, formal-standards block alliances differ from their market-based analogues in structural form. Virtually all of the block alliances engaged in the ITC standards and prestandards environments are organized as corporations with officers, boards of directors, and dues-paying members. Market-based alliances are characterized as star (firm centered) or clique (dyadic relationships) (Gomes-Casseres, 1994; Vanhaverbeke & Noorderhaven, 2001). Potential motivations for this difference are discussed below.

Supplemental Alliances

Supplemental alliances form within the context of a specific standardization effort and are associated with an SDO. One of the key attributes of the supplemental alliances generally is that both market-building and technical-specification roles are executed. While responsibility for formal specification may rest with an SDO, the supplemental alliance can coordinate testing and experimentation outside the normal operations of the institution, achieving additional efficiency and presumably faster time to market.

The 10Gigabit Ethernet Alliance (10GEA), a supplement to the operation of the IEEE 802.3 committee, may provide a good illustration. In January 2000, the committee authorized the establishment of the 10Gigabit task force, the working group operating under the 802.3ae standard. At almost the same time, the 10Gigabit Ethernet Alliance announced its formation. Although these are independent bodies, the 10GEA had the objective of driving consensus on and contribution to the developing standard. As one alliance executive noted, "an industry consortium like Gigabit [Ethernet Alliance] or 10GEA facilitates meetings outside of IEEE meetings…[T]he alliance could be reducing the standards efforts anywhere from six months to 18 months" (Hochmuth, 2001a). This can happen because there is substantial overlap between the task force and the 10GEA: Many members participate in both organizations.

The 10GEA carries on programs to facilitate interoperability testing (which helps clarify aspects of the standard) and provides public technology demonstrations as market-supporting activities. The purpose of these activities is not only to help define key issues in the emerging standard, but to build customer confidence about the technology (Hochmuth, 2001b).

Another aspect of the linkage to a particular standard is that the supplemental alliances are generally self-limiting. Emergent networks or those that exhibit interdependencies such as the need for agreement on standardization or the existence of a common threat by environmental change would terminate sooner than would those created by an engineered process. This earlier mortality was attributed to the premise that emergent coalitions draw together firms with similar interests, generally from the same industry. When the need for standardization or the environmental threat is resolved, the motivation to collaborate diminishes and the alliances terminate. This may well be because alliance activities impose additional governance cost on the innovation process in the form of dues to the alliance and the commitment of staff to the process. Since the purpose of alliance activities is the joint creation of value (Nooteboom, Berger, & Noorderhaven, 1997), the cessation of relevant activities generating value means that only governance costs are incurred and there is no longer a reason to make the expenditures.

Prop 1a: Supplemental alliances will emerge when:
1.   there are perceived time constraints and
2.   there is significant within-SDO technological competition.

Prop 1b: Supplemental alliances will terminate when the standard is finalized.

Developmental Alliances

In the longer run, proponents of a particular technology stream face two problems: how to develop generations of the product from the stream and, equally, how to fend off or manage developments in other technologies. In networking technologies, for instance, specifications are generated to adapt a particular technology to transmission over various media such as twisted copper pair or optical fiber. Consumers may demand various transmission speeds for applications from desktops to data backbones. In order to broaden the appeal of the technology, developers may also consider various converters or interfaces that permit communication between different types of transmission systems such as a gateway between an Ethernet local network and an ATM backbone.

Developmental block alliances arise to manage research and the threat of competition, and are distinct from supplemental alliances in that developmental alliances are not standard centered but technology centered. Organizations such as the ATM Forum, Optical Internetworking Forum, and the DSL Forum are oriented toward the cooperative development of new extensions of focal technologies. The outputs of these groups are not standards per se, but rather publicly available specifications. For instance, the ATM Forum develops specifications that are transmitted to the ITU, ETSI, and the IETF. Similarly, the OIF submits specifications to the ITU and IEEE, among others. As such, the alliances have substantial control over what technologies actually make it to the standards body. The developmental alliances can be longer lived than the supplemental forms. The ATM Forum has been in operation since 1991, and the DSL Forum and IMTC since 1994.

These alliances also have marketing objectives as they manage the threat of competition. The DSL Forum has half a dozen marketing groups covering issues such as "mind share," tradeshows, and public relations in addition to the six technical work groups. As DSL faces direct competition from technologies such as cable modems, marketing is a key issue. On the other hand, the International Multimedia Telecommunications Consortium is much more heavily weighted toward technical groups as it focuses on the reconciliation and harmonization of multiple standards and establishing interoperability testing. However, it still coordinates marketing efforts for the harmonization process through several working groups.

Prop 2a: Developmental alliances will emerge when:
1.   intertechnology competition is high and
2.   the need for coordination or bridging across technologies is high.

Prop 2b: Developmental alliances will persist as long as innovations are being developed within the technology stream.

Breakaway Alliances

Alliances sometimes emerge as an alternative to the SDO process. Here again, the prospect of institutional failure can drive the process, but there are several other conditions that need to be obtained. If firms resort to formal standardization to avoid market failure, one reason must be that there is concern over the power that alliances or firms have in the market. That is, the blocks do not have sufficient market power to force the issue. For a block of firms to secede from the formal process implies the opposite: that collectively, they have sufficient power to generate a market standard.

An example is the emergence of the standard for cable modems in the United States. Although the original responsibility for developing the standard was placed with an IEEE committee in 1994, 2 years later no standard was evidently forthcoming and the threat of DSL as an alternative was strengthening. At this point, a consortium of cable firms already involved in the standards process broke away to develop a simple standard of their own. The consortium introduced DOCSIS in 1997. Although this was not a group recognized by any formal standards body, the consortium submitted its architecture to the ITU, which formally standardized it in March 1998. The IEEE committee was disbanded shortly after (Hranac, 2001).

Table 2. Factors affecting the formation of specific block alliances

Developmental	Supplemental	Breakaway
Strong intertechnology competition	Strong intrastandard competition	Strong intrastandard competition
Need for intertechnology coordination	Time constraints or perceived need for speed to market	Time constraints or perceived need for speed to market
		High level of control over user base

The consortium, called the Multimedia Cable Network System, was composed of the major cable-service providers in the United States, including Cox Communications and Time-Warner Cable. As these are fundamentally local monopolies, they could control the choices subscribers made about modem technology through both supply (leasing modems to consumers) and the decisions about what technology the cable suppliers would support. That is, given any decision on modem technology, users would have no choice but to adapt or seek an alternative technology. No other choice within the cable infrastructure would be available. This gave substantial market power to the cable companies.

Prop 3: Breakaway alliances will share the same environmental drivers (time constraints, high within-SDO technological competition) with supplemental alliances but differ in that alliance members will exhibit significant market power.

Table 2 lists the environmental characteristics argued to drive the emergence of block alliances in their three forms in the formal standards process, summarizing the propositions.

DISCUSSION

Block alliances are a common feature of formal standard setting in the ICT industry, but their presence raises some interesting issues. First, why is the structure of these alliances different from alliances in market-based competition? Second, given a structure of alliances, what are the connections between the alliances and the implications thereof?

Alliance Structure

With very few exceptions, block alliances in the ICT standards process are organized around a corporate or board model. This structure differs from the star vs. network or clique characterization of Gomes-Casseres (1994) or Vanhaverbeke and Noorderhaven (2001) found in market-based block alliances. A star structure requires a central firm with the key technology (e.g., JVC or Mips) and is usually a function of multiple bilateral alliances. At least, the members of the star network are connected through the central firm. However, as noted above, in many networking or communications technologies, no single firm controls all of the key components. Therefore, no single firm can claim the center of the star. Neither is the alliance structure an emergent one composed of dyadic linkages. The prevailing alliance governance form here is a formal structure, with articles of incorporation, dues-paying members, officers of the alliance corporation, and a board of directors.

An open question is specifically why the block alliances have so frequently chosen this form. One can argue for efficiency causes: The corporate model is likely to be more effective than the clique or network model at coordinating activities such as interoperability testing and interactions with standards bodies. As Vanhaverbeke and Noorderhaven (2001) observe, without a central firm, alliances generally rely on dense linkages to achieve coordination and, in a clique configuration, this can become prohibitively costly. This is particularly pertinent in the case of coordinating interdependencies such as sequential steps in the value chain. The corporate model may offer other operational advantages: Organizing as a mutual-benefit corporation is seen as a necessary step toward qualifying as a tax-exempt organization in the United States. Funding for these organizations typically comes from member dues, not from commercial activity. Operating as a tax-exempt organization provides more efficient use of the income.

A more important role may be a safe harbor in terms of antitrust prosecution. Some alliances, such as the ATM Forum, organize under the 1993 National Cooperative Research Production Act of the United States (NCRPA), which permits the development and testing of technologies as well as the exchange of production information. The act specifically prohibits certain activities such as market allocation, production agreements, or pricing discussions, but also offers several advantages. First, collective activities are not automatically regarded as anticompetitive but must be judged under a rule of reason that speaks to an economic analysis of alleged restraint of trade. Second, damages awarded in civil court to successful claimants are limited to actual damages, not the treble damages available under the Clayton Act (Hamphill, 1997).

Many of the alliances are not organized under NCRPA but as mutual-benefit bodies. Even so, this may still offer some protection, particularly under U.S. federal guidelines regarding collaboration among competitors. If collaboration results in procompetitive efficiencies, such as quicker time to market, then this is weighed against any anticompetitive behaviors that might be charged. With respect to anticompetitive behavior, the guidelines assess carefully the nature of the agreement itself, particularly the business purpose (Federal Trade

Commission, 2000). Given this, it is noteworthy that the alliances without exception publish a specific statement of purpose along with proscriptions against anticompetitive actions such as sharing information about pricing or production. More inquiry is needed into the motivations for alliance structure choice and if and how it might be extended to market-based block alliances.

A Network of Networks?

Another aspect of the alliance structure of this industry that deserves exploration is the connection between developmental alliances and SDOs, and between developmental and supplemental alliances. For instance, in the first Ethernet standardization process, the alliance of Digital Electronics, Intel, and Xerox had not only opened proceedings with the IEEE, but also had begun work with ECMA in parallel (Genschel, 1997; Sirbu & Hughes, 1986). Genschel observes that the application to ECMA by the consortium helped revitalize the IEEE process, which had earlier come to a halt. More currently, the ATM Forum has had liaison and collaborative efforts underway with 35 organizations around the world, some of which are SDOs (Berendt, 1998). It would be interesting to understand if alliances operating across multiple SDOs are more or less successful at advocating their key issues (i.e., generating an acceptable standard) than more focused alliances. This may well be moderated by other developmental-alliance linkages to the extent that multiple alliances operate within the same SDO, providing an opportunity for politicking.

Even more interesting is the examination of the vertical linkages between developmental and supplemental groups. What is clear from inspection of the memberships is that these are not mutually exclusive bodies. Many firms participate in multiple alliances, more or less as a function of the markets they are in. In particular, there is overlap in the vertical stages of standard development, that is, the same firms may belong to a developmental alliance and subsequently to the relevant supplemental alliance. So, firms may be a part of a group that submits a publicly available specification to an SDO as a sort of protostandard; the firms then initiate a supplemental alliance to coordinate and expedite the process. This is intuitively appealing from an efficiency perspective as the parties to the standards-focused alliance already have a clear understanding of the technology and the implications of the specification, and it would be useful to understand if they indeed deliver that efficiency.

Finally, from an industry-structure perspective, it would be interesting to understand if there are consistent groups of firms allying over time and across markets. This is not quite a strategic group issue as allying firms may represent noncompeting parts of the overall industry. However, there are signs that long-term coalitions of firms might exist and define themselves as one alliance or another depending on need. If a firm controls all of the resources or capabilities needed to compete, do the longer term coalitions reflect a complete set of such resources? This has been observed in other industries (Gulati & Gargiulo, 1999), but it has also been observed that in high-technology industries, the effect of prior firm relationships may not be as strong. Thus, the alliance structure may be an emergent property of deeper industry characteristics and power arrangements. In this case, some of the emerging network-analysis tools could usefully be brought to bear.

These questions highlight the fundamental issue with block alliances in standard setting: Are technology adopters helped or damaged by alliance actions? Prior work has argued that the alliances are successful at bringing standards to market fairly quickly due to constrained membership, how the work is funded, and how property rights are treated (Krechmer, 2000). Together with the coordination benefits described above, alliances may be more efficient than SDOs. This should open markets more rapidly. On the other hand, some of these alliance attributes in conjunction with the development of standards vetted by formal SDOs could raise issues of collusion and anticompetitive behavior. Krechmer has pointed out that the formation of consortia by industry leaders may set objectives that benefit those firms, and other firms have little option but to accede. This has several implications.

First, this may lead to charges of anticompetitive actions. Balto (2001) cites the Addamax vs. Open Software Foundation case as an example of how a plaintiff could argue, albeit unsuccessfully in that case, that the structure of the alliance implied excessive market power. However, even a failed suit can be costly, and litigation might chill the alliance and fast standardization process. Second, and perhaps more important, is the problem of innovation itself. Anderson and Tushman (1990) have argued that discontinuous or radical innovations are very rarely the work of incumbents in an industry but rather the contribution of outsiders. To the extent that industry leaders control the standardization process, consumers may not receive the benefits of new approaches.

Alliances as Technology Sources

Acquisition is a well-known method of sourcing technology in the ITC industry (Chaudhuri & Tabrizi, 1999; Deogun, 2000; Hagedoorn & Duysters, 2002), and the potential role of block alliances as a marketplace for buyers and sellers warrants consideration. Acquirers and targets are often members of the same alliance, and the purchases are frequently consummated well before the standard has been approved. This raises several questions. First, why do firms that lack a core technology enter these alliances? Second, why do firms go to the expense of acquisition before the outcome of the standards process is clear?

One approach, particularly in the case of early or prestandard acquisition, is to consider real-options logic (ROL) as an explanation. ROL argues that many investments firms make are analogous to financial options in that they convey the right but not the obligation to make subsequent investments. Specifically, a staged investment under uncertainty can yield information that later permits managers to decide whether to proceed further with the project. This is known as a deferral option. While ROL is most often used for intrafirm decisions, it has been applied to relations between firms in the context of joint ventures (Kogut & Kulatilaka, 1994) and equity alliances (Bowman & Hurry, 1993; Hurry, Miller, & Bowman, 1992). In these cases, options are construed less as a right than as preferential access to purchasing opportunities.

More recently, researchers have begun to focus on the value of growth options or the choice to invest rather than defer under uncertainty. These growth options are particularly important in high-technology industries in order to be able to participate in future development (Folta & O'Brien, 2004; Leiblein, 2003), and are predicated on the pace of innovation and the advantages accruing to early movers. This has particular relevance to industries where standards are common. The uncertainty around technology investments can be decomposed into technical (pertaining to the likelihood of technical success) and market or input cost (exogenous market factors; Dixit & Pindyck, 1994; McGrath, 1997). However, McGrath (1997) also argues for a third form: factors outside the firm but subject to strategic action such that a firm may influence outcomes to its advantage. Certainly, standards and block alliances are one manifestation of this.

However, alliances rarely convey strong property rights, so the actions a firm takes on behalf of a particular technology owned by another firm may not pay off. Block alliances can provide a context within which to test implications of real-options logic regarding staged investment, preferential access, information accrual, and the decision to acquire. In other words, even firms lacking a relevant technology may undertake the relatively low level of investment required to participate in block alliances to help overcome information asymmetries regarding the technological merit of potential target firms. Block alliances may serve not only as coordinating mechanisms but as information markets.

CONCLUSION

In this chapter, I have developed the dual roles of block alliances in standard setting, particularly with respect to how these roles are distinctively different in formal standard environments as opposed to market competition. I have also developed a classification scheme based on environmental drivers and alliance objectives. The discussion of alliance structure has augmented the block-alliance investigation through the distinction of a separate corporate governance form that differs from both star and clique structures. This may result from both efficiency motivations and from legal or antitrust considerations. Direction for future research, especially with respect to the analysis of the alliance networks, was explored.

ACKNOWLEDGMENTS

An early version of this chapter was presented at the Midwest Academy of Management meeting, April 2002. The author gratefully acknowledges the help of Jay Barney, Michael Leiblein, Konstantina Kiousis, Doug Ruby, Bob Grow, Ray Hansen, Bruce Tolley, and Jonathan Thatcher, as well as the insights of several anonymous reviewers.

REFERENCES

Anderson, P., & Tushman, M. L. (1990). Technological discontinuities and dominant designs: A cyclical model of technological change. *Administrative Science Quarterly, 35*, 604-633.

Arthur, B. (1989). Competing technologies, increasing returns and lock-in by historical events. *The Economic Journal, 99*, 116-131.

Balto, D. A. (2001). Standard setting in the 21st century network economy. *Computer and Internet Lawyer, 18*(6), 5-19.

Barney, J. (1991). Firm resources and sustained competitive advantage. *Journal of Management, 17*(1), 99-120.

Berendt, A. (1998). Setting standards for a new era. *Telecommunications, 32*(7), 26-27.

Besen, S. M., & Farrell, J. (1994). Choosing how to compete: Strategies and tactics in standardization. *The Journal of Economic Perspectives, 8*(2), 117-131.

Bikhchandani, S., Hirshleifer, D., & Welch, I. (1998). Learning from the behavior of others: Conformity, fads, and informational cascades. *The Journal of Economic Perspectives, 12*(3), 151-170.

Bowman, E. H., & Hurry, D. (1993). Strategy through the options lens: An integrated view of resource investments and the incremental-choice process. *Academy of Management Review, 18*(4), 760-782.

Carayannis, E., & Alexander, J. (1999). The wealth of knowledge: Converting intellectual property to intellectual capital in co-operative research and technology management settings. *International Journal of Technology Management, 18*(3/4), 326-352.

Chaudhuri, S., & Tabrizi, B. (1999). Capturing the real value in high-tech acquisitions. *Harvard Business Review, 77*(5), 123-130.

Cowan, R. (1991). Tortoises and hares: Choices among technologies of unknown merit. *The Economic Journal, 101*(407), 801-814.

Cusumano, M., Mylonadis, Y., & Rosenbloom, R. (1992). Strategic maneuvering and mass-market dynamics: The triumph of VHS over Beta. *Business History Review, 66*(1), 51-95.

David, P., & Greenstein, D. (1990). The economics of compatibility of standards: A survey. *Economics of Innovation and New Technology, 1,* 3-41.

David, P. A., & Shurmer, M. (1996). Formal standards-setting for global telecommunications and information services. *Telecommunications Policy, 20*(10), 789-815.

De Laat, P. B. (1999). Systemic innovation and the virtues of going virtual: The case of the digital video disc. *Technology Analysis and Strategic Management, 11*(2), 159-180.

Deogun, N. (2000, January 3). Year end review of markets and finance 1999: Europe catches merger fever as global volume sets record. *Wall Street Journal* (Eastern ed.), p. R8.

Didio, L. (1993). Vendors squabble over Fast Ethernet standard. *LAN Times, 10*(3), 1.

Dixit, A. K., & Pindyck, R. S. (1994). *Investment under uncertainty.* Princeton, NJ: Princeton University Press.

Doz, Y. L., Olk, P. M., & Ring, P. S. (2000). Formation processes of R&D consortia: Which path to take? Where does it lead? *Strategic Management Journal, 21*(3), 239-266.

Eisenhardt, K. M., & Schoonhoven, C. B. (1996). Resource-based view of strategic alliance formation: Strategic and social effects in entrepreneurial firms. *Organization Science, 7*(2), 136-150.

Farrell, J., Monroe, H. K., & Saloner, G. (1994). *The vertical organization of industry: Systems competition versus component competition.* Standford University Graduate School of Business, Research Paper No 1329.

Farrell, J., & Saloner, G. (1985). Standardization, compatibility and innovation. *RAND Journal of Economics, 16*(1), 70-83.

Farrell, J., & Saloner, G. (1986). Installed base and compatibility: Innovation, product preannouncements, and predation. *The American Economic Review, 76*(5), 940-955.

Farrell, J., & Saloner, G. (1988). Coordination through committees and markets. *RAND Journal of Economics, 19*(2), 235-252.

Farrell, J., & Saloner, G. (1992). Converters, compatibility, and the control of interfaces. *Journal of Industrial Economics, 40*(1), 9-35.

Federal Trade Commission & Dept. of Justice (2000). *Antitrust guidelines for collaborations among competitors.* Washington, DC: Federal Trade Commission.

Folta, T. B., & O'Brien, J. P. (2004). Entry in the presence of dueling options. *Strategic Management Journal, 25,* 121-138.

Genschel, P. (1997). How fragmentation can improve coordination: Setting standards in international telecommunications. *Organization Studies, 18*(4), 603-622.

Gold, D. (1993). Will Fast Ethernet be here fast? *LAN Times, 10*(6), 40.

Gomes-Casseres, B. (1994). Group versus group: How alliance networks compete. *Harvard Business Review, 72*(4), 62-71.

Gulati, R., & Gargiulo, M. (1999). Where do interorganizational networks come from? *American Journal of Sociology, 104*(5), 1439-1493.

Hagedoorn, J., & Duysters, G. (2002). External sources of innovative capabilities: The preference for strategic alliances or mergers and acquisitions. *Journal of Management Studies, 39*, 167-188.

Hamphill, T. A. (1997). U.S. technology policy, intraindustry joint ventures and the National Cooperative Research and Production Act of 1993. *Business Economics, 32*(4), 48-54.

Hawkins, R. (1999). The rise of consortia in the information and communication technology industries: Emerging implications for policy. *Telecommunications Policy, 23*, 159-173.

Hill, C. W. L. (1997). Establishing a standard: Competitive strategy and technological standards in winner-take-all industries. *Academy of Management Executive, 11*(2), 7-25.

Hiltzik, M. (1999). *Dealers of lightning: XEROX PARC and the dawn of the computer age.* New York: HarperCollins.

Hochmuth, P. (2001a). 10G Ethernet gains momentum. *Network World, 18*(5), 21-24.

Hochmuth, P. (2001b). 10G Ethernet standard hits a snag. *Network World, 18*(30), 14.

Hranac, R. (2001). Cable modem 101: Covering the nuts and bolts. *Communications Technology.*

Hurry, D., Miller, A. T., & Bowman, E. H. (1992). Calls on high-technology: Japanese exploration of venture capital investments in the United States. *Strategic Management Journal, 13*(2), 85-101.

Islas, J. (1997). Getting round the lock-in in electricity generating systems: The example of the gas turbine. *Research Policy, 16*, 49-66.

Josifovska, S. (1997, March 19). Major modem makers turn up the heat in bid to win 56K crown. *Electronics Weekly*, 3.

Katz, M. L., & Shapiro, C. (1986). Technology adoption in the presence of network externalities. *The Journal of Political Economy, 94*(4), 822-841.

Katz, M. L., & Shapiro, C. (1992). Product introduction with network externalities. *Journal of Industrial Economics, 40*(1), 55-83.

Katz, M. L., & Shapiro, C. (1994). Systems competition and network effects. *The Journal of Economic Perspectives, 8*(2), 93-115.

Kogut, B., & Kulatilaka, N. (1994). Options thinking and platform investments: Investing in opportunity. *California Management Review, 36*(2), 52-71.

Korzeniowski, P. (1998, March 2). Finally, 56-Kbps modems hit the fast lane. *Telepath*, p. T25.

Krechmer, K. (2000). Market driven standardization: Everyone can win. *Standards Engineering, 52*(4), 15-19.

Kristiansen, E. G. (1998). R&D in the presence of network externalities: Timing and compatibility. *RAND Journal of Economics, 29*(3), 531-547.

Leiblein, M. J. (2003). The choice of organizational governance form and performance: Predictions from transaction cost, resource-based and real options theories. *Journal of Management, 29*(6), 937-961.

McCarron, S. P. (1998). Standards are made, not born. *UNIX Review, 7*(9), 14-21.

McGrath, R. G. (1997). A real options logic for initiating technology positioning investments. *AMR, 22*(4), 974-996.

Mookherjee, D., & Ray, D. (1991). Collusive market structure under learning by doing and increasing returns. *Review of Economic Studies, 58*, 993-1009.

Nooteboom, B., Berger, H., & Noorderhaven, N. G. (1997). Effects of trust and governance on relational risk. *Academy of Management Journal, 40*(2), 308-338.

Peteraf, M. A. (1993). The cornerstones of competitive advantage: A resource-based view. *Strategic Management Journal, 14*, 179-191.

Rafter, M. V. (1997, December 15). ITU backs compromise for 56K modem standard. *Internet World*.

Sansom, R., & Juliano, M. (1993). Architecture not the sole ATM question. *Network World, 10*(2), 23-24.

Schilling, M. A. (1998). Technological lockout: An integrative model of the economic and strategic factors driving technological success and failure. *Academy of Management Review, 23*(2), 267-284.

Schilling, M. A. (1999). Winning the standards race: Building installed base and the availability of complementary goods. *European Management Journal, 17*(3), 265-274.

Schilling, M. A. (2002). Technology success and failure in winner-take-all markets: The impact of learning orientation, timing, and network externalities. *Academy of Management Journal, 45*, 387-398.

Shy, O. (1996). Technology revolutions in the presence of network externalities. *International Journal of Industrial Organization, 14*, 785-800.

Sirbu, M., & Hughes, K. (1986). *Standardization of local area networks.* Paper presented at the Telecommunications Policy Research Conference.

Vanhaverbeke, W., & Noorderhaven, N. G. (2001). Competition between alliance blocks: The case of the RISC microprocessor technology. *Organization Studies, 22*(1), 1-30.

Wade, J. (1995). Dynamics of organizational communities and technological bandwagons: An empirical investigation of community evolution in the microprocessor market. *Strategic Management Journal, 16*, 111-133.

Chapter IV

Standardization and Other Coordination Mechanisms in Open Source Software

Tineke M. Egyedi,
Delft University of Technology, The Netherlands

Ruben van Wendel de Joode,
Delft University of Technology, The Netherlands

ABSTRACT

Open-source software (OSS) offers programmers the opportunity to elaborate and adapt source code. It is an opportunity to diverge. We would therefore expect incompatible strains of software to develop, and, consequently, a demand for standardization to arise. However, this is only partly the case. In this chapter we explore which other coordinative mechanisms are at work next to committee standardization. We identify four other categories of coordinative mechanisms and illustrate their relevance in OSS development. They complement committee standardization, can be used in standardization, and are sometimes an alternative to standardization.

INTRODUCTION

"What if, when the software doesn't work or isn't powerful enough, you can have it improved or even fix it yourself? What if the software was still maintained even if the company that produced it went out of business?" (Perens, 1999, p. 171)

Perens (1999) refers to the possibility to adapt and improve open-source software (OSS) to specific user needs. Open-source code[1] means that anyone with programming skills can understand how the software works. Moreover, the user licenses that govern OSS explicitly allow modification and redistribution of the modified source code. Most OSS are freely available, and they are easily accessible on and downloadable from the Web sites of open-source communities. These communities primarily consist of hobbyists. Many of them are sizeable. For example, the Apache community roughly comprises 630 contributors of which about 90 belong to the core developer group (http://www.apache.org/~jim/committers.html). In the communities, a variety of mature software programs is developed and maintained (e.g., Linux and Apache).

In effect, anyone with access to the Internet can modify the source code and start a new branch thereof. To estimate the number of people that might do so, let us imagine that 1,000 people download a software program—a much too modest figure.[2] When assuming that 10% has programmer skills, of which again 10% actually feels the need or urge to modify software (e.g., they found a bug), this still amounts to at least 10 modifications to the original software (van Wendel de Joode, de Bruijn, & van Eeten, 2003).

There is no entrance fee for participating in OSS development. In most cases, one does not even need to register to download software and modify its source code. Low entry barriers ease the emergence of diversity in OSS (Franck & Jungwirth, 2003; Sharma et al., 2002). The absence of any significant entry barrier—apart from engineering expertise—also explains the variety among those actively involved in OSS development: "[T]here are many different types of contributors to open-source projects: organizations and individual initiators and those who help with other people's projects; hobbyists and professionals" (Markus, Manville, & Agres, 2000, p. 15).

The different types of involvement and diverse backgrounds imply different requirements with regard to the software being developed. With both motive and opportunity to modify code, one would expect many branches (forks) and variants in open-source software to emerge. Such diversity more likely leads to incompatibility and fragmentation.

As Kogut and Metiu (2001, p. 257) note, "[t]he danger of open-source development is the potential for fragmenting the design into competing versions." A classic example of fragmentation is UNIX, a multiuser operating system and open-source initiative avant la lettre. It was a de facto standard in the late 1970s. Different UNIX variants developed, which fragmented the market.

An important means to counter diversifying fragmentation and the incompatibility that this brings is standardization (e.g., Linux Standards Base [LSB]; Egyedi & van Wendel de Joode, 2003). However, standardization alone does not explain the relative coherence in certain areas of OSS where an OSS has acquired a dominant market share (e.g., 90% of all e-mail programs are based on Sendmail, and Apache has a 65% share in the Web server market[3]) or a very promising position in the market (e.g., the Linux kernel is supported by a number of large and highly respected software houses such as IBM). Divergence and potential incompatibility also seem to be tempered by other forms of coordination, which brings us to our main research questions.

Which coordination mechanisms next to standardization are at work in OSS development? What is their significance for standards policy?

The chapter is organized as follows. To start with, coordination is a theme that standards literature discusses in some detail. A scan thereof points to four categories of coordination mechanisms next to committee standardization. These categories help to identify coordination mechanisms in OSS development, which is the core of this chapter. Our focus is on the Apache and the Linux case. The two cases highlight a number of interesting similarities and differences. We address their possible contribution to committee standardization in the closing section.

The empirical data from which we draw has been gathered by Ruben van Wendel de Joode. His data is based among other things on content analyses of Web sites and mailing lists, and, in particular, face-to-face (semistructured) interviews with OSS experts, 47 in all, from the Netherlands, Germany, and the United States. Most interviews took place from October 2001 to August 2002.

THEORIES OF COORDINATION

Some form of coordination is required to maintain consistency in software development in order to safeguard the internal compatibility of software. This can be done by preventing or governing divergence, or by achieving technical convergence retrospectively. Standardization literature addresses several such coordination mechanisms. A number of classic studies stem from economic literature. Therein two types of coordination mechanisms are compared (e.g., Farrell & Saloner, 1988; Katz & Shapiro, 1985): committee standardization and coordination by the market. Those who view committee standardization as a market process will feel somewhat uncomfortable with this distinction. There are other objections as well as we will discuss further on. However, we use the distinction here to organize some main theoretical concepts in a recognizable way. The concepts are later used as a stepping stone toward identifying coordination mechanisms other than committee standardization in OSS development.

Committee Standardization

The need for technical coordination follows from the interrelatedness of components and products, that is, from the network character of many technical artifacts (Schmidt & Werle, 1991). The more complex the system is in terms of different components, the more in number and the more diverse the actors involved. The higher the number of actor dependencies, the higher the need to coordinate market strategies and define common technical specifications for the development of a technical field. Committee standards are such specifications. They serve to create technical compatibility among products and services, and direct the actions and orientation of other market players. Standards "help [in] coordinating the actions of those who design, produce or use technical components as parts of more comprehensive technical systems" (Schmidt & Werle, p. 6).

Standards efforts in OSS (e.g., Unix and Linux) indicate that standardization is very relevant to OSS development. However, our research question focuses on coordination mechanisms in OSS that complement committee standardization. These more implicit mechanisms are discussed below.

Market Coordination

Standardization literature frequently refers to two concepts related to market coordination: the bandwagon mechanism and positive network externalities.[4] Farrell & Saloner (1988, p. 236) describe the bandwagon mechanism as follows: "If an important agent makes a unilateral public commitment to one standard, others then know that if they follow that lead, they will be compatible at least with the first mover, and plausibly also with later movers."

The bandwagon mechanism is a descriptive concept. It captures a leadership-induced response by the market in situations where a technical alternative exists. Technical convergence (i.e., reduced divergence) is a by-product of this response. The notion is also relevant for OSS in situations where a fork in OSS development arises.[5]

There is still little insight into how the OSS market works. Prognoses are uncertain. In these situations, marketing and sociocognitive elements such as consumer expectations and software-developer reputation become all the more influential in coordinating the OSS market. They trigger what Katz and Shapiro (1985) refer to as the psychological bandwagon mechanism. Reputation and expectation management can sway the market toward one or the other OSS alternative.

The notion of positive network externalities (or consumption externalities) is of an analytic kind. Katz and Shapiro (1985, p. 424) circumscribe it as follows: "There are many products for which the utility that a user derives from consumption of the good increases with the number of other agents consuming the good."[6] They identify three main sources of network externalities:
• the direct physical effect of the number of purchasers on the quality of the product (e.g., the number of connected households increases the value of a connection to the telephone network),

- indirect effects (e.g., if more computers are sold of a certain kind, more software will be developed for this computer), and
- the quality and availability of a postpurchase service for the good depend on the experience and size of the service network, which may in turn vary with the number of units of the good that have been sold (e.g., the service network of a popular car brand is denser and possibly better since the same set of servicing skills are needed).[7]

The notion of network externalities provides a rationale for coordinative responses, and, in extremis, for the emergence of monopolies (McGowan & Lemley, 1998). It provides a rationale for these responses but is itself not a mechanism of coordination. For, although the notion is always treated under the headings of the market, network externalities explain committee standardization as well as other market-based responses to coordination needs. To illustrate this, in complex markets, the externalities of compatibility are inherently high. Where this is the case, network externalities create a pull toward compatibility and trigger coordination mechanisms.

Are the concepts of the bandwagon mechanism, reputation and expectation management, and network-externalities-induced coordination mechanisms applicable to OSS development? The open-source movement is a network rather than a market. Wegberg & Berends (2000a, 2000b) argue that, like a market, a network is a way to cooperate and coordinate activities. However, in markets a sharp distinction exists between contributors (suppliers) and users (buyers), whereas it is characteristic of networks that all participants make a contribution (Wegberg & Berends, 2000a, 2000b).[8] Because of this difference, other coordination mechanisms may be relevant in OSS settings besides those discussed above.

For these other mechanisms, we turn to Egyedi (2000). Under the headings of compatibility-enhancing strategies, Egyedi discusses a set of coordinative measures that Sun Microsystems took with regard to the development of the Java programming environment. These measures are listed in Table 1. Some are coercive, that is, their coordinative effect depends on the implicit guidance that they provide and on voluntary compliance to them (e.g., instructional books and standardization, respectively). Other measures more forcefully impose coordination. Typically, coordination is then enforced by administrative requirements (e.g., membership registration) and legal means (e.g., licenses and contracts). In Sun's case, measures such as licensing, test suites, and formalizing the Java community process were used to prevent divergence, and coordinate and safeguard the compatible development of Java.

Regulatory Coordination

Reviewing the above measures, a subset can be captured by the term regulatory coordination thereby government regulation and company rules

Table 1. Coordinative strategies used in respect to Java development (Egyedi, 2000)

Coercive Strategies	Forceful Strategies
Open-Source Code	Community Process
Instructional Books	Participation Agreement
Certified Training Programs	Technology License and Distribution Agreement
Distribution of the Software-Development Kit (SDK)	Community Source-Licensing Model
ANSI/JTC1 Standardization	Reference Implementations
JTC1 PAS Process	Test Suites
ECMA/JTC1 Fast-Track Procedure	Compatibility Logo

enforce coordinative behavior. Contracts, licenses, intellectual property rights (IPRs), and participation agreements (e.g., membership registration) are typically used in this manner.

For example, the Technology License and Distribution Agreement for Java (March 1996) specifies how Microsoft may use the Java technology. It states, for example, that Microsoft's products are to incorporate the latest updates of the Java platform, and that Microsoft may use the Java-compatible logo for marketing purposes if its products pass Sun's test suites (free of costs). If Microsoft does not comply, among other things, it has to pay a fine.

Operational Coordination Mechanisms

Another subset of coordinative strategies we refer to as operational coordination mechanisms because they address operational aspects of technology development and use. Coordination, here, is technology specific. The table mentions instructional books on Java, certified Java training programs, reference implementations, and testing and distribution of the Java software-development kit.

Briefly elaborating, the instructional books that Sun and other authors write on Java, and the training programs that lead to certified Java developers contribute to a coordinated development of the Java trajectory. Those who follow the training program will learn the tricks of the trade (heuristics), are introduced to the Java philosophy, and so forth. A shared outlook increases the likelihood of a coordinated outcome, although it is no guarantee. In this respect, a more effective coordinative strategy is the distribution of the Java SDK. The SDK contains a full suite of development tools. SDK-based programs run on a standard Java platform.

Coordination by Authority

Farrell and Saloner (1988) mention a third category of coordination mechanisms: coordination by authority.[9] Coordination by authority is at stake "if one player is nominated in advance to choose [among alternatives], and others then defer to him" (e.g., gatekeepers; Farrell & Saloner, p. 250). We broaden their definition to include coordination based on informal authority. The source of (informal) authority can reside in technical expertise, economic success, a formal position in the organization, and so forth.

For example, when Sun first formalized the Java community process in The Java Community Process Program Manual, the formal procedures for using the Java specification-development process (December 1998), it was heavily criticized because in the proposed procedures, Sun would still veto decisions regarding the eligibility of specification requests. It was perceived to be a gated community process.

In summary, we discern four main categories of coordination next to committee standardization:[10]

- market coordination,
- regulatory coordination,
- operational coordination, and
- coordination by authority.

They help identify more specific mechanisms of coordination in open-source communities.

COORDINATION MECHANISMS IN OSS COMMUNITIES

In the following, we examine the mechanisms of coordination in the development of Apache and Linux. Apache is a Web server, the primary task of which is to host content, like Web sites and databases. Rob McCool laid the foundation for the Apache Web server, work that was later elaborated by a small—but over time increasingly larger—group of geographically dispersed people. Currently, the Apache Web server hosts more than 65% of the active Internet Web sites.

Linux is an operating system. Linus Torvalds started the development of the Linux kernel in 1991. By placing the source code on the Internet, Torvalds was able to attract and involve many people. Together they created an operating system that is said to be the only real threat to the hegemony of Microsoft Windows (Wayner, 2000).

Bandwagon Mechanism

Popular open-source projects tend to become even more popular. As one of the respondents remarks: "It is like a chain reaction, popularity will lead to more users, more users will lead to more developers, to more applications and thus to more users, and so forth." At stake is a self-reinforcing process similar to that of the bandwagon effect. Below we discuss two causes thereof: the reputation of OSS developers and the level of activity of OSS projects.

First, a main reason to get involved in OSS is the enhanced reputation that one receives when making high-level contributions to an open-source project (e.g., Benkler, 2002; Ljungberg, 2000; Markus et al., 2000; McGowan, 2001; O'Mahony, 2003; Selz, 1999; von Hippel, 2001). One effect of reputation is that to earn a good reputation, developers will choose open-source projects that provide them with a large audience (Lerner & Tirole, 2002). The involvement of other reputable developers will, in turn, attract more developers to that community. According to the former project leader of the Debian community, "Reputation does help: when you see a good name, you will take a look." Obviously, a community that consists of highly skilled developers is more likely to produce software that is of high quality.

A second mechanism that reinforces the bandwagon effect is the level of activity of the OSS project. Many OSS Web sites list statistics that indicate how active the different projects are. One such site is Freshmeat. It contains a so-called popularity index that is based on the number of subscriptions, URL (uniform resource locator) hits, and record hits of an OSS project (http://software.freshmeat.net). Another site is SourceForge (http://sourceforge.net). SourceForge hosts a large number of open-source projects. The site uses a number of statistics to measure the activity in projects (e.g., the number of downloads). These statistics affect people's choices, as one respondent explains: "If, for the same purpose, there are two different versions of a library to choose from, and one has been downloaded 10,000 times and another 20 times, it is clear which one you choose first."

Thus, the level of activity can be a criterion for developers to get involved in a certain project. This leads to positive feedback: The new developer will also download software and by doing so will automatically increase the level of activity. Developers may choose projects that have a higher level of activity because it implies a quicker response to the questions asked, because problems are likely to be discovered and fixed sooner, and because it suggests higher quality software.

Regulatory Coordination Mechanisms

Legislation is an important source of coordination in open-source communities. The coordinative effects of intellectual property rights and contractual law are particularly relevant, as will be discussed in detail below. Regarding IPRs, copyrights are important because they are the core of open-source licenses (e.g., Bonaccorsi & Rossi, 2003; Dalle & Jullien, 2003; Markus et al., 2000; McGowan, 2002; Perens, 1999; Wayner, 2000), while trademarks are relevant because most open-source communities have trademarked the names of their software.

Regarding contractual law, a number of communities have contracts that newcomers have to sign before they are allowed to add source code directly in the CVS (concurrent versioning systems).

As noted earlier, there are a large number of different open-source licenses. The Open Source Initiative hosts a Web site where they collect and provide copies of licenses they consider to be open-source licenses (http://www.opensource.org/licenses/). Currently they list 43 licenses. These have in common the fact that the source code is open and that others are allowed to change the code. Therefore, each of these licenses essentially facilitates further divergence. Still, certain licenses do so more than others. To illustrate this we analyze two of the most popular licenses in more detail: the General Public License (GPL) and the Berkeley Software Distribution (BSD) license.[11] Most other open-source licenses are elaborations of one of them (van Wendel de Joode et al., 2003).

The two licenses are compared on three aspects in Table 2. The first aspect is whether or not the licensed software can be mixed with closed-source software. As the table shows, software licensed under the GPL cannot be mixed with non-OSS. The license explicitly states that software that is mixed with GPL software becomes licensed under the GPL as well.[13] Software licensed under the BSD can be mixed with non-OSS.

The second licensing aspect of interest is whether OSS can become a proprietary distribution, that is, whether software is permitted to be distributed as closed-source code. As Table 2 indicates, GPL software cannot be distributed in a proprietary version, while software under the BSD license can.

The third characteristic listed above is whether relicensing is allowed. Both licenses forbid others to relicense the software.

How do these licensing features affect diversity in OSS development? The GPL tends to reduce divergence more than other licenses do. According to Perens (1999, p. 185), "If you want to get the source code for modifications back from people who make them, apply a license that mandates this. The GPL... would be [a] good choice(s)." However, the GPL does not require developers to send back their modifications to the original author of the software. It only obliges those who distribute modified GPL software to disclose the source code. Still, the GPL does stimulate some level of convergence since others have the

Table 2. A summary of two open-source licenses (Perens, 1999)[12]

License	Can be mixed with non-OSS	Proprietary distribution is possible	Can be relicensed by anyone
GPL	No	No	No
BSD	Yes	Yes	No

possibility to integrate modifications into the original software. This is different from the BSD license. Software licensed under the BSD can be taken privately, which means that the modified source code does not need to be made available. Therefore, software licensed under the BSD is more prone to divergence.

Trademarks are symbols related to a specific product. They are used to differentiate the product from competing products. For example, Linux, Apache, and Debian have trademarked their names for this purpose (O'Mahony, 2003). On a number of occasions, the Apache Software Foundation (ASF) had to protect its trademark against others who called their product, for example, WinApache.[14] Trademarks diminish divergence for two reasons. First, well-known trademarks like Apache and Linux attract new developers, thereby reinforcing the bandwagon effect. Second, trademarks make it less likely that developers create a different version of, for instance, Apache. Since the Apache trademark is well known, it will be difficult to start a project based on the Apache software that achieves the same level of popularity using a name that in no way resembles that of Apache.

Apache requires that developers sign a participation agreement before they may contribute directly to existing Apache source code.[15] The primary reason is to prevent them from contributing software that includes material copyrighted by a company. The coordinating effect of having people sign a participation agreement can go both ways. The agreement works as a boundary: People have to make an effort before they can become a part of the Apache community. If people are not willing to make this effort, the agreement may increase divergence. Although, they can still download the software and modify it, they cannot upload their modifications themselves. They may make their modified version available on another Web site and by doing so create divergence.

However, for those who have crossed the boundary, boundaries increase the chances that developers get to know each other, which will reduce uncertainty and has a coordinating effect (e.g., Ostrom, 1990). As a member of the community, adherence to one software version is more likely.

Operational Coordinating Mechanisms

The third set of coordinating mechanisms discussed here are the operational coordinating mechanisms. Developers in open-source communities use several tools to support their activities. These tools focus and structure their work. The interviews indicate that there are many tools. In this chapter we highlight four examples of operational coordination mechanisms, namely the use of CVSs, manuals, to-do lists, and orphanages.

The CVS is a client-server repository. It is an automated system that allows remote access to source code and enables multiple developers to work on the same version of the source code simultaneously (http://www.cvshome.org). A board member of the ASF describes the CVS-based software-development process as follows.

A typical software-development process in CVS starts with a first version of the software. Developers in the community download this version and try to understand how it works. Once they understand the software, they are likely to have comments, suggestions, and new ideas. Some of them will write source code and add it to the existing source code in the CVS. Other developers will test the modifications, and may comment on or even remove the newly added source code if they disagree with it. Obviously, opinions among developers will differ about whether or not the source code is good enough. If they cannot reach agreement, some of them may decide to take the prior version from the CVS and modify that version. In this manner, two different versions are created. At a later stage, the community will choose which application is to be taken as a starting point for further development.

The CVS can be said to increase coordination because it allows a restricted degree of divergence. Those who disagree with certain modifications are challenged to prove that their own suggestions are better—within the confinement of the community. The community then decides whose adaptation is best. Without the option of in-house divergence, dissidents would be forced to prove their point outside of the community, which could lead to a new, competing version of the software. By allowing divergence, the CVS keeps the community together thereby easing convergence.

Another operational coordinating mechanism is the use of manuals. The Debian community, for example, which is an open-source community that maintains the open-source Debian distribution,[16] has several manuals (e.g., a policy manual and a package manual). They all aim to support the coordinated development of code. One maintainer in the Debian community explains:

The policy manual enforces...integration in Debian. [It indicates] the way in which we put packages together. These are very simple rules saying that we will comply with the Linux file system hierarchy standard...Next to the policy manual there is a package manual that provides some technical details. For instance if you have binaries, which require one or more libraries then how do I make sure that the library is indeed installed on the computer? [It takes care of] dependencies between packages.

One of the things the manuals prescribe is that each maintainer must classify the software. There are three classifications, namely, *provides*, *depends*, and *conflicts*. The classification *provides* should be used to explain what the software is supposed to do, for example, Netscape provides Internet browser functionality. The classification *depends* is used to explain which other packages are needed to make the software work. The classification *conflicts* indicates that the software will not work properly if a certain package is already installed on the computer. According to a respondent, "In this system you assign relationships and that is what makes the system as a whole work." In other words, it institutionalizes coordination.[17]

A third mechanism is the to-do list. Many developers have new and innovative ideas; however, sometimes they lack the time to actually convert ideas into code. For this purpose, so-called to-do lists are created. As the name implies, a to-do list is a list or inventory of things that one or more people consider to be important or want to have. A community that has adopted a very structured to-do-list approach is the PostgreSQL community. Typically, an idea is transferred to this list when someone has sent "an e-mail with a suggestion."[18] The Apache community also has such lists, but they appear to follow a less systematic approach.

The to-do list works as a coordinative mechanism because it signals to developers what issues other members of the community find important. Participants do not have to discuss and explain to others why they think so. They can just put the item on the to-do list. Others can judge for themselves what issues they find interesting and want to work on. In a way, the to-do list serves as a marketplace in which demand, for example, a certain software feature, meets supply, namely developers with the knowledge, time, and motivation to develop the feature.

A fourth mechanism is the so-called orphanage. In most OSS communities, software projects are headed by a project leader or a maintainer. This person is responsible for collecting changes and integrating them in the software. What happens when such a person is no longer interested in maintaining the software? In the Debian OSS community, members have created an orphanage.[19] The orphanage lists the projects that have been abandoned and are in need of a new maintainer. Every project in the orphanage has two links. The first link is to the official message in which either (a) the maintainer indicates that he or she wants to cease his or her activities, or (b) someone else reports that the maintainer is missing in action.[20] The second link is to (a) the most recent stable version of the software, (b) the latest unstable development version of the software, and (c) the latest testing release of the software. The links help to find the software and to take over its maintenance. By a number of fairly simple actions, the orphan is removed from the Web site and the new maintainer can start his or her work.

The orphanage has a coordinative effect since the likelihood that existing projects are continued increases and the likelihood of the creation of rival software that leads to divergence decreases. The orphanage eases the transfer between maintainers. The orphanage approach has, however, not been adopted by either the Apache or the Linux communities.

Coordination by Authority

There are several examples of coordination by authority in open-source communities. Two main ones are addressed and illustrated here, that is, the coordinative effect of distributions (e.g., Debian) and the role of gatekeepers in OSS communities (i.e., Linux and Apache).

The Debian distribution has an important converging effect. As a respondent points out, "[t]here are only a few users who choose among software versions...Usually, the decisions are made by the distributions." Debian and companies like Red Hat and SuSE, who develop, maintain, and ship commercial distributions of the Linux operating system and a selection of applications, decide which applications and which version of the Linux kernel to include in their distribution. These three parties have a considerable market share, which lends authority to their distributions. If all three decide to include a certain application among competing ones, the chosen one is likely to become the most popular one.

There is some controversy about Linus Torvalds' role in the Linux community. Some people argue that he clearly has leadership (e.g., Lerner & Tirole, 2002). Others argue the opposite. They view his role as a rather limited one, one with no formal authority over the Linux community (e.g., van Wendel de Joode et al., 2003). Either way, his presence stimulates convergence. He maintains the Linux kernel and is the only person with the prerogative to include additional pieces of source code. In principle, people can ignore his decisions and build another version of the Linux kernel. However, most developers respect Linus Torvalds and his decisions, and use his version. This obviously creates a lot of convergence.[21]

The structure of the Apache community differs from that of the Linux community. Apache developers work in a decentralized setting. They can directly contribute source code into the CVS—if they have signed the participation agreement. However, bodies within Apache like the Apache Software Foundation and the project management committees (PMCs) theoretically have the authority and power to impose decisions on the community. Whether they actually exercise their power or not, the members of these bodies are well known and highly visible, and will be listened to more carefully than other Apache developers. As such, they are the gatekeepers of the Apache community.

Comparing the two communities, Linux and Apache only partly apply the same coordination mechanisms (Table 3). The list of features of each does not present a clear picture of a more coordinated and a less coordinated OSS project. However, the Apache community could be argued to use more elaborate and influential internal coordination mechanisms than Linux does (e.g., participation agreements and CVSs). Well-institutionalized mechanisms for intracommunity coordination can largely prevent unnecessary divergence.

One hypothesis is that communities in which such intracommunity mechanisms are lacking, external coordination mechanisms must compensate for this lack. External mechanisms like standards initiatives need to be invoked to reduce

Table 3. Linux and Apache compared

Coordination Mechanisms[1]	Apache	Linux
Licenses	Apache Software License (BSD-like)	GPL
Trademarks	Yes	Yes
Participation Agreements	Yes	No
CVSs	Yes	No[2]
Manuals	Coding Style Guide	Coding Style Guide
To-Do lists	Yes	No
Orphanages	No	No
Gatekeepers	ASF & PMC	Linus Torvalds

[1] *The mechanisms' reputation, level of activity, and distribution have not been included in this table because they cannot be ascribed to one particular community. Reputation, for instance, is tied to individuals rather than communities. A distribution can include Linux or Apache, but this is also not necessarily tied to a particular community. Finally, levels of activity are used by outsiders to compare communities (projects).*
[2] *Currently, there is controversy in the Linux community about whether or not to adopt BitKeeper (http://kt.zork.net/kernel-traffic/).*

divergence ex post. The standardization effort of the Linux Standards Base illustrates the argument, an effort that seems to be less necessary for Apache.

CONCLUSION

The tension between the incentive to change open-source code (divergence) and the need for compatible open-source software (convergence) can be addressed by different coordination mechanisms. We identified four categories of coordination mechanisms next to committee standardization, that is, market coordination, regulatory coordination, operational coordination, and coordination by authority, and we have discussed their workings. These mechanisms complement standardization, can partly be applied by standardizers, and partly form an alternative to standardization.

- **Complementary to Standardization:** Coordinative efforts other than standardization complement and support standardization. Stricter coordination within OS communities (internal coordination) appears to make committee standardization, which is a form of external coordination, superfluous. Formulated differently, the more open the OSS development process, the more likely standardization will be required to recreate compatibility at a later stage.
- **Useful in Standardization:** Not all the mechanisms discussed are new to standardization. Sometimes similar mechanisms are at work within stan-

dardization (e.g., formal and informal gatekeepers, rules on IPR, and psychological bandwagon effects). For example, the essence of CVS is comparable to the approach used for Internet standardization (IETF) where the procedures allow for the development of competing alternatives (i.e., in the expectation that the market will select the most useful one). Further research is required to determine whether the standards setting could also benefit by using one or more of the other coordinative mechanisms. One line of investigation would be whether it is possible to actively invoke the help of other coordination mechanisms in order to ease negotiations during standards development. That is, should standards bodies more actively encourage prestandardization? Should their policy aim for fortification of these other coordination mechanisms?

- **Alternative to Standardization:** The coordination mechanisms explored in this chapter are partly also an alternative to standardization. This raises the question of whether, in view of achieving compatibility, it matters which coordination mechanisms are used. For example, is standardization an equally desirable mechanism of coordination as, for example, the CVS? On what situational aspects would the answer to this question depend?

In this chapter we focused on OSS communities. These communities share many characteristics with network communities. As noted earlier, networks differ from markets. In networks, all participants make a contribution, whereas in markets, contributors and users are distinct groups. Can our insights be generalized to markets? Or does the difference between markets and networks mean that other coordination mechanisms are at stake? What are the theoretical implications? We recommend further research along this line of enquiry.

ACKNOWLEDGMENTS

This chapter is a revised version of Egyedi and van Wendel de Joode (2004), which in turn was a reworking of Egyedi and van Wendel de Joode (2003). We would like to express our gratitude for the insightful comments on the latter article made by three anonymous reviewers of and participants to the Third IEEE Conference on Standardization and Innovation in IT (SIIT2003).

REFERENCES

Benkler, Y. (2002). Coase's penguin, or, Linux and the nature of the firm. *Yale Law Journal, 4*(3), 396-446.
Bonaccorsi, A., & Rossi, C. (2003). Why open source software can succeed. *Research Policy, 32*(7), 1243-1258.

Broersma, M. (2002, July 2). *Linux standard gets the go-ahead.* Zdnet.com. Retrieved from http://techupdate.zdnet.com/techupdate/stories/main/0,14179,2873230,00.html

Dalle, J. -M., & Jullien, N. (2003). *"Libre" software: Turning fads into institutions* [Working paper]. Retrieved from http://opensource.mit.edu/papers/dalle2.pdf

Edwards, K. (2001). *Epistemic communities, situated learning and open source software development.* Paper presented at the Epistemic Cultures and the Practice of Interdisciplinary Workshop, NTNU, Trondheim.

Egyedi, T. M. (2000). Compatibility strategies in licensing, open sourcing and standardization: The case of JavaTM. *Fifth Helsinki Workshop on Standardization and Networks* (pp. 5-34).

Egyedi, T. (2002a). Strategies for de facto compatibility: Standardization, proprietary and open source approaches to Java. *Knowledge, Technology, and Policy, 14*(2), 113-128.

Egyedi, T. (2002b). *Tension between OSS-divergence and user needs for interoperability.* Contribution to the Workshop Advancing the Research Agenda on Free/Open Source Software, Brussels, Belgium. Retrieved from http://www.infonomics.nl/FLOSS/workshop/papers.htm

Egyedi, T. M. (2002c). *Trendrapport standaardisatie: Oplossingsrichtingen voor problemen van IT-interoperabiliteit.* Delft, The Netherlands: Ministerie van Verkeer en Waterstaat, Rijkswaterstaat/ Meetkundige Dienst.

Egyedi, T. M., & Hudson, J. (2001). Maintaining the integrity of standards: The Java case. *Standards Compatibility and Infrastructure Development: Proceedings of the Sixth EURAS Workshop* (pp. 83-98).

Egyedi, T. M., & van Wendel de Joode, R. (2003). Standards and coordination in open source software. *SIIT 2003 Proceedings: Third IEEE Conference on Standardization and Innovation in Information Technology* (pp. 85-98).

Egyedi, T. M., & van Wendel de Joode, R. (2004). Standardization and other coordination mechanisms in open source software. *International Journal of IT Standards & Standardization Research, 2*(2), 1-17.

Farrell, J., & Saloner, G. (1988). Coordination through committees and markets. *RAND Journal of Economics, 29*(2), 235-252.

Franck, E., & Jungwirth, C. (2003). Reconciling rent-seekers and donators: The governance structure of open source. *Journal of Management and Governance, 7*(4), 401-421.

Katz, M., & Shapiro, C. (1985). Network externalities, competition and compatibility. *American Economic Review, 75*(3), 424-440.

Kogut, B., & Metiu, A. (2003). Open-source software development and distributed innovation. *Oxford Review of Economic Policy, 17*(2), 248-264.

Lerner, J., & Tirole, J. (2002). Some simple economics of open source. *Journal of Industrial Economics, 50*(2), 197-234.

Liebowitz, S. J., & Margolis, S. E. (1998). *Network externalities (effects).* Retrieved from http://www.utdallas.edu/~liebowit/palgrave/network.html

Ljungberg, J. (2000). Open source movements as a model for organising. *European Journal of Information Systems, 9*(4), 208-216.

Markus, M. L., Manville, B., & Agres, C. E. (2000). What makes a virtual organization work? *Sloan Management Review, 42*(1), 13-26.

McGowan, D. (2001). Legal implications of open-source software. *University of Illinois Review, 241*(1), 241-304.

McGowan, D. (2002). Developing a research agenda for open-source licensing practices. *Proceedings of the NSF Workshop on Open-Source Software,* Arlington, VA.

McGowan, D., & Lemley, M. A. (1998). Could Java change everything? The competitive propriety of a proprietary standard. *Antitrust Bulletin, 43,* 715-773.

O'Mahony, S. C. (2003). How community managed software projects protect their work. *Research Policy, 32*(7), 1179-1198.

Ostrom, E. (1990). *Governing the commons: The evolution of institutions for collective action, the political economy of institutions and decisions.* Cambridge, MA: Cambridge University Press.

Perens, B. (1999). The open source definition. In C. DiBona, S. Ockman, & M. Stone (Eds.), *Open sources voices from the open source revolution* (pp. 171-189). Sebastopol, CA: O'Reilly & Associates.

Schmidt, S. K., & Werle, R. (1991). *The development of compatibility standards in telecommunications: Conceptual framework and theoretical perspective.* Koeln, Germany: Max-Planck-Institut fuer Gesellschaftsforschung.

Schmidt, S. K., & Werle, R. (1992). *Technical discourse in international standardization.* Paper presented at 4S/EASST, Goteborg.

Schmidt, S. K., & Werle, R. (1998). *Co-ordinating technology: Studies in the international standardization of telecommunications.* Cambridge, MA: MIT Press.

Selz, D. (1999). *Value webs: Emerging forms of fluid and flexible organizations.* St. Gallen: University of St. Gallen.

Sharma, S., Sugumaran, V., & Rajagopalan, B. (2002). A framework for creating hybrid-open source software communities. *Information Systems Journal, 12*(1), 7-25.

Van Wendel de Joode, R., de Bruijn, J. A., & van Eeten, M. J. G. (2003). *Protecting the virtual commons: Self-organizing open source communities and innovative intellectual property regimes* (Information technology & law series). The Hague, the Netherlands: T. M. C. Asser Press.

Von Hippel, E. (2001). Innovation by user communities: Learning from open-source software. *Sloan Management Review, 42*(4), 82-86.

Wayner, P. (2000). *Free for all: How Linux and the free software movement undercut the high-tech titans.* New York: HarperBusiness.

Wegberg, M. van, & Berends, P. (2000a). *Competing communities of users and developers of computer software: Competition between open source software and commercial software.* NIBOR working paper, NIBOR/RM/00/001. Retrieved from http://edata.ub.unimaas.nl/www-edocs

Wegberg, M. van, & Berends, P. (2000b). The open source code movement: The Linux operating system. In S. Schruijer (Ed.), *Multi-organizational partnerships and cooperative strategy* (chap. 46, pp. 283-287). Tilburg, The Netherlands: Dutch University Press.

ENDNOTES

[1]   Source code is the human-readable part of software (Edwards, 2001).

[2]   The Freshmeat Web site (http://freshmeat.net) publishes a top-20 popularity index of OSS. If this index in any way reflects the number of downloads, then our example in the running text of 1,000 downloads is a very modest one.

[3]   http://news.netcraft.com/archives/2003/04/13/april_2003_web_server_survey.html

[4]   Game theory is frequently used to model coordination by committees and markets (e.g., the battle-of-the-sexes game, war of attrition, grab-the-dollar game; Farrell & Saloner, 1998). The assumptions needed to reduce the complexity of the coordination problem (e.g., that players have a choice between two incompatible new technologies or systems that are championed by different companies) make game theory less suitable for understanding OSS development. However, the motives mentioned in this literature for choosing one standard above others may very well be applicable to the choice between OSS software (e.g., technical superiority, vested interests, expectations, and company reputation).

[5]   According to the Jargon Files (http://jargon.watson-net.com/jargon.asp?w=fork), a fork is "[w]hat occurs when two (or more) versions of a software package's source code are being developed in parallel which once shared a common code base, and these multiple versions of the source code have irreconcilable differences between them."

[6]   Liebowitz and Margolis (1998, p. 1) specify positive consumption externalities by distinguishing two components to consumer value: the autarky value, which is "the value generated by the product even if there are no other users," and the synchronization value, the additional value "derived from being able to interact with other users of the product...[I]t is this latter value that is the essence of network effects."

7   Product information is also more easily available for popular brands (Katz & Shapiro, 1985).

8   Von Hippel (2001) argues that users are the primary developers of OSS.

9   This is similar to what Schmidt and Werle (1991) address as coordination by hierarchy.

10 The categories emphasize different features of coordination. They are largely, but not fully, mutually exclusive.

11 According to the Freshmeat statistics, the GPL is by far the most used license (i.e., 65% of the projects). The second most popular license is the Lesser GPL (LGPL), which differs only slightly from the GPL. The third most popular license is the BSD license, which is used in 5.3% of the projects analysed by Freshmeat (http://software.freshmeat.net/stats/).

12 The table is based on a more elaborated discussion of open-source licenses by Perens (1999).

13 This is the case when GPL and non-OSS are mixed to become one program. Only then will non-OSS become licensed under GPL. It is not entirely clear, however, what the definition is of one program.

14 This example is based on an interview with a board member of the Apache Software Foundation.

15 For example, http://jakarta.apache.org/site/agreement.html

16 The Debian distribution is based on the Linux kernel and many tools from the GNU project. It comes with more than 8,700 packages. It is the only significant noncommercial Linux distribution (http://www.debian.org/doc/manuals/project-history/ch-intro.htmlp.s1.1).

17 The Apache and the Linux communities do not have manuals, but they have coding style guides that indicate what the source code should look like, and, for example, how comments should be included. See http://www.apache.org/dev/styleguide.html and http://www.linuxjournal.com/article.php?sid=5780

18 This is cited from an interview with a member of the steering committee in the PostgreSQL community.

19 The term orphan was used by a Debian developer in one of the interviews and on the following Web site: http://www.debian.org/devel/wnpp/orphaned

20 For a description of the way a maintainer can be reported as missing in action, see the discussion at http://lists.debian.org/debian-qa/2002/02/msg00111.html

21 Linus Torvalds does not have any formal authority, which raises the conceptual question of what difference there is between the bandwagon mechanism and coordination by authority. In the latter case, the bandwagon mechanism is triggered by the (perceived?) authority of a person or an institution. Does this differ from the way the bandwagon mechanism works in other situations? Coordination by authority can be created and influ-

enced. This would appear to be more difficult for other situations where bandwagon mechanisms may arise, a difference that has direct implications for organizing coordination in a field of technology.

Chapter V

Beyond Consortia, Beyond Standardization?
Redefining the Consortium Problem

Tineke M. Egyedi, Delft University of Technology, The Netherlands

ABSTRACT

This chapter analyzes the rhetoric that surrounds the problem of consortia, that is, the supposed lack of democratic procedures. The social shaping of the standardization approach is applied. Two cases are used to illustrate what is at stake in consortium standardization, namely, the standardization of Java in ECMA, and XML in W3C. The findings show inaccuracies and inconsistencies in the way the consortium problem is defined: The dominant rhetoric underestimates the openness of most industry consortia and overestimates the practical implications of formal democratic procedures. This unbalanced portrayal and sustained ambiguity about what is meant by democracy are part of the meaning of negotiation at work in the actor network. Implicitly, the European network still predominantly defines standardization as an instrument of regulatory governance. This marginalizes the role of consortia. The chapter offers several suggestions to redefine the consortium problem.

INTRODUCTION

In Article 14 of the "Council Resolution of 28 October 1999 on the Role of Standardization in Europe" (Council of the European Union [EU], 2000), the European Commission is requested to examine how the European Union should deal with specifications that do not have the status of formal standards. The council recognizes "an increasing tendency of interested parties to elaborate technical specifications outside recognized standardization infrastructures" (p. 1, Article 7). The resolution distinguishes between standards developed by official standards bodies such as the International Organization for Standardization (ISO), those on the European level, for example, the Comité Européen de Normalisation (CEN), and those from other sources. In the resolution context, another main source is the standards consortium.[1] The council's feeling is, apparently, that there may be a need to deal differently with consortium standards than with formal ones, a feeling which, for example, the U.S. government shares (Center for Regulatory Effectiveness [CRE], 2000).

No accepted definition exists as yet for the term standards consortium. In practice, it can cover a variety of alliances. Some standards consortia focus solely on the development of technical standards or specifications (*specification groups*; Updegrove, 1995). These may be R & D- (research and development) oriented and precompetitive (*research consortia*; Updegrove; *proof-of-technology consortia*; Weiss & Cargill, 1992). They may focus on heightening the usability of existing standards (*implementation and application consortia*; Weiss & Cargill). Or, their goal may be to formalize dominant existing practices and de facto standards. Again, other consortia may foremost promote the adoption of a certain technology (*strategic consortia*; Updegrove), organize educational activities for users of standards (Hawkins, 1999), or combine these activities with specification development. In this chapter, a standards consortium refers to an alliance of companies and organizations financed by membership fees, the aims of which include developing publicly available, multiparty industry standards or technical specifications. In practice, mostly large companies are members of these consortia.

The council's resolution is but one example that there has been some unease and discussion about the role of standards consortia in the network of actors involved in standardization. The actor network appears to be caught up in a polarized discussion about what type of organization best serves the market for democratic and timely standards: standards consortia or the traditional formal standards bodies. The general feeling is that standards consortia work more effectively, but that they have restrictive membership rules and are undemocratic. The latter is a cause of concern for governments that require democratic accountability of the standards process if they are to refer to such standards in a regulatory context.

Much depends on the democratic rhetoric, as it is called in the following, for it is this rhetoric that largely determines who is part of the European actor network on standardization and who is not. Those who are in may have good arguments to emphasize the need for democratic standardization. However, after having gone uncontested for many decades, maybe it is time to question the obvious to avoid a situation where standards consortia are being marginalized for the wrong reasons.

In the following, the assumptions and arguments from which the current actor network largely draws its legitimacy are analyzed. The social shaping approach is used to do so. Two cases of consortium standardization are discussed to uncover possible inconsistencies and asymmetries in the reasoning applied. Indeed, in the last sections of this chapter, a redefinition of the consortium problem is argued for.

METHODOLOGY

This chapter uses data based on research funded by the European Commission. The latter's request was, in short, to gather up-to-date case material on consortium standardization, and to develop a perspective that offers new leads for policy. This required a reexamination of the standards consortium problem and of the underlying assumptions. Does the way the problem of standards consortia is defined in dominant rhetoric accurately describe what is at stake? An explorative, qualitative approach was applied. Apart from a study of the literature and content analysis of documents, two in-depth case studies took place: Java standardization in ECMA, an international industry association for standardizing information and communication systems, and standardization of the extensible markup language (XML) in the World Wide Web Consortium (W3C). The standards were selected because of their high profile and the sufficient amount of information this implied. Data were gathered foremost by means of participant observation (i.e., attending ECMA standard committee meetings), interviews with committee participants (face to face and by e-mail), and, with respect to W3C standardization, content analysis of electronic contributions to and e-mails about the standards process. The research resulted in the report Beyond Consortia, Beyond Standardization? New Case Material and Policy Threads (Egyedi, 2001a).

In this chapter the case findings are used to confirm or, when counterexamples arise, correct commonsense assumptions (i.e., falsify the dominant rhetoric, if you will). They serve to illustrate, develop, and argue a different line of reasoning. In the section that argues for a redefinition of the consortium problem, a participative social shaping approach is taken, one that interprets the research findings with the explicit aim of contributing to the debate.

CONCEPTUAL FRAMEWORK

The essence of the social shaping approach—or social constructivism as the body of theory is also called—is straightforward. It emphasizes that everyday knowledge and scientific knowledge are social constructions (Berger & Luckmann, 1966; Bloor, 1976). The sense of objectivity is based on a shared perception thereof. That is, facts, artifacts, and events can be interpreted in different ways. This is captured by the term *interpretative flexibility* (Collins, 1981): Their meanings thus become negotiable. Groups negotiate these meanings. In the process of acquiring consensus, the interpretative flexibility is reduced. This is called *closure* (Pinch & Bijker, 1984). Closure leads to the stabilization of a meaning. Callon (1980), the developer of the actor network theory, speaks of the dominant problem in an actor network when referring to this process.

Social constructivists inspired Bijker (1990) to develop the social construction of technology (SCOT) model. They also inspired the conceptual framework of the social shaping of standardization (Egyedi, 1996). The framework draws together and builds forth on notions from the sociology of knowledge and social constructivism (Berger & Luckmann, 1966; Bloor, 1976; Collins, 1981; Kuhn, 1970), social studies of technology theories (Bijker, 1990; Callon, 1980; Dosi, 1982; Pinch & Bijker, 1984), institutional theory (Jepperson, 1991; March & Olsen, 1989; Powell & DiMaggio, 1991), and theories from social psychology (Berkowitz, 1980).

Table 1 summarizes the main line of reasoning in the social shaping of standardization approach. For the purpose of this chapter, only part of the framework is relevant. In particular, understanding the last two columns of Table

Table 1. Main elements in the social shaping approach to standardization (Egyedi, 1996)

Social-Actor Characteristics	Standards Committee		Standards Organization	Actor Network on Standardization
Social Shaping by (Specific Actor)	interest groups	practitioner communities	institutional provisions	corporate actors, interest groups, etc.
1. Attribute	interests	paradigms	standards ideology	interests
Social Shaping Through (Process)	negotiating problem definition & standards solutions		structuring negotiation	Negotiating problem definition of standardization
Social Shaping of (Target)	standards		standards process	role of standardization
Role of Standards Organization	internally oriented (institutional context of committee)		mediates standards ideology via institutional provisions	externally oriented (as actor agency)

1 suffices to locate the consortium problem and contextualize actor-network issues that are analyzed further on. First, the defining concepts in Table 1 are explained. Next, the levels of social aggregation in standardization are separately discussed, in particular, the level of the organization and of the actor network.

Social Actors and Attributes

Social actors are the main unit of analysis in the framework. The term refers to a set of individuals (or organizations). Table 1 distinguishes three social actors: the standards committee (i.e., working group, etc.), the standards organization (i.e., standards body, consortium, etc.), and the actor network on standardization (i.e., consisting of standards organizations, interest groups, actor agencies such as the European Commission, World Trade Organization, etc.).

These social actors vary in what binds them together (*attributes*). That is, they differ with regard to what makes them social sources of influence in the field of standardization. Possible attributes include common interests, shared problems, shared beliefs, common activities, a common profession, and so forth. For example, a *practitioner community* is a group characterized by shared knowledge, heuristics, methods, and professional beliefs (Kuhn, 1970).

Social shaping occurs at all three levels relevant in standardization. However, the shaping process and the target of shaping differ. For example, at the standards-committee level, whether analyzed in terms of interest groups (the attribute is interest) or practitioner communities (the attribute is technological paradigm), groups negotiate the standards content (the target is the standard). At the level of the standards organization, however, the source of social shaping are the standards procedures. The values and beliefs embedded in the procedures (the attribute is standards ideology) structure the negotiation process (the target is the standards process).

In the following, the last two columns of Table 1, that is, the level of the standards organization and that of the actor network, are discussed in detail.

Standards Organization

Moving a decision process from one arena to another with different structural features changes its outcome (Besen, 1990; March & Olsen, 1989). Analogously, the institutional context of standards committees is highly relevant to the outcome of the standards process (Bonino & Spring, 1991; Genschel, 1993). Organizational procedures embed ideas on how standardization should proceed, beliefs on what is important in the process of establishing standards and why it is important, assumptions about the standards environment, and so forth. These shared ideas, assumptions, values, and beliefs are captured by the term *standardization ideology* (Egyedi, 1996). For example, the procedures of the formal standards bodies embed democratic values and reflect the desirability of a technically and politically neutral standards process. The ideal of decision making by consensus reflects serious consideration of minority viewpoints. Together, such features make up the tissue of the formal standards ideology. Once ideas are institutionalized, they acquire a taken-for-granted quality and are not easily dismissed or changed (e.g., March & Olsen, 1989). The institutionalized ideology of the formal standards bodies thus fosters continuity in their standardization approach. It indicates the role that the formal standards bodies aim to play, and clarifies the direction in which institutional provisions influence current standards work. Standards procedures regulate the committee process. Rules are laid down on how the negotiation of insights and interests should proceed. As the rules of a game affect its outcome, institutional provisions for formal standardization affect standards. The same reasoning applies, of course, to the institutional provisions of the standards consortia. In both cases, the social shaping process at stake is a delayed and mediated form of social construction. In the wordings of Table 1, the standards organization shapes the standards process by means of institutional structures and procedures. The latter mediate ideas and beliefs about how standardization should occur.

Actor Network on Standardization

An actor network is, loosely speaking, a society of organizations (Jepperson, 1991). To paraphrase Mulder (1992), in the field of standardization, it is the ensemble of relationships that serve the objective of developing, maintaining, and applying standards. It typically comprises standards organizations, interest groups (industry and users), and government agencies. It is a bounded network. Participants must acknowledge each other's roles in respect to standardization. Again paraphrasing Mulder, a relationship between actors exists if they agree on the roles that each of them will fulfill with regard to standardization. However, the relationship between network constituents need not be voluntary. Actors may unwillingly be forced to recognize other actors. They are interdependent for they cannot alter their role without renegotiating their relations with other participants in the network (Mulder, 1992).

There are no clear-cut network boundaries. Who is in and who is out is negotiable and is the cause of boundary skirmishes. Certain actors may occupy a pivotal position and have a disproportionate influence on how the network defines the dominant problem. Where significant changes in the network are handled through one actor, the actor is viewed as an *obligatory point of passage* (i.e., a kind of gatekeeper; Callon, 1986). For example, in the period leading up to the European common market in 1992, the European Commission played this role in regard to European standardization (Egyedi, 2000).

Actors negotiate the authoritative assignment of values and meanings (Albert de la Bruhèze, 1992). At the actor-network level, negotiations often concern the significance of standardization in respect to other political issues. This can be done, for example, by the purposeful problematization of an aspect of standardization or by the coupling of standardization to other established political priorities (e.g., a faltering regional industry). The meaning on which the actor network temporarily settles affects the position of the standards organizations in the network, and thereby the conditions under which formal and consortium standardization take place. Thus, actors negotiate about:

a. the problem of standardization (its role, i.e., its significance and purpose) and thereby the role of the actor network and

b. their position in the network and the network boundaries (who is included or excluded).

ANALYSIS OF THE PROBLEM DEFINITION

According to the dominant rhetoric, why are standards consortia a problem? The answer lies in the way governments sometimes use formal standards. To avoid frequent revision of European legislation in reaction to technical changes and standards developments, the European Union drew up the New Approach (European Commission, 1985). This approach entails that, where necessary, the European Union refers in legislation to the technical specifications developed and/or accredited by the European standards bodies. If for this purpose a new standard is needed, the government can mandate (fund) a standards body to develop one. In other words, if a reference to standards is needed for EU legislation or public procurement, then the standards referred to will preferably be formal standards, or *voluntary consensus standards* as they are also called. The latter are developed in institutions with a democratic decision culture, one that matches the political democratic context in which regulation is developed. A politically accountable, democratic development process is required. There is reciprocal dependence and, consequently, a strong alliance between the European Union and the European formal standards bodies. Together these two parties own the consortium problem.

From the point of view of the formal standards bodies, the problem is that standards consortia are rivals (CEN/ISSS, 2000). The reasoning is that they attract expertise from the group of large industrial companies, which often contributes significantly in terms of organizational support and technical contributions. Consortia can produce specifications quicker than the formal bodies do because they need not bother with a lengthy democratic and open process that demands consensual decision making, well-balanced representation of interest groups, and so forth (Rada, 2000). The CRE report (2000), a report that will be referred to more often because of its high significance for this chapter, explicitly mentions that consortia deliver nonconsensual specifications. They need not compromise on standards content as much as the formal standards bodies do. Because of their speed, time-sensitive technology development is addressed there. That is, they are a competitive but undemocratic standards environment (e.g., CRE).

At least, this is the widely shared view of those who compare the formal standards bodies with standards consortia.[2] If this view is correct, the government's problem with consortia would be that, although consortia address a very relevant area of technology in an efficient way, a politically correct alliance cannot be set up that allows reference to consortium standards. However, the problem as defined is not as obvious as may seem. Recent literature already indicates that some underlying assumptions can be questioned. For example, doubts are raised about the speed of consortium standardization (Hawkins, 1999; Krechmer, 2000); findings differ with respect to whether the rise of standards consortia has led to a decline of industry participation in formal standardization (Cargill, 2000; Hawkins). Moreover, the politically correct aims of the formal process are often not met, which makes a comparison difficult (Cargill, 1990).

In the following sections, the social shaping approach is used to analyze two cases of consortia. The findings partly support some of the doubts raised in standardization literature and raise new ones.

STANDARDS PROCEDURES OF ECMA AND W3C

Standards bodies influence committee processes by means of procedures and structures. They mediate a standards ideology. The ideology of the formal standards bodies has remained rather stable during the 20th century—apart from a slight shift during the last decade. For example, participation in the international standards process, which was and is still based on national membership (national standards bodies), has become a less principled issue. The direct membership of interest parties and companies (e.g., in the European Telecommunications Standards Institute [ETSI]) poses less of a problem.

But other ideological elements still apply or are striven for, such as the consensus principle; the voluntary application of standards; the quality of standards; the broad constituency of national delegations; democratic working methods by means of a well-balanced influence of national members in the management bodies of standards organizations and an open, democratic process of decision making; impartial, and politically and financially independent organizations and procedures; widely used, and thus in principle, international standards; fair competition and fair trade; openness in information; and a rational, technical discussion (Egyedi, 1996).

The dominant rhetoric on the consortium problem is usually not very specific about what is meant by democracy. Reading between the lines, the most pressing problems appear to concern the openness of membership procedures (i.e., the diversity of participants and the inclusion or exclusion of groups) and decisions-making procedures (i.e., consensus and minority treatment). In the following sections, these two categories of procedures are examined with regard to ECMA and W3C as they appear in writing and, subsequently, as they work out in practice. The ECMA, an international Europe-based industry association for standardizing information and communication systems, was founded in 1961 and is one of the oldest consortia. W3C was founded in 1994 and is one of the younger ones.

Membership and Decision Procedures

Membership. ECMA and W3C demand membership fees. The fee struc-ture roughly takes company profit into account: the more profit, the higher the fee. Small and medium-sized enterprises (SMEs) can participate, and so can governmental and educational institutions—although ECMA procedures are less explicit on this account. Thus, membership is in principle open. There are some practical and organizational exclusion mechanisms (fees, participation in com-mittees, etc.).

ECMA and W3C differ in respect to what the membership categories mean. Only ordinary ECMA members (i.e., fully paying 70,000 Swiss Franc in 2001) have a vote in the General Assembly (GA), whereas W3C bestows no extra voting benefits on full members ($50,000 per year).

Decision Procedures. The overall decision structure of ECMA and W3C differs strongly. In ECMA, the ultimate power lies in the hands of the General Assembly (inclusive), in which only ordinary members have a vote (exclusive). In other words, there is full democracy among members that pay the ordinary membership fee (large companies). In W3C, the ultimate power lies in the hands of the director, who formally has the role of a benevolent dictator. There is no democracy at this level. Indeed, if it is necessary to move on, the chair of W3C technical committees (TCs) also has far-going powers. However, the W3C standards process is consensus oriented, strives for a vendor-neutral solution, and is an open process, according to the procedures. Some procedures give room to minority standpoints. W3C's mix of these democratic ideological features with dictatorship leads to interesting procedural directions (W3C, 2001). ECMA procedures are internally, and ideologically, consistent. They are consensus oriented and give room to minority standpoints as well.

Whereas W3C's director appoints committee chairs, ECMA committee members elect the chair. If there is a deadlock in an ECMA TC, voting takes place by a simple majority of the members present at the meeting. In both W3C and ECMA, each member may assign several representatives to participate in the standards work, but the company only has one vote.

All in all, these two cases indicate that the procedures of (some) consortia allow them to be reigned in an autocratic manner. But at the committee level, they largely share the democratic values of the formal standards ideology and strive for consensus, address minority viewpoints, and so forth.

Effectuation of Procedures

Consortia. Consortia are valued by some for their lack of bureaucracy and for the high pace of standards development in which this leads to. Others criticize them for lacking elaborate procedures (e.g., intellectual property rights), or conversely, for adopting the bureaucratic procedures of the formal standards bodies.[3]

In either case, it may not be wise to overly focus on consortium-standard-ization-based procedures. Indeed, Hawkins (2000) argues that an informal manner of handling procedures is typical for consortia. This casts doubts on whether consortium procedures are applied.

Formal Bodies. Analogously, we should question whether formal standards bodies handle their procedures in accordance with the underlying ideology. There, too, practice diverges from theory. For example, in practice, the formal standards process is an exclusive one. The participation of end users and SMEs is very low (Jakobs, 2000). Moreover, while formal bodies usually restrict access to committee drafts to participants—and seek consensus within this group— consortia more often actively seek comments from outside and post their drafts on the Web (Rada, 2000). Furthermore, formal procedures are sometimes strategically exploited in less-than-democratic ways. Examples are staging a voting during the Christmas holiday (Egyedi, 2001b), and the dominant presence of U.S. multinationals in European national delegations (Cargill, 1999). Whereas governments emphasize that consortia adhere to democratic procedures, the lack of diversity and well-balanced participation in formal standardization is well known. At present, it is as difficult for the formal bodies to amend this situation as it would be for consortia to meet such objectives. Both settings have much in common.[4]

To sum up, the two case studies indicate that formal standards bodies and standards consortia include and exclude the same constituency. Consortia more explicitly target industrial parties. Like the formal bodies, they also strive for consensus, address minority viewpoints, and so forth; that is, they share the values of the formal ideology at the committee level. However, unlike the formal bodies, there is much variation in who is the ultimate gatekeeper: an individual director, like in W3C, or a collective, like the GA in ECMA. This is the only real difference. The rest is foremost persuasive rhetoric.

CONSORTIA IN EUROPE

A reciprocal dependence exists between the European Commission and the European formal standards bodies (e.g., EU mandate standards development).[5]

In the period leading up to the European common market in 1992, the ties were strengthened. From the Commission's point of view, the actor network on standardization served the purpose of regional governance (i.e., breaking down technical barriers to trade, and harmonizing regulation among member states). In the framework of the New Approach (1985), which arranged reference to standards in legislation and regulation, the European standards bodies were asked to play a preeminent role in the harmonization process. They willingly complied, but by doing so adopted a heavily regulation-oriented interpretation of standardization. The commission became an obligatory point of passage in the actor network: It determined which organization was to be part of the actor network and which one was not. Standards consortia that were accredited by the formal bodies fortified the formal standards setting and were included (Genschel, 1993). Nonaccredited consortia were usually excluded. Little has changed since then except that the number of influential standards specifications developed by nonaccredited standards consortia has increased exponentially.

The formal standards bodies have always embraced democratic ideals. This makes them particularly suitable as a government instrument of political and regulatory governance. By prioritizing democracy as a core value in standardization, as the Commission does, the latter strengthens the position of formal standards bodies (e.g., ISO and CEN) in the actor network. Consortium accreditation is part of an inclusion process. However, the rhetoric that defines consortia as a democratic problem is part of a negotiation process about who has a central position in the actor network. It marginalizes the role of consortia. The rhetoric derives part of its persuasiveness from keeping the meaning of the term democracy vague as well as the assumptions that define its value. Their interpretative flexibility and negotiability are essential to dominant actors in the network.

The problem which consortia have is that they do not easily fit into a context which defines standardization as an instrument of political and regulatory governance. Their way of operating is best captured by defining standardization as a means to structure and self-govern market developments (CRE, 2000; Hawkins, 1998). Not acknowledging consortia would place consortium standardization outside the Commission's sphere of influence. It would estrange the actor network from an important source of standards. However, were the Commission to acknowledge consortia that do not seek democratic certification as serious players, this would destabilize its relationship with the formal standards bodies. It would undermine the latter's role as standards-developing organizations, and compromise the importance of the democratic standardization ideology, which many people still think of as a primary asset.

REDEFINITION OF THE PROBLEM

Recapitulating, the rhetoric emphasizes that consortia are a problem because they are undemocratic and therefore unfit for use in a network used as an instrument of regulatory governance. Does this well describe what is at stake? As previous sections showed, there are several questionable aspects to the consortium problem as it is defined. They are listed in Table 2 and require further thought. Below, the necessity of their redefinition is argued and illustrated.

Democratic Standardization (1, 2)? The problem as defined says that consortia procedures are unfit for use in a regulatory-governance context. However, whether the standardization setting should at all be interpreted as an area of regulatory governance is a question hardly addressed. In the current political and economic climate (e.g., privatization, neoliberalism, multinationalism), imposing the democratic-governance ideal on the standardization setting seems to be an anachronism. Except for de jure situations, where government legislation refers to standards, the democratic process should not be prescribed as a criterion for accepting standards. Instead, democratic specification development should be qualified as a voluntary, marketable product asset (e.g., in a similar way has high quality). Consumers will then decide whether its democratic origin is an extra asset (i.e., market democracy). This would apply to the vast majority of technical standards. In exceptional cases, where government reference to standards is at stake (an estimated 30% of, e.g., the ETSI standards), regarding compatibility standards, the democracy criterion should be worded more realis-

Table 2. Aspects of the consortium problem as it is defined in the dominant rhetoric and the proposed redefinition

	Aspect of the Consortium Problem	as Defined	as Redefined
1	standardization setting characterized as	area of (democratic) regulatory governance	fair market competition
2	democratic standards process 1. compatibility standards	consensus, well-balanced participation	1. multiparty & simple majority voting (if no judicial status, no criterion)
	1. health/safety/environment/etc. standards		2. consensus, well-balanced participation
3	approach to consortia	general	differentiated
4	aim of technical coordination	standardization	compatibility
5	stage of standardization emphasized	standards development	standards diffusion and implementation
6	stage of technology development emphasized	technology research & development	technology diffusion in market

tically in terms of multiparty participation and simple majority voting. Regarding standards for health, safety, and so forth, the democracy criterion of consensual decision making and a well-balanced participation in standards development are very appropriate. However, in these cases the democratic procedures of the formal standards bodies should be followed with more care. They should be monitored more systematically.

Scope of Consortia Activities (3). As noted in the introduction, there are different kinds of consortia. For example, some try to involve as many actors as possible (inclusive), while others consist of a select number of like-minded participants (exclusive). Their activities can strongly differ (e.g., standards-development or implementation oriented). Therefore, a differentiated approach is needed to deal with the consortium problem.

Coordination of Technology: Compatibility (4). The consortium problem is defined as a standards-development problem. Standards development, whether by means of formal standardization (technical committees), hybrid workshops (e.g., CEN/ISSS workshops), or specification consortia, is a means to coordinate technology development. However, the ultimate aim is not a standard but compatibility. Compatibility can also be achieved by other means than standardization. Other compatibility strategies are, for example, pure proprietary or proprietary-led multiparty specification development, and the community-source approach to software development (Egyedi, 2001b).

Many issues that seem very important from the standardization standpoint take on a different meaning if compatibility is centered on. For example, the distinction between specifications and standards becomes unimportant. Moreover, at times, proprietary specifications can be at least as effective in fostering de facto compatibility as standardization strategies are. In comparison, multiparty, nonconsensual consortium standards would seem preferable if seen from the democratic viewpoint.

Most importantly, shifting the focus from standardization toward compatibility largely dissolves the consortium problem as defined.

Market Coordination by Specification and Strategic Consortia (5, 6). Most specification consortia are also implementation oriented. They aim at coordinating technology development in a multivendor environment. They succeed if companies implement the specifications consistently. The result is, ideally, a coordinated market segment, an outcome that requires the support of a business community (Hawkins, 1999). Strategic consortia focus on developing such communities.

As the previous section showed, the consortium problem as defined foremost focuses on standards development and R&D questions instead of standards implementation and technology diffusion in the market, respectively. A redefinition of the problem should take place along these lines.

CONCLUSION AND RECOMMENDATIONS

The democratic rhetoric holds, in a nutshell, that standards consortia lack democratic procedures, which makes them a problematic source of standards for Europe. In the previous, this rhetoric was examined more closely on two levels.

At the organizational level, the standards procedures and workings of two consortia were studied to falsify, as it were, beliefs and assumptions. Indeed, there are some inaccuracies and inconsistencies in the way the consortium problem is defined. The cases indicate that dominant rhetoric underestimates the openness of most industry consortia and overestimates the practical implications of the formal democratic procedures. The findings show that, according to paper procedures, the committees of the formal standards bodies and standards consortia roughly work in the same way. Consortia, too, strive for consensus, address minority viewpoints, and so forth. Although the latter more explicitly targets industrial parties, in practice both settings include and exclude the same constituencies. One real difference is that, formally, there is much variation among consortia about who is the ultimate gatekeeper: an individual or a collective. In practice, this difference may be nominal as seems to be the case with W3C. But, to be sure, more extensive research is needed that examines how standards procedures (of consortia and formal standards bodies) are followed up in practice.

The other level of analysis was that of the actor network on standardization. At this level, the democratic rhetoric is interpreted as part of a negotiation process. It is a tool in shaping, first, the dominant meaning of standardization and the purpose of the actor network. The main tenet in the democratic rhetoric is, implicitly, that standardization is mainly an instrument of regulatory governance. Standards that are used in this context require political accountability: democracy and openness. A weakness in the rhetoric line of reasoning is the underlying assumption that most standards are used in a regulatory context, which is not the case, and the idea that democracy has an intrinsic value also in nonregulatory environments.

Second, the democratic rhetoric sets the network boundaries and determines the position of actors in the network. Part of the shaping process is how the term democracy is defined, namely, only vaguely by inference. This interpretative flexibility prevents too-easy compliance to democratic procedures by unwelcome parties. The outcome of past negotiations is a network currently dominated by the formal and accredited standards bodies, a regulation-oriented network that marginalizes consortia. Would the dominant meaning of standardization and the purpose of the actor network be redefined, for example, as a means to structure and self-govern market developments, this would immediately favorably affect the position of standards consortia.

The way the consortium problem is defined in dominant rhetoric does not match well with developments in standardization. Therefore, a redefinition is recommended along the following lines:

- Except for de jure situations, a democratic standards process should not be imposed as a criterion for accepting standards. The significance of democratic development should be a voluntary product asset.
- In de jure situations, where democratic legitimacy is still important, procedures on paper do not suffice. The factual standards process should be monitored.
- Since consortia activities and procedures can differ widely, a differentiated approach may also be needed to deal with the redefined consortium problem.
- The ultimate aim of (compatibility) standardization is compatibility. The latter can also be achieved by other means than standardization. A focus on compatibility instead of standardization largely dissolves the consortium problem as defined.
- The current actor network, and the consortium problem as defined, focuses on standards development and R&D questions. Instead, the network should shift its focus toward issues of standards implementation and technology diffusion, respectively.

In conclusion, a standards policy is recommended that bypasses possible rivalry between standardization settings and goes beyond the inclusion of consortium standardization. Based on the shifts in emphasis proposed above, policy should, on the one hand, reflect a more pragmatic view for the majority of standards; on the other hand, it should give more substance to the aim of democratic accountability required in de jure contexts.

POSTSCRIPTUM

Since the research for this chapter was done (Egyedi, 2001a, 2003), an equilibrium seems to be emerging between formal standards bodies and standards consortia. Not only do prospective standards participants in most cases think it irrelevant which type of standards body is at stake (NO-REST TNO/RWTH, 2005), but also the number of consortia seems to be declining (Blind & Gauch, 2005). The decline in the number of standards published by ISO/IEC JTC1, which can partly be explained by the rise of consortia in the late 1990s, seems to have stopped (Egyedi & Heijnen, 2005). Of interest is, of course, whether this is due to a more restricted and focused application of the democratic ideal or whether those concerned are proceeding beyond standardization toward compatibility and interoperability. This is a matter for further research.

ACKNOWLEDGMENTS

The research on which this chapter is based was funded by the European Commission (DG Enterprise) and Verdonck Holding B. V. I gratefully acknowledge their contribution, but must add that they are in no way responsible for and need not agree with its content. I further thank the anonymous reviewers for their valuable suggestions.

REFERENCES

Albert de la Bruhèze, A. A. (1992). *Political construction of technology: Nuclear waste disposal in the United States.* Unpublished doctoral dissertation, University of Twente, Enschede, The Netherlands.

Berger, P., & Luckmann, T. (1966). *The social construction of reality.* Harmondsworth, England: Penguin Books.

Berkowitz, L. (1980). *A survey of social psychology* (2nd ed.). New York: Holt, Rinehart and Winston.

Besen, S. M. (1990). The European Telecommunications Standards Institute: A preliminary analysis. *Telecommunications Policy, 14*(6), 521-530.

Bijker, W. E. (1990). *The social construction of technology.* Unpublished doctoral dissertation, University of Twente, Enschede, The Netherlands.

Blind, K., & Gauch, S. (2005, September 21-23). Trends in ICT standards in European standardization bodies and standards consortia. In T. M. Egyedi, & M. H. Sherif (Eds.), *Proceedigns of the Fourth International Conference on Standardization and Innovation in Information Technology,* Geneva, Switzerland (pp. 29-40).

Bloor, D. (1976). *Knowledge and social imagery.* London: Routledge and Kegan Paul.

Bonino, M. J., & Spring, M. B. (1991). Standards as change agents in the information technology market. *Computer Standards & Interfaces, 12,* 97-107.

Callon, M. (1980). Struggles and negotiations to define what is problematic and what is not. In K. D. Knorr, R. Krohn, & R. Whitley (Eds.), *The social process of scientific investigation: Sociology of sciences yearbook* (pp. 197-219). Dordrecht, The Netherlands: Reidel.

Callon, M. (1986). The sociology of an actor-network: The case of the electric vehicle. In M. Callon, J. Law, & A. Rip (Eds.), *Mapping the dynamics of science and technology* (pp. 19-34). London: MacMillan.

Cargill, C. (1999). Consortia and the evolution of information technology standardization. *SIIT '99 Proceedings* (pp. 37-42).

Cargill, C. (2000). *Evolutionary pressures in standardization: Considerations on Ansi's national standards strategy.* Paper presented at The

Role of Standards in Today's Society and in the Future, Subcommittee on Technology of the Committee on Science, U.S. House of Representatives. Retrieved from http://www.house.gov/science/cargill_091300.htm

CEN/ISSS. (2000). *CEN/ISSS survey of standards-related fora and consortia* (4th ed.). Retrieved from http://www.cenorm.be/isss

Center for Regulatory Effectiveness (CRE). (2000). *Market-driven consortia: Implications for the FCC's cable access proceeding.* Retrieved from http://www.thecre.com/pdf/whitepaper.pdf

Collins, H. M. (1981). Stages in the empirical programme of relativism. *Social Studies of Science, 11*, 3-10.

Council of the European Union. (2000). Council resolution of 28 October 1999 on the role of standardization in Europe. *Official Journal of the European Communities*, 2000/C 141, 1- 4.

Dosi, G. (1982). Technological paradigms and technological trajectories. *Research Policy, 11*, 147-162.

Egyedi, T. M. (1996). *Shaping standardization: A study of standards processes and standards policies in the field of telematic services.* Dissertation. Delft, The Netherlands: Delft University Press.

Egyedi, T. M. (2000). Institutional dilemma in ICT standardization: Coordinating the diffusion of technology? In K. Jakobs (Ed.), *IT standards and standardization: A global perspective* (pp. 48-62). Hershey, PA: Idea Group Publishing.

Egyedi, T. M. (2001a). *Beyond consortia, beyond standardization? New case material and policy threads* (Final report for the European Commission). Delft, the Netherlands: Delft University of Technology.

Egyedi, T. M. (2001b). Why Java™ was not standardized twice. *Computer Standards & Interfaces, 23*(4), 253-265.

Egyedi, T. M. (2003). Consortium problem redefined: Negotiating "democracy" in the actor network on standardization. *International Journal of IT Standards & Standardization Research, 1*(2), 22-38.

Egyedi, T. M., & Heijnen, P. (in press). Scale of standards dynamics: Change in formal, international IT standards. In S. Bolin (Ed.), *The standards edge: Future generation.* Ann Arbor, MI: Bolin Communications.

European Commission (1985). Retrieved from www.newapproach.org

Genschel, P. (1993). *Institutioneller wandel in der standardisierung van informationstechnik.* Unpublished doctoral dissertation, University of Köln, Köln, Germany.

Hawkins, R. (1998). *Standardization and industrial consortia: Implications for European firms and policy* (Working Paper 38). Sussex, UK: ACTS/SPRU.

Hawkins, R. (1999). The rise of consortia in the information and communication technology industries: Emerging implications for policy. *Telecommunications Policy, 23*, 159-173.

Hawkins, R. (2000). *Study of the standards-related information requirements of users in the information society.* Final report to CEN/ISSS,
February 14. Retrieved from www.cenorm.be/isss

Jakobs, K. (2000). *Standardization processes in IT: Impact, problems and
benefits of user participation.* Braunschweig/Wiesbaden, Germany:
Vieweg.

Jepperson, R. L. (1991). Institutions, institutional effects and institutionalism. In
W. W. Powell, & P. J. DiMaggio (Eds.), *The new institutionalism in
organizational analysis* (pp. 143-163). Chicago: University of Chicago
Press.

Krechmer, K. (2000). Market driven standardization: Everyone can win. *Standards Engineering, 52*(4), 15-19.

Kuhn, T. S. (1970). *The structure of scientific revolutions* (2nd ed.). Chicago:
University of Chicago Press.

March, J. G., & Olsen, J. P. (1989). *Rediscovering institutions: The organizational basis of politics.* London: Macmillan.

Mulder, K. F. (1992). *Choosing the corporate future: Technology networks
and choice concerning the creation of high performance fiber technology.* Unpublished doctoral dissertation, University of Groningen,
Groningen, the Netherlands.

NO-REST TNO/RWTH. (2005). D05/6: Report on demand factors and on the
supply side for standards for networked organisations (EU/NO-REST
project; Contract No. 507 626). Brussels, Belgium: European Commission.

Pinch, T. J., & Bijker, W. E. (1984). The social construction of facts and
artefacts: Or how the sociology of science and the sociology of technology
might benefit each other. *Social Studies of Science, 14*, 399-441.

Powell, W. W., & DiMaggio, P. J. (1991). Introduction. In W. W. Powell, & P.
J. DiMaggio (Eds.), *The new institutionalism in organizational analysis*
(pp. 1-40). Chicago: University of Chicago Press.

Rada, R. (2000). Consensus versus speed. In K. Jakobs (Ed.), *IT standards and
standardization: A global perspective* (pp. 19-34). Hershey, PA: Idea
Group Publishing.

Updegrove, A. (1995). Consortia and the role of the government in standard
setting. In B. Kahin, & J. Abbate (Eds.), *Standards policy for information infrastructure* (pp. 321-348). Cambridge, MA: MIT Press.

Weiss, M., & Cargill, C. (1992). Consortia in the standards development
process. *Journal of the American Society for Information Science,
43*(8), 559-565.

World Wide Web Consortium (W3C). (2001). *World Wide Web Consortium
process document.* Retrieved from http://www.w3.org/Consortium/Pro-
cess-20010208/

ENDNOTES

[1]   The CEN Web site www.cenorm.be/isss includes an elaborate list of consortia.

[2]   U.S. sources comment that consortia are generally more under control of their members, have more money to maoeuvre, and their specs are accepted in practice in U.S. public procurement (Cargill, 1999). The benefit of the formal bodies is, for example, less chance of anit-trust claims (CRE, 2000).

[3]   One reason why consortia sometimes adopt the formal procedures is that consortium standardization is at times a stepping-stone for formalization. In those cases, consortia must conform to the formal ground rules.

[4]   Interestingly, both formal standars bodies and standards consortia have been accused of administrative rubberstamping. W3C has been accused of rubberstamping the products of major vendors because of the "member submission process" (Rada, 2000, p. 22). This process makes it possible to consider proposals developed outside of W3C. The W3C rules explicitly state that this process is "*not* a means by which Members ask for 'ratification' of these documents as W3C Recommendations" (W3C, 2001).

[5]   In U.S. legislation (i.e., the OMB circular), allegedly, no preference is given to formal (consensus) standards *vis a vis* (non-consensus) consortium standards (CRE report, 2000).

Section III

The Issue of Speed

Chapter VI

Standardization and Competing Consortia:
The Trade-Off between Speed and Compatibility

Marc van Wegberg, University of Maastricht, The Netherlands

ABSTRACT

The consortia movement in the standardization world has led to a fragmentation of standardization processes. This fragmentation is partly of a competitive nature, where rival coalitions support competing technologies. A critique on this movement is that it fragments technologies and multiplies the number of standards. The aim of supporting competing technologies may reflect experimentation with different technological paths. It may also, however, reflect differences in intellectual property rights of firms. From a user's perspective, the competing technologies may represent spurious differences that increase uncertainty, and create transaction costs. The consortia do have a function for end users: Established industry-wide standard development organizations (SDOs) may be slow to act, bureaucratic, and inflexible to changes in users' needs and new opportunities; consortia speed up the process of standardization. This chapter argues that consortia do indeed tend to correct these coordination failures of the official SDOs. They do so at a cost, however, and because of this, industry-wide SDOs still have a role to play.

INTRODUCTION

The standardization landscape in the information and communication technology (ICT) industries is fragmented in many different standardization bodies, industry consortia, and alliances. Some of these coalitions cooperate with each other, while others compete. The consortia movement is a major cause of competitive fragmentation of standardization. Practitioners and analysts argue with each other about whether this fragmentation leads to coordination failures. The existence of competing standardization coalitions may prevent coordination on a common standard. The argument in this chapter is that consortia exist for a reason. A better understanding of why companies have standardization strategies that give rise to fragmentation may show the possible advantages of fragmentation.

An important form of innovation in the ICT industries consists of developing new combinations of components. The ability to connect devices can increase their utility to end-users. An end user can increase the utility of a product by connecting it to complementary products. Hardware and software are examples of this. Connecting different devices may also benefit users by enabling them to communicate with each other by voice or data communication. In both examples, *network externalities* are realized. These are defined as situations where the utility of a product (or a service) to a user increases when more users use the same product or compatible technologies. *Compatibility standards* set specifications for components that make it possible to connect these components to each other. By improving the connectability of products, compatibility standards make it possible to realize network externalities. They create value for the end users or for their suppliers. The standardization process can therefore be an important value-generating process. How this process is organized affects the outcome of the standardization process. Standardization processes are partly organized in coalitions. How many coalitions there are, and how many members each has, is known as the *coalition structure* of the standardization process (Bloch, 1995). An important aspect of the coalition structure is the *level of centralization*, defined here as the extent to which decision making about standards is concentrated in one or more coalitions.

The most centralized coalition structure is the *grand coalition*: a coalition that includes all participants in the standardizing process. In the case of a standard that affects an industry, this will be an industry-wide coalition. It may take the form of an official standards development organization (SDO). A grand coalition has access to the widest number of players and their information. A consensual decision-making process means that specifications are accepted only if no one (or at most a sufficiently small minority) holds out against them. The consensus provides legitimacy to its specifications. Due to its comprehensive membership, information about the new standards is widely available in the field. The comprehensive coordination makes it possible to convene on a single standard. This is an important step towards ensuring that technologies in use are fully compatible, and positive network externalities can be realized.

While there are quite a few grand coalitions, many standardization processes are highly fragmented (Genschel, 1997). A special case of a fragmented process occurs when multiple coalitions compete with each other in a standardization process. If competing coalitions set different, incompatible standards, some network externalities will not be realized. How centralized the emerging coalition structure will be depends on the pros and cons of the various possible coalition structures. The choice of coalition structure faces trade-offs, one of which is between the speed of decision making and the level of compatibility that can be achieved. A grand coalition may be slow to act. It may comprise participants with different backgrounds and antagonistic objectives. Antagonism may lead to intransigent behavior by firms, which slows down decision making the more consensus is valued. If a new standard substitutes for existing technologies, some firms may participate to slow down standardization (Lint & Pennings, 2003).

One way to speed up decision making is for (potential) members to split up to create faster consortia (David & Shurmer, 1996; Warner, 2003). Participants with different or opposite interests may, for instance, be excluded from the coalition. Those who are excluded may go on to form their own committee (Axelrod, Mitchell, Thomas, Bennett, & Bruderer, 1995; Belleflamme, 1998; Bloch, 1995; Economides & Flyer, 1998; Greenlee & Cassiman, 1999). For instance, Bloch (1995) argues that the more intense competition is in the product market, the more firms are tempted to exclude rivals from their coalition. The better substitutes their products are, the less likely that a grand coalition will appear, and the more likely that rivals will establish competing committees. If competing committees are formed, they may accept different, incompatible technologies as a standard.

These arguments suggest that there is a *common view* within the literature on the disadvantages of a grand coalition. This common view implies a circumscribed defense of fragmentation. A grand coalition has a better chance of ensuring compatibility between the technologies used in an industry than competing coalitions. The higher the degree of compatibility between the technologies actually adopted by service providers, the more service providers can realize positive externalities. A grand coalition, however, may also take more time to arrive at a decision than competing coalitions. It has more opposite interests to accommodate. This delay represents an intra-coalition coordination failure. The smaller size of competing coalitions, and the competition between them, tend to speed up their decision making. This greater speed does come at the possible risk of selecting incompatible technologies, which generates an inter-coalition coordination failure (see Figure 1).

Figure 1. Common view on coalition structure

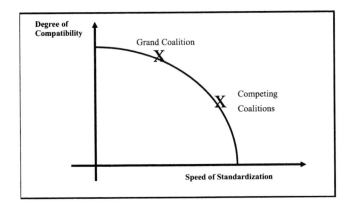

Figure 1 shows the two propositions inherent in the common view. Firstly, speeding up decision making about standard setting tends to reduce the compatibility between the technologies that end users will adopt. Secondly, while a grand coalition is better at achieving conditions of compatibility, competing coalitions are better suited to select standards quickly.

This chapter subjects the common view to a critical and theoretical analysis. The next section presents some examples of grand coalitions and competing coalitions. It illustrates that standard setting is a conflicted, political process. Because of this, standard setting tends to occur in coalitions. Exposed to conflicting political pressures, coalitions can both experience consolidation as well as fragmentation. The subsequent section of this chapter asks: Is there support from the literature that grand coalitions are reduced in their speed of decision making by internal antagonism? And do competing coalitions arrive at standards quicker than a grand coalition would, at a cost of less compatibility? The focus in the review is on insights about the decision-making and interaction process in standardization. It argues that there are sound, basic theoretic reasons to believe that a trade-off between speed and compatibility exists. It does suggest, however, that cases exist where a grand coalition can decide more quickly about a standard than competing coalitions. While a fragmented standardization landscape may have its benefits, a balanced view should acknowledge that there are cases where a grand coalition is quick to overcome competing interests.

Examples of Grand Coalitions and Competing Coalitions

Two examples illustrate the push and pull between a grand coalition and competing coalitions. The following quote from the IT website *The Register* on the Open Mobile Alliance (OMA) may illustrate the common view:

The OMA is intended to harmonize a barrage of mobile industry standards for 3G, hitherto a barrier to creating a seamless mobile Internet...But wide participation will slow OMA and put global standards out of reach. Let's face it: The more players that get involved, the more time it will take for them to agree on anything at all—and many will participate only to slow down development and control the market. (Forrester, in *The Register*, 13 June 2002)

The OMA (http://www.openmobile alliance.org/) develops and supports standards for mobile telecommunication and data communication services. It resulted from a merger between different industry consortia, including the Open Mobile Architecture, the WAP Forum, the Location Interoperability Forum, and the MMS Interoperability (MMS-IOP) Group. Fragmentation may apparently go too far, in which case consolidation occurs. With more then 200 members, the OMA is however very large. Within the fragmented standardization context, the OMA intends to play a central role: "The Open Mobile Alliance is designed to be the center of all mobile application standardization work" (OMA, 28 October 2003, http://www.openmobilealliance.org/faq.html).

The DVD Forum (http://www.dvd forum.org) is another example of a grand coalition. It is an industry-wide consortium that develops and promotes the DVD format as a standard. As of October 2003, it had 212 members. According to the association, "The geographic distribution of our members, as of March 2003, was as follows: 38% Japanese, 29% Asian, 20% American, and 13% European." Characteristic of the politically charged standardization process of DVD and derivative standards, the DVD Forum notes: "Forum Members are not required to support the DVD Format to the exclusion of other formats" (DVD Forum, http://www.dvdforum.org/about-mission.htm, 27 October, 2003). The website of the forum publishes antitrust guidelines to ensure that its activities do not develop into a cartel. Discussions in the DVD Forum may not refer to prices or to costs. These antitrust requirements help to create a non-commercial, technical atmosphere that may reduce the level of conflicts between the representatives of the companies within the Forum. Conflicts between the DVD Forum members may exist. That they do exist is apparent from the competing technologies for DVD-rewritable.

There are currently three technologies for DVD-rewritable: DVD-RAM, DVD+RW, and DVD-RW. Each of these has a coalition of supporters. For example, Sony, Ricoh, Hewlett-Packard, Philips, Mitsubishi, and Yamaha support DVD+RW. This coalition is called the DVD+RW Alliance (http://www.dvd rw.com/). The DVD-RAM Promotion Group members are Hitachi, Hitachi-LG Data Storage, Hitachi Maxell, LG Electronics, Matsushita Electric Industrial, Samsung Electronics, TEAC, Toshiba, and JVC (CNET, 21 August 2003, http://news.com.com/2110-1041-5066673 .html). Apart from that, there is also the

DVD-R recordable technology. It is supported by the recordable DVD Council (http://www.rdvdc.org/english/index.html).

The DVD Forum did not select a priory a standard for DVD-rewritable. It does note on its homepage: "Please note that the '+RW' format, also known as DVD+RW' was neither developed nor approved by the DVD Forum. The approved recordable formats are DVD-R, DVD-RW, and DVD-RAM" (DVD Forum, 28 October 2003, http://www.dvd forum.org/forum.shtml). The developers of the DVD+RW format chose to keep their work outside of the DVD Forum. The other coalitions liaise with the DVD Forum by means of the Forum's working groups. There is one working group for DVD-RAM, and one for both DVD-R and DVD-RW.

The DVD example shows that a grand coalition can exist that endorses a unified standard, notably the DVD standard. It also shows cases with competing coalitions. The grand coalition is, with more than 200 members, very large indeed. The political nature of their cooperation can hamper the activities of the Forum. Having competing coalitions, however, leads to incompatibility, which may harm end users and slow down their adoption of the new formats. The OMA reacted to the fragmentation of mobile standard setting by consolidating various initiatives and coalitions. The next section explores a trade-off that may explain the push and pull between, on the one hand, consolidation into a grand coalition and, on the other hand, fragmentation into competing coalitions.

Trade-Off between Timing and Compatibility

This part of the review asks if a trade-off exists between timing and compatibility. Is a coalition structure that is geared to achieving compatibility ill-suited to select a standard quickly? The proposition we wish to explore is that the higher the level of centralization of a coalition structure, the higher the chance that it generates compatibility at the industry level and the longer standardization tends to take. The more inclusive an individual coalition is, the longer it takes to agree on a standard. By implication, the larger a coalition is, the slower its decision making tends to be. In a setting with a given set of firms, this implies that a grand coalition will decide slower than competing coalitions would. We discuss various arguments that support this idea. Both the process of forming a coalition and the decision-making process within a coalition take time.

Forming a coalition is a time-consuming process. There is an initialization phase to a partnership (Zajac & Olsen, 1993). In this initialization phase, firms communicate, negotiate, analyze feasibility studies, and forge relational exchange norms. Which factors can facilitate this process? One factor will be the size of the coalition. The more members in a committee, the more alternative technologies there may be to choose between. Communication takes time. The more people are involved, the more time it takes to communicate with them. If participants have different backgrounds, it can take time to translate concepts between them. They will need to develop a common vocabulary. In the Internet tradition, for example, participants in a standardization process will often need to define concepts first, using a document type called a Request for Comments, RFC. An example is RFC 2119 of the Internet Engineering Task Force (IETF), which defines among other things the meaning of concepts such as must, must not, shall, etc., when used in standards.[1]

There are factors that may shorten the initialization phase. Prior contacts between the partners can facilitate the set-up of a partnership. This may explain an insight about partner choice in strategic alliances: firms with prior contacts are more likely to become partners in an alliance (Gulati, 1995). Firms may form a private consortium in order to work with long-time allies, while excluding potential rivals and outsiders. A consortium consisting of insiders may be formed quickly, which underlies the speed of decision making by small competing coalitions. A permanent SDO can shorten the initialization phase by adopting standard procedures and routines for new workgroups. This offers some compensating speed advantage to SDOs that act as a grand coalition.

Decision making in a coalition is a political process. Time is both a consequence of the process and an instrument in it. Important players in the process are the sponsors. A *sponsor* is a firm that actively supports a particular specification for a standard through a standardization process. It is likely to be a firm that expects that this specification will bring it more revenues than any other. It is, for example, the innovator who developed the technology that can be standardized. A sponsor may insist on the coalition standardizing on its preferred specification. If other organizations are equally intransigent, the combination of their efforts slows down the process of creating consensus and selecting a standard.

A technology sponsor may try to win over the grand coalition by actively participating in its activities. Organizations may invest in influencing activities to influence the coalition's decision making (Besen & Farrell, 1994). In the ICT industries, for instance, sponsors influence official standardization bodies by means of the contributions they make to the work of these bodies. An interesting paper observes that large companies increased the number of staff they dispatch to meetings of official organizations, such as the IETF and the IEEE (the Institute of Electrical and Electronics Engineers) (Heywood, Jander, Roberts, & Saunders, 1997). They hire people who have gained influence and reputations in official standard-setting organizations. They also try to influence who will chair workgroups of standardization organizations. As Heywood et al. (1997) show, standardization organizations are aware of these possibilities, and try to design rules to suppress them.

Rules and procedures in standardization coalitions may suppress politicking behavior of individual participants. They may also diminish the influence of these participants. As a result, these firms may abandon the coalition and set up competing coalitions. In a decentralized coalition structure with multiple competing coalitions, an individual firm has a larger chance of influencing its particular coalition. A grand coalition may want to prevent defection by being very responsive to the interests of its members. Too much responsiveness may bog down the standardization process (Sherif, 2003b). The more contingencies a specification needs to be tailored to, the more complicated the specification will be. Hence, a grand coalition faces some tension between speeding up decision making and ensuring compatibility of new technologies.

In terms of the speed of decision making, competing coalitions have several advantages over a grand coalition. The competition may provide incentives to speed up. If one coalition sets a standard before another one, the latter may be too late to get its standard adopted in the market. Each coalition tends to select its members carefully. It can exclude members, unlike the grand coalition, which needs to open up to all potential members. The ability to exclude potentially obstructive or intransigent members reduces the tensions within a coalition. Competing coalitions can thus be less inflicted by tensions than a grand coalition. With lower levels of politicking, they can act faster. The sheer size of the coalition may also help: a smaller coalition can act quicker than a larger one. Fewer participants need to participate in decision making. A smaller and more focused group may have a shorter initialization stage than a grand coalition. The participants may already be familiar with each other, which reduces the effort needed for initialization. These are compelling arguments to suggest that a grand coalition tends to take more time to select a standard than competing coalitions.

A second impact of coalition structure is that the more centralized it is, the more likely that the market will adopt compatible technologies. In particular, a grand coalition is more likely to adopt a single standard than competing coalitions. A grand coalition can select a standard from among competing technologies. It may also try to combine different technologies into a compromise standard. The DVD standard is the result of such a compromise.[2] The DVD consortium (the precursor of the DVD Forum) combined the multimedia CD coalition of Philips, Sony, and 3M with the super-density CD of Toshiba and Time Warner. The combination gave rise to the DVD specification, albeit after disagreements in the DVD Forum about the specifications and licensing schemes delayed the introduction of DVD products (van Wegberg, 2003). A grand coalition or industry-wide standard development organization can also design specifications to reduce incompatibility. For example, the IEEE has developed several standards for wireless data communication. Some of these technologies use the same unlicensed frequency band. As a consequence, wireless systems can interfere. Interference can diminish the quality of the signal. The IEEE developed standards for Wireless Local Area Networks and Wireless Personal Area Networks that reduce the disadvantages of interference.[3]

The effort in a grand coalition to create compatibility is another reason why the process may take more time than competing coalitions. The more different parties are taken on board in a standardization coalition, the more complicated it is to develop specifications for a standard that all can accept and can be compatible with. One of the solutions for this problem is to develop standards with options (Egyedi & Dahanayake, 2003). The different options reflect different interests and views of the parties combined in the SDO or coalition. This allows technology providers and adopters to switch these options on or off, depending on which interests, views, and preferences they have. The advantage of including options in the standard is that it enables the SDO to forge a compromise that is acceptable to its members. A disadvantage is that in the implementation phase, differences come to light in the technologies used by the standard's adopters, due to their different choices of which options to (de)activate. The need to compromise between its many members may thus reduce the ability of a grand coalition to create de facto interoperability (compatibility) in the implementation of the standard.

Grand coalitions such as formal standard-setting bodies know they need to speed up decision making (David & Shurmer, 1996). One of the solutions is to produce incomplete standardization by means of meta-standards. A *meta-standard* establishes some conditions and aspects of a standard, without specifying the standard itself in full detail. Settling details of a standard can take a lot of time. Setting a meta-standard avoids the need to fine-tune the standard, and thus speeds up decision making. The advantage of having a meta-standard in an early stage is that it preempts companies that might otherwise commit to incompatible technologies. A disadvantage of setting a meta-standard is that private companies adopt technologies that are only partially standardized. They may, inadvertently or otherwise, implement technologies that are partially incompatible. This solution therefore gives up some compatibility in order to speed up decision making in a grand coalition.

A grand coalition will not always be able to ensure compatibility when its members simultaneously play a game of *de facto* standardization. A participant in the grand coalition may try to create a fait-a-compli to force the coalition to bend to its wishes. It may start an installed base with its technology in an attempt to strengthen its bargaining power in the coalition. This leads to *hybrid standardization*: a standardization process where firms pursue standardization by two simultaneous paths, using both the market mechanism and negation in a coalition (Axelrod et al., 1995; Farrell & Saloner, 1988; Funk & Methe, 2001). In a hybrid standardization process, a grand coalition cannot guarantee compatibility. The best way to prevent incompatibility in this case is for the grand coalition to agree on a standard fast, before some of its members have committed themselves to a de facto standard. A hybrid standardization process may thus speed up standardization, but at a cost (Farrell & Saloner, 1988). The cost may be that companies adopt incompatible technologies.

There are thus various reasons why a grand coalition may not be able to guarantee compatibility. Political compromises lead to ambiguous standards, with options that can be activated or deactivated at will, thus creating compatibility. A meta-standard can speed up standardization, at a cost of leading to incompatible technologies in use. Hybrid standardization increases the bargaining power of coalition members that have a strong position in their product markets. Their early pre-emptive moves in the product market build up an installed base for a technology that the coalition may be hard-pressed to ignore. The presence of competing specifications in technologies reduces the level of compatibility, however.

While the grand coalition may not guarantee compatibility, the presence of competing coalitions need not per se lead to incompatibility. If one committee adopts a standard quickly, the members of a competing coalition may adopt that standard (Genschel, 1997). In this case firms abandon the slowest committee. Furthermore, competing committees make choices that do not have to be entirely incompatible. They may adopt partially overlapping technologies, leading to partial compatibility. In his book about Bill Gates, Wallace (1997) gives a more controversial example, when he attributes to Microsoft the strategy of *embrace and extend*. This strategy confronts a successful technology of a rival, but not by developing an incompatible alternative. Instead, the embrace and extend strategy is to adopt the technology, and then to extend it with proprietary extensions.[4] Once users adopt these extensions, control over the technology shifts to Microsoft. Competing coalitions may have the same motive to adopt partially or wholly a rival committee's technology. As a result, in a setting of competing coalitions, committee standards can be hybrids that combine elements from competing technologies (for an example, see Mangematin & Callon, 1995).

If standardization coalitions or organizations cannot avoid incompatible technologies, they can design standards such as to make it possible to have gateways between incompatible technologies. For example, the IETF and the ITU (the International Telecommunications Union) cooperated to develop the Megaco/H.248 gateway protocol to act as a gateway between dissimilar networks.[5]

To conclude the discussion: it is likely, but by no means certain, that a grand coalition takes more time to conclude a standard than smaller competing coalitions would. It is, moreover, not self-evident that a grand coalition ensures a higher degree of compatibility than competing coalitions would. Only when a grand coalition compensates for a longer duration of decision making, by increasing the expected degree of compatibility, do firms face the trade-off between speed of decision making and compatibility. The literature review thus suggests caution with respect to the view that there is a trade-off between compatibility and speed of decision making.

Reasons for the Consortia Movement

The consortia movement has led to fragmentation of standardization. It has compromised the ability of SDOs to act as a grand coalition. There may be many reasons for this process (see, e.g., Shapiro & Varian, 1999). Figure 1 offers a good framework to understand one set of reasons. Figure 1 suggests that if the preference that firms have for a high level of compatibility would decrease, and their preference for speedy standardization would increase, they would want to switch from a grand coalition to competing coalitions. A shift in preferences from compatibility to time-to-market will explain fragmentation.

An important step in exploring firms' preferences for coalition structures is the value they attach to compatibility. An open industry-wide standard can be a platform for new services. This applies to *anticipatory standards*, which are standards for new technologies that will develop new services and associate markets (Sherif, 2003b). There are direct network externalities if new services enable users to communicate with each other. The more users can communicate with each other, using standards-compliant equipment, the more benefits they derive from participating in the new service. This is how a standard can create value. The higher the degree of compatibility between the technologies used by vendors, the greater the value created in the product market (Katz & Shapiro, 1985).

On the demand side of the market, compatibility may not be very important. If users tend to communicate in small communities, the sheer size of the network of users may not increase their benefit. As long as a standard dominates within their particular community, they may not care about standards used by other communities. A study by Cowan and Miller (1998) offers support for this intuition. It studies a case with local externalities, where neighbors are potential adopters of a standard to communicate or cooperate with each other. It finds that users may adopt incompatible technologies. Since they communicate locally, they are not aware (and do not care) of far-away users adopting a different technology.

Even if there is no standard, service providers may realize network externalities for their end users. They can start gateway services that connect users of incompatible technologies. Converters link the users of otherwise incompatible technologies. They help these users to achieve positive externalities (Choi, 1996).

On the supply side of the market too, there may be insufficient preferences for compatibility. Innovators may benefit if their innovation is accepted as a standard. They want to earn revenues from their intellectual property rights on the innovation. They are concerned about the *appropriability* of revenue streams. They may care more about their share of the revenue stream from a standard than about the absolute size of revenues created by the standard. Patents are a case in point; they give the innovator some control over revenue flows generated by the innovation. If a standard increases the value of the intellectual property right on a technology, the firm may be more interested in supporting one particular technology than in having an industry-wide standard per se. A concern with intellectual property rights on technologies may lead to fragmentation of the standardization process. Blind (2001) finds empirical support for the argument that patent protection may make companies reluctant to set standards via national or international standard-setting organizations (SDOs).

The value of compatibility may thus be limited. The importance of time-to-market, on the other hand, is increasingly emphasized. If there are first-move advantages in competition, firms are likely to disagree with a time-consuming decision process in a committee. If, however, there are second-move advantages, there is a benefit to waiting, and speeding up standardization may not be a priority at all. If technology improves continually, for example, users may switch to an incompatible technology if its quality is sufficiently higher than the established technology (Katz & Shapiro, 1992; Shy, 1996). Shy (1996) shows among others that if new technology is backward-compatible with the old technology, users are more likely to switch to the new technology. This reduces the lifetime of the older technology. The lifetime of a standard thus depends on the willingness of users to wait for better technology to appear.

Speeding up standardization will be valuable if the benefits from the standard are time-dependent. The standard is a specification for a technology. There are expectations about when the technology will be superseded by a superior technology. Technologies have a lifecycle; the more time in this cycle absorbed in the standardization process, the less time remains for using it in marketable products. A standard may be a platform for new or improved services. These services themselves have a product lifecycle (Sherif, 2003a). Delaying the product introduction delays the start of revenues. Due to the time preference of the potential vendors, they are likely to want to speed up market introduction.

Figure 2 summarizes the relationships discussed so far.

A shift in the objectives of firms may have occurred, away from stressing network externalities, and toward greater emphasis on pre-emptive moves, first-move advantages, and intellectual property rights. This shift itself would explain the fragmentation created by the consortia movement (see Figures 1 and 2).

Solutions and New Roads

The situation in Figure 1 illustrates the quandary that both grand coalitions and sets of competing coalitions find themselves in. The problem for a grand coalition is that its main value added is to enhance compatibility between the

Figure 2. Choice of coalition structure in the trade-off between speed and compatibility

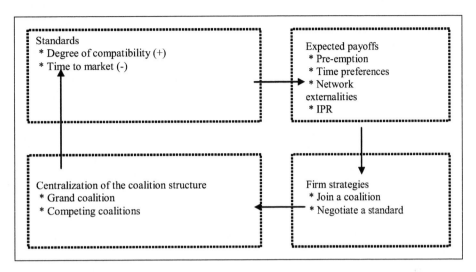

technologies that adopters will be using. The more it works toward achieving this goal, the more it will be perceived as a slow mover. We discussed some moves to speed up decision making of a grand coalition. It may establish options in a standard or generate a meta-standard. These may increase the speed of decision making, but they do jeopardize the compatibility and interoperability of technologies used. A grand coalition that does not give in to pressures to move faster may see its members defect to private consortia. Many grand coalitions, and notably official SDOs, try to break the trade-off by learning from the consortia movement (Krechmer, 2003).

Faced with the trade-off between speed of decision making and compatibility, a grand coalition may try to have its cake and eat it too. Some official SDOs have established workshops for different standards. The European SDO CEN/ISSS uses workshops to attract new members and develop consensus quickly.[6] CEN/ISSS encourages private companies to participate in these workshops. Another way to involve private companies is by creating workgroups and giving these a fast-track access to standards approval. These workgroups may become substitutes for private consortia. There may be some fragmentation left, but this will take the form of multiple workgroups or workshops within the common framework of an SDO or grand coalition.

For competing coalitions, Figure 1 illustrates that their problems lie in a different field. Their main added value is timely standardization within a focused group of companies. The problem they run into is fragmentation, and an associated loss of coordination at the level of the industry. They may overcome this problem by consolidation. The DVD example showed how two competing coalitions merged into the DVD Consortium. The resulting coordination may fall short of creating compatibility. The DVD Forum, for instance, hosts two workgroups that work on incompatible specifications for DVD-rewritable: the DVD-RAM and DVD-RW specifications.

These solutions suggest that some convergence occurs between SDOs and consortia. Some consortia, like the DVD Forum and the OMA, acquire SDO-like characteristics. They are large, inclusive, industry wide, and open to new members. Some SDOs become more consortium-like, by providing fast-track approval procedures, and opening workgroups or workshops to private companies. However, differences in priorities and objectives are likely to persist between private consortia and official SDOs (Krechmer, 2003).

Where fragmentation is unavoidable, coordination can be improved by creating liaisons between SDOs and consortia. Several SDOs have established MoUs (Memorandums of Understanding) about their relationships with one another and with private consortia. While these new forms of coordination may reduce the unintentional levels of incompatibility that may result from fragmentation, they may also increase the time and effort needed to develop new standards. They therefore represent new, intermediate forms of coalitional standardization processes.

CONCLUSION

This chapter argues that having an industry-wide standardization coalition may slow down standardization, compared to cases where smaller coalitions of firms compete with each other. The conflicts of interest between the participants in the grand coalition lead to politicking, which in turn can stall decision making. Competing technology sponsors hold the grand coalition to ransom, in order to get their preferred technologies selected as standard. Competing coalitions have a greater incentive to speed up their decision making. The presence of competing coalitions may, however, lead to incompatible technologies being adopted in the marketplace. This may exacerbate a conflict of interest between technology sponsors, on the one hand, and technology adopters on the other hand. Technology adopters may have a greater preference for compatibility and interoperability. Their fear for incompatibility may hold back standardization in competing committees. Technology adopters give in to politicking themselves, in order to avoid an unwanted level of incompatibility. As a result, in certain situations a grand coalition can decide quicker than competing committees. It satisfies the greater preference that technology adopters tend to have for compatibility. This result sheds doubt on the common view that industry-wide standardization coalitions are slow.

In choosing a coalition structure, firms balance two types of coordination failure: the failure to select an industry-wide standard (when competing committees support competing technologies) and the failure to decide in a timely manner (when negotiations in a grand coalition lead to stalemate or when competing committees hold back, fearing incompatibility). According to the common view, the need to reach an agreement between competitors within a grand coalition leads to lengthy negotiations. If sponsors of rival technologies set up their own coalitions, this can speed up standardization. Competition between standardization coalitions does, however, squander some benefits of having a standard (the so-called network externalities). The new insight of this chapter is that a fear for resulting incompatibility affects the timing of decision making in a negative way. Technology adopters may slow down the coalition they participate in, in order to keep track of what a rival coalition is up to. This does slow down the standardization process with competing coalitions. When that happens, a grand coalition can select a standard faster than competing coalitions can.

REFERENCES

Axelrod, R., Mitchell, W., Thomas, R. E., Bennett, D. S., & Bruderer, E. (1995). Coalition formation in standard-setting alliances. *Management Science, 41*(9), 1493-1508.

Belleflamme, P. (1998). Adoption of network technologies in oligopolies. *International Journal of Industrial Organization, 16*, 415-444.

Besen, S.M., & Farrell, J. (1994). Choosing how to compete: Strategies and tactics in standardization. *Journal of Economic Perspectives, 8*(2), 117-131.

Blind, K. (2001). *The impact of intellectual property rights on the propensity to standardize at standardization development organizations: An empirical international cross-section analysis.* Unpublished manuscript.

Bloch, F. (1995). Endogenous structures of association in oligopolies. *RAND Journal of Economics, 26*(3), 537-556.

Choi, J. P. (1996). Do converters facilitate the transition to a new incompatible technology? A dynamic analysis of converters. *International Journal of Industrial Organization, 14*(6), 825-835.

Cowan, R., & Miller, J. H. (1998). Technological standards with local externalities and decentralized behavior. *Journal of Evolutionary Economics, 8*, 285-296.

David, P. A., & Shurmer, M. (1996). Formal standards-setting for global telecommunications and information services: Towards an institutional regime transformation? *Telecommunications Policy, 20*(10), 789-815.

Economides, N., & Flyer, F. (1998). Equilibrium coalition structures in markets for network goods. *Annales d'Economie et de Statistique,* (49/50), 361-380.

Egyedi, T. M., & Dahanayake, A. (2003). *Difficulties implementing standards.* Paper presented at the 3rd IEEE Conference on Standardization and Innovation in Information Technology (SIIT 2003), Delft, October.

Farrell, J., & Saloner, G. (1988). Coordination through committees and markets. *RAND Journal of Economics, 19*(2), 235-252.

Funk, J. L., & Methe, D. T. (2001). Market- and committee-based mechanisms in the creation and diffusion of global industry standards: The case of mobile communication. *Research Policy, 30,* 589-610.

Genschel, P. (1997). How fragmentation can improve coordination: Setting standards in international telecommunications. *Organization Studies, 18*(4), 603-622.

Greenlee, P., & Cassiman, B. (1999). Product market objectives and the formation of research joint ventures. *Managerial and Decision Economics, 20*(3), 115-130.

Gulati, R. (1995). Social structure and alliance formation patterns: A longitudinal analysis. *Administrative Science Quarterly, 40,* 619-652.

Heywood, P., Jander, M., Roberts, E., & Saunders, S. (1997). Standards, the inside story: Do vendors have too much influence on the way industry specs are written and ratified? *Data Communications,* 59-72.

Katz, M. L., & Shapiro, C. (1985). Network externalities, competition, and compatibility. *American Economic Review, 75*(3), 424-440.

Katz, M. L., & Shapiro, C. (1992). Product introduction with network externalities. *The Journal of Industrial Economics, 40*(1), 55-83.

Krechmer, K. (2003, October). *Face the FACS.* Paper presented at the 3rd IEEE Conference on Standardization and Innovation in Information Technology (SIIT 2003), Delft.

Lint, O., & Pennings, E. (2003). The recently chosen digital video standard: Playing the game within the game. *Technovation, 23*(4), 297-306.

Mangematin, V., & Callon, M. (1995). Technological competition, strategies of the firms and the choice of the first users: The case of road guidance technologies. *Research Policy, 24*(3), 441-458.

Shapiro, C., & Varian, H. R. (1999). *Information rules: A strategic guide to the network economy.* Boston: Harvard Business School Press.

Sherif, M. H. (2003a). *Technology substitution and standardization in telecommunications services.* Paper presented at the 3rd IEEE Conference on Standardization and Innovation in Information Technology (SIIT 2003), Delft, October.

Sherif, M. H. (2003b). When is standardization slow? *Journal of IT Standards & Standardization Research, 1*(1), 19-32.

Shy, O. (1996). Technology revolutions in the presence of network externalities. *International Journal of Industrial Organization, 14*(6), 785-800.

van Wegberg, M. v. (2004). Compatibility choice by multi-market firms. *Information Economics and Policy,* forthcoming.

Wallace, J. (1997). *Overdrive: Bill Gates and the race to control cyberspace.* New York: John Wiley & Sons.

Warner, A. G. (2003). Block alliances in formal standard setting environments. *Journal of IT Standards & Standardization Research, 1*(1), 1-18.

Zajac, E. J., & Olsen, C. P. (1993). From transaction cost to transactional value analysis: Implications for the study of interorganizational strategies. *Journal of Management Studies, 30*(1), 131-145.

ENDNOTES

[1]   See http://www.ietf.org/rfc/rfc2119.txt for this RFC (Nov. 2002). Note that an RFC is a standard, in the language of the IETF.

[2]   Aguilar, R. (1996). Philips, Sony team to hurry up DVD, CNET, http://news.com.com/2102-1023-219879.html, August 2, 1996.

[3]   IEEE—http://www.ieee.org. Shellhammer, S. (2001). IEEE 802.15.2 Clause 5.1 Description of the Interference Problem, July 2, 2001.

[4]   There is debate about whether this strategy is legal. Judge Motz of the U.S. district court in Baltimore ordered Microsoft to include Sun Microsystems' version of Java with the Windows operating system (CNET, http://news.com.com/2102-1001-978786.html, December 23, 2002). The verdict is a response to Sun's suing Microsoft "for allegedly violating antitrust law in dropping Sun's version of Java and including its own version, which Sun alleges to be incompatible with its technology."

[5]   IETF, 18-2-2002, http://www.ietf.org/html.charters/megaco-charter.html, and IETF, Request for Comments: 2805, April 2000, http://www.ietf.org/rfc/rfc2805.txt.

[6]   Lecture by John Ketchell, Director of CEN/ISSS, SIIT 2003 Conference, Delft.

This chapter was previously published in the Journal of IT Standards & Standardization Research, 2(2), 18-33, July-December, Copyright © 2004.

Chapter VII

When is
Standardization Slow?

M. H. Sherif, AT&T, USA

ABSTRACT

This chapter presents an analysis of the political, marketing, and technical aspects associated with the perception that standardization is slow. A framework is defined to evaluate the rate of standardization in terms of meeting users' needs.

INTRODUCTION

It is widely believed that formal standard bodies are less responsive to market needs than industrial associations or consortia. This belief has had at least two consequences. First, there has been an unprecedented increase in the number of ad hoc groups to promote specific technologies. Second, some traditional standards-developing organizations have fallen out of favor and have reduced their activities. Yet the foundation of such a perception has never been fully investigated.

While technical specifications provide the basis for the development of products and services, they also reflect the influence of participants and their power to shape events. When similar technologies offer comparable performances, parties may adopt divergent or competing paradigms, and their disagreements would not be resolved by proof because the various groups "practice their trades in different worlds" (Kuhn, 1970). Thus, any analysis of the responsiveness of standards to market needs must take into account the merits of the technologies involved, the standardization strategies of the participants, as well as any unstated assumption, whether collective or individual (Sherif, 1997a).

In this chapter, we propose a way to assess the optimum speed of standardization by separating the technical issues from immediate business needs and/or ideological persuasion. The analysis excludes the social and interpersonal aspects of the standardization process on the basis that they relate mostly to the management of human resources. The chapter briefly reviews the current political context in which claims on the relative responsiveness of standardization organizations are stated. Then it summarizes the marketing viewpoint with respect to the need for standards. While this view assists in defining marketing strategies, it overlooks the nature of the innovation and the technological complexity of equipment and services. To overcome these shortcomings, the chapter then introduces a framework to assess the impact of an innovation by combining the notion of value-chain discontinuity (Christensen, 1997) and the technology s-curve (Betz, 1993). Next, the framework is applied to illustrative examples from wireless digital telecommunications services. The final section summarizes the main conclusions of this chapter.

THE POLITICAL CONTEXT OF INFORMATION AND COMMUNICATION TECHNOLOGIES STANDARDIZATION

It is well known that the production of technical specifications intertwines technical considerations with business, social, interpersonal, political, and ideological aspects. This is especially true in the case of ICT that is embedded in a wide variety of products and services. ICT attempts to fulfill a multiplicity of needs exacerbate the inevitable tension among the different requirements, particularly because the time horizon of manufacturers (of terminal and of network equipment) is often much shorter than that of network operators (McCalla & Whitt, 2002; Sherif, 1997b). Furthermore, ICT services do not involve equipment only, but include network-management systems and operation-support systems (OSSs; fault management, performance management, billing, security, etc.), as well as proprietary methods and procedures (M&Ps) specific to a given network provider.

It is often asserted that the new breed of standard organizations and consortia is capable of turning around specifications at a faster rate. Many espouse the view that the Internet standardization process is "fast and democratic...light years away from the protracted and plodding activities of the official standards bodies" (Turner, 1996). Mr. Anthony Rutkowski (1995, p. 597), previously the executive director of the Internet Society, wrote:

the Internet standards development process is by far the best in the business...The traditional bodies struggle to evolve within a standards market place that finds their products largely unacceptable, yet still run processes that incur collective costs of tens of millions of dollars per year.

At the same time, the Internet is "a distributed collaboration and innovation engine that has produced a thriving new field of electronic communication and a ten-billion-dollar global marketplace growing faster than any communications technology yet devised" (Rutkowski, 1995, p. 597).

Mr. Kowack (1997, p. 55), a member of the board of directors of the Commercial Internet eXchange Association, the industry association for Internet-service providers and businesses, stated:

the general direction of the Internet and the activities of its global community of users, consistent with Paine's concept of "natural individual propensity for society," are both philosophically legitimate and arguably superior to jurisdictional claims by nation-states...Internet advocates may not have a dominant argument in any territory with a government, but they will have philosophical legitimacy on their side.

Complaints about slowness of standardization may be less relevant for technologies with long lead times for development and deployment, such as for infrastructure projects. In these instances, agreements on shaky technical foundations for the sake of producing documents may not be very helpful in the long run. Furthermore, from a user's viewpoint, it is more likely that the quality of the final product or service counts more than the time to define a formal standard. For example, the performance of speech-compression algorithms needs to be assessed under a wide range of network conditions and for as many languages as possible to satisfy users worldwide. This is why in the past, official standards bodies concerned with the telecommunications infrastructure were deliberately slow. The International Telecommunications Union's Telecommunications Sector (ITU-T), previously known as the CCITT (Comité Consultatif International Télégraphique et Téléphonique), used to approve its recommendations every 4 years. This pace, however, while more than adequate for the mature technologies of telephony, could not track the successive waves of innovations in data communications or accommodate instabilities associated with the deregulation and privatization of telecommunications. Faced with growing criticism, the ITU revised its mode of operation first by shortening the cycle to

2 years, then by abolishing it entirely. Today, ITU-T recommendations are adopted as soon as they are ready and, in some cases, the time interval from start to finish could be compressed to 18 months. Nevertheless, the ITU is still being considered a slow organization.

In reality, many consortia based their work on existing formal standards. For example, the hypertext markup language (HTML) and extensible markup language (XML) descend from the standardized general markup language (SGML) that ISO defined in 1986 on a contribution from IBM. Furthermore, after a flurry of some initial successes, new standardization organizations often experience schedule delays. The ATM Forum needed several years to revise its Private Network to Network Interface (PNNI). Likewise, it took nearly 10 years for the Object Management Group (OMG) to define the object-management procedures for the Common Object Request Broker Architecture (CORBA) of CORBA 2.2 (1998), which had to be extensively revised to issue CORBA 3 in 2000. Finally, the backlog of delayed or canceled standards by the IETF (Internet Engineering Task Force) is increasing. One cannot escape the conclusion that statements like "Formal standardization is too slow," which are contradicted by objective facts, have political and ideological motivations.

Political considerations can appear in many other areas as well, such as in membership selection, in the decision-making process, and in the policy on intellectual property rights (IPRs; Kipnis, 2000). One clear example of political bias is the description of the IETF as an open and democratic forum (Huitema, 1995) even though the basic Internet standards were developed by a closed group of contractors when access to the Internet was restricted by the U.S. Department of Defense (DoD; Malkin as cited in Lehr, 1995). Many Internet enthusiasts perceive it as a challenge to established legal and commercial principles such as the jurisdiction of the nation states and have attributed the success of the IETF to the merits of a free initiative (Kowack, 1997; "The Second Annual Roundtable on the State of the Internet," 1996). In reality, the Internet was funded, initiated, and controlled by public funds through the U.S. DoD, and in 1998 the U.S. Department of Commerce authorized the Internet Corporation for Assigned Names and Numbers (ICANN) to oversee the domain-name assignment and registration process. Incidentally, ICANN has been accused of "ignoring all the lessons about checks and balances, representation, and procedure that can be drawn from centuries of experience with governance institutions" (Mueller, 2002, p. 16).

In short, criticisms of established standards associations, particularly those with government involvement, parallel sustained attempts to replace political and democratic control of institutions with technocratic and financial procedures and to privatize the public space. The confluence of all these phenomena can be interpreted through the observation that unfettered markets must be engineered by weakening or dissolving all intermediary social institutions between individuals and economic entities that may challenge market-driven mechanisms (Gray, 1998).

THE PRESSURE TO
STANDARDIZE AND INNOVATE

Leaving ideological considerations aside, slowness can be considered from the point of view of specific user communities, in particular, the marketing professionals and the technologists. The latter group, in turn, has several camps, such as those that are involved in the design of equipment, those that deal with software products, and those that are involved in services. In the case of telecommunication services, technology competence covers configuration, fault management, accounting and billing, security, customer care, and so forth. Innovations in telecommunication services take place in one or more of several areas: networking technologies (transmission and switching), operations-support systems, and methods and procedures in the network operations center (NOC). Each of these areas operates on a different schedule even though standardization has traditionally stopped at that of network elements. Another common conceptual error is to assume that all types of innovations have the same characteristics and are operating under the same time constraints.

The urgency of a standard varies on whether the innovation is incremental, architectural, or radical, or a platform innovation. Incremental innovations occur in mature technologies to increase profitability by reducing costs and improving efficiencies. It is estimated that half of the economic benefit of a new technology comes from process improvements after the technology has been commercially established (Christensen, 1997). With incremental innovations, the growth in the revenues from telecommunications services and the corresponding demand for supporting networking resources takes place 4 to 10 years after the peak sale of the network equipment (the time lag depends on the types of service; McCalla & Whitt, 2002). This creates the following contention among standards users: Service providers rely on incremental innovation to increase revenues while the profits of manufacturers are mostly associated with platform innovations.

In the rest of the chapter, we restrict our attention to the technological aspects only. (The service aspects are discussed in the chapter entitled "Standards for Telecommunication Services"). We use the evolution of facsimile to illustrate the point as shown in Figure 1. The transition from purely analog facsimile was a platform innovation. Two platforms were in competition: a purely digital platform on ISDN (G4 fax) and a hybrid analog and digital platform (G3 fax). With the advent of the Internet, transporting facsimile using the TCP/IP protocol stack was a purely architectural innovation.

Figure 1. Evolution of facsimile over three of the four innovation quadrants

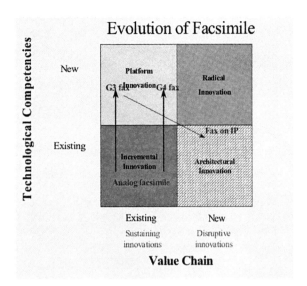

We are now ready to relate the characteristics of the standards (speed and details) to the type of innovation and the technology life cycle. This is the subject of the following section.

STANDARDIZATION, INNOVATION, AND THE TECHNOLOGY LIFE CYCLE

The purpose of this section is to offer an objective evaluation of the speed of standardization as it relates to the technology life cycle and the type of innovation. Figure 2 illustrates how the type of standards needed at each phase of the technology life cycle depends on the technology maturity (Betz, 1993; Weiss & Cargill, 1992).

Anticipatory standards are needed to specify the production system of the new technology provided they allow for errors. The associated specifications cannot be as robust as when the technology is mature and the market for the service is well defined. Given the uncertainties, small-group discussions unencumbered by procedural rules are more appropriate. Accordingly, proof-of-concept consortia seem to offer a suitable environment for competing firms to share knowledge. For example, the development of CORBA, Bluetooth, and WAP protocols were done by proof-of-concept consortia. Similarly, the development of the TCP/IP protocol stack in the early phases of the Internet (late 1970s and early 1980s) occurred in an environment akin to that of proof-of-

Figure 2. Timing of standards in relation to the technology s-curve

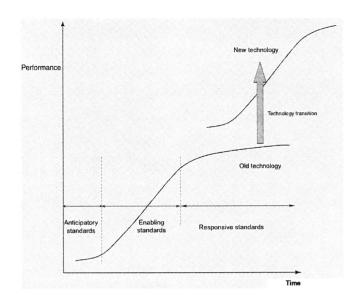

concept consortia. Formal standard bodies may also participate provided that they allow experimentation, as shown in the case of the IEEE 802.11 series of specifications. In fact, anticipatory standards may be able to benefit the most from a hybrid mode of standardization in which the tasks are divided among formal standards organizations and industry-driven consortia (Vercoulen & Wegberg, 2000).

Enabling standards relate to refinements in the production system as well as improved embodiment of the technology. They are usually generated while the performance of the innovation is improving exponentially and initial products are commercialized. Standardization in this phase provides an avenue to the dissemination of knowledge about the technology. Formal standards organizations or implementation consortia can equally develop enabling standards: formal specifications on the one hand, and on the other, development tools, conformance verification processes, and/or the promotion of the technology.

Responsive standards relate to the manifestation of the technology in a service system, that is, a "completed and connected set of transformational technology systems used in communicating and transacting operations within and between producer/customer/supplier networks" (Betz, 1993, p. 361). To define a responsive standard, the initial innovators may have to release technical information earlier than anticipated or shift product differentiation to areas such as quality, customer support, or maintenance services. Therefore, if there is a concern (based on unfulfilled needs and current technological progression) that a technology discontinuity may happen within 5 years, then a responsive standard may not be worth the effort. For example, with the transition to the third-generation wireless networks being actively pursued, we should not expect many responsive standards for wireless communications.

Application consortia and formal standard organizations can provide a neutral ground for defining users' requirements and deployment architectures of responsive standards. Note that application consortia deal mostly with responsive standards.

Example: Wireless Telecommunications

Restricting ourselves to digital wireless communications, it is clear that GSM was a platform innovation, and the development of its specifications was an enabling standard for both the service operators and the manufacturers. Once the initial platform was in place, subscriber growth was achieved through acquisitions. This required some architecture innovations to integrate different operations-support systems and to improve efficiencies through electronic data communications. Such innovations do not usually call for much standardization because they are kept proprietary.

With the growth of subscribers saturating, the emphasis has been to persuade existing customers to subscribe to additional data services to increase revenues or to compensate for the reduction in the price of voice calls. Thus, for service providers, GPRS and SMS are responsive standards in the sense that the market was already established and the performance of the GSM system had reached a saturation level. In contrast, manufacturers would tend to see these standards as anticipatory because they are needed for the design and deployment of the equipment. Clearly, there is a tension among manufacturers and service providers in terms of the urgency of coming up with a specification.

As explained earlier, WAP is an architecture innovation to offer mobile information services through a common specification defined by a consortium of companies (i.e., it is an anticipatory standard). The intent of this architecture was the support of a large number of available and planned wireless networks, for example, GSM, CDMA, and so forth, and various portable devices (palmtops, phones, computers, personal digital assistants). It used a new variant of HTML, wireless markup languages (WMLs), to display data with a telephone terminal equipped with a microbrowser. Its commercial failure shows the importance of market input for the success of architecture innovations and artificial standards.

Transition to the third-generation wireless devices may be considered as a discontinuity in both the value chain and in the technology competencies. The UMTS (Universal Mobile Telecommunications Service) is an anticipatory standard for new services for voice and data (144 to 344 Kb/s outdoors, 2 Mb/s indoors). It is still not obvious what services should be offered and what configurations of customers, competitors, and suppliers will emerge in the new environment. Some standardization is of course needed to achieve interoperability or to ensure adequate service levels (e.g., to ensure roaming); nevertheless, the whole transition is fraught with the myriad of problems associated with radical innovations. For such an innovation, standardization should proceed very cautiously and focus on nothing but the basic interoperability among the various platforms.

SUMMARY AND CONCLUSION

There are two main conclusions to the chapter. First, unqualified statements on the benefit of one type of standardization over another are essentially ideological statements that, in the current context, tend to favor deregulation, privatization, and the establishment of unfettered markets. Second, it is possible to evaluate on more objective terms the importance of the speed of standardization for a given situation.

While standardization can take place anytime, successful standards meet the expected needs of the technology users. By relating the type of standards to the technology s-curve, it may be also possible to evaluate whether the standardization of a technology makes any sense. Depending on the complexities involved, it may take anywhere from a year to 3 to 4 years, or even 8 to 10 years for a standard to be fully and successfully implemented. If the consensus is that the performance of a given technology can reach its maximum in less than a year, then an enabling standard may not be feasible. Very few firms would voluntarily devote precious resources to that effort.

By matching the timing of standards to the intrinsic capabilities of the technology as well as the market situations, it is possible to analyze the adequate speed of standardization. Standard professionals may build on the process outlined in this chapter to audit the work in various standard bodies and plan their efforts accordingly. Hopefully, this chapter will encourage additional contributions on this topic.

REFERENCES

Abernathy, W., & Clark, K. N. (1985). Innovation: Mapping the winds of creative destruction. *Research Policy, 14*, 3-22.
Betz, F. (1993). *Strategic technology management.* New York: Mc-Graw Hill.
Christensen, C. M. (1997). *The innovator's dilemma: When new technologies cause great firms to fail.* Boston: Harvard Business School Press.
Gray, J. (1998). *False dawns: The delusions of global capitalism.* London: Granta Books.
Huitema, C. (1995). *Le routage dans l'Internet.* Paris: Eyrolles. (French version).

Huitema, C. (1995). *Routing in the Internet.* Englewood Cliffs, NJ: Prentice Hall. (English version).

Kipnis, J. (2000). Beating the system: Abuses of the standards adoption process. *IEEE Communications Magazine, 38*(7), 102-105.

Kowack, G. (1997). Internet governance and the emergence of global civil society. *IEEE Communications Magazine, 35*(5), 52-57.

Kuhn, T. S. (1970). *The structure of scientific revolutions* (2nd ed.). Chicago: The University of Chicago Press.

Lehr, W. (1995). Compatibility standards and interoperability. In B. Kahn, & J. Abbate (Eds.), *Standards policy for information infrastructure* (pp. 123-147). Cambridge, MA: MIT Press.

McCalla, C., & Whitt, W. (2002). A time-dependent queueing network model to describe the life-cycle dynamics of private-line telecommunication services. *Telecommunications Systems, 19*(1), 9-38.

Mueller, M. (2002). Why ICANN can't. *IEEE Spectrum, 39*(7), 15-16.

Rutkowski, A. M. (1995). Today's cooperative standards environment and the Internet standards-making model. In B. Kahn, & J. Abbate (Eds.), *Standards policy for information infrastructure* (pp. 594-600). Cambridge, MA: MIT Press. (A slightly modified version is available online at http://www.isoc.org/internet/standards/papers/amr-on-standards.shtml.)

The Second Annual Roundtable on the State of the Internet. (1996). *IEEE Spectrum, 33*(9), 46-55.

Sherif, M. H. (1997a). *L'Internet, la société de spectacles et la contestation.* Paris: Université Paris-Dauphine, Institut de Recherches Internationales, Groupe de Recherche Économique et Sociale.

Sherif, M. H. (1997b). Study Group 16 and the interface between the Internet and public networks. *Proceedings of Telecom Interactive '97,* Geneva, Switzerland.

Turner, A. (1996, March). A question of protocol. *Communications International,* 74-78.

Vercoulen, F., & Wegberg, M. v. (2000). *Standard selection modes in dynamic, complex industries: Creating hybrids between market selection and negotiating selection of standards.* Universiteit Maastricht. Retrieved from http://www-edocs.unimaas.nl/files/nib98006.pdf

Weiss, M., & Cargill, C. (1992). Consortia in the standards development process. *Journal of the American Society for Information Science, 43*(8), 559-565.

Chapter VIII

Ready – Set – Slow:
A View from Inside a CEN Working Group

Jürgen Wehnert, urbane ressourcen, Germany

ABSTRACT

There is one thing most people seem to know about standardisation: It is slow. While innovation rates and product cycles grow shorter the principal mechanisms for making standards has not changed a lot over the years. This is at least true for the official standardisation bodies which are based on voluntary work and consensus building. Industry initiatives, however, can work with different targets, timing, and funding. The chapter investigates the standard making from within the very working groups which actually develop standards. The example from within Europe's CEN shows why standard making is slow and what could be done about it. A lot boils down to management issues and funding, but also to industry interest and EU policies. One thing is for sure: The members of standardisation working groups do not delay standard making on purpose, but are part of a system which requires a certain degree of revision and innovation.

INTRODUCTION

Standards are an accepted and necessary part of our industrial culture: They provide vast societal benefits. Standardisation is an essential enabling mechanism and a tool for many products, services, administrations, and business relations. Originally, there were isolated national standardisation bodies. Later, they were organising themselves first regionally and eventually on a global scale. In Europe, both national and international standardisation bodies provide powerful vessels for ever more and increasingly complex normalisation. The national governments recognise the importance of the work, as does the European Union (EU). Today, the latter has to be considered the most important official body in Europe to monitor standardisation, to interact with its governing bodies, to partly sponsor it, and to influence it.

Currently, the main concerns voiced about standardisation are costs and the slow going of the work. M. H. Sherif (2003) has analysed the perceived slowness of standardisation on various principal levels. This contribution adds a grassroots perspective with a particular view to IT standardisation. It first describes the framework and conditions under which working groups operate within CEN. This may help those not active in the practical standardisation work to understand why standardisation is often difficult, slow, expensive, and not very responsive. Maybe, though, there is scope for faster and more efficient work: Changing a few working principles and improving working conditions may help a lot. After all, if official standardisation has a role in this world, it has to improve and position itself better against growing industry standardisation.

RELEVANT STRUCTURES WITHIN CEN

Policy makers are aware of the importance of standardisation, the industry requires it, and the necessary organisations and structures are all in place. Before we investigate how the "doing" level is set up, we will be briefly looking at some relevant elements of the CEN organisation, the European body that governs most of the work related to information technology. CEN is organised in a number of technical committees (TCs), each of which is dedicated to a specific subject; for example, TC48 is dedicated to domestic gas-fired water heaters. There are about 220 committees, of which about 10 are more or less concerned with IT matters.

The TCs receive their guidance from CEN's Technical Board (BT): "The Technical Board of CEN is responsible for the development of technical policies and for the overall management of technical activities to guarantee coherence and consistency of the CEN standardization activities system-wise" (CEN Web site, http://www.cenorm.be/standardization/bt.htm).

The BT has a number of means to fulfill these responsibilities: by setting up task forces or by mandating specific investigations.

The BT exclusively serves as a policy- and decision-making body, while the TCs are related to putting the work into practice:

CEN Technical Committees (CEN/TCs) are responsible for the programming and planning of the technical work in the form of a Business Plan, for the monitoring and the execution of the work in accordance with the agreed Business Plan and for the management of the standards making process, including the respect of CEN's policies and the consensus building amongst all interested parties represented through the CEN national members, the CEN Associates and the Affiliates. (CEN Web site, http://www.cenorm.be/ standardization/tech_bodies.htm)

Each TC has a number of working groups (WGs) in which the actual standardisation work is being carried out. Each of these groups is dedicated to a more specific aspect of the overall theme of the TC it is allocated to. An example would be TC224, Machine Readable Cards (i.e., credit-card-sized magnetic stripe cards, smart cards, etc.), with currently four working groups, one of them being WG11, Surface Transport Applications. The number of working groups allocated to the TCs varies. Also, the life cycle of a WG may be shorter than that of its TC, as it may be dissolved after completing its specific work or because its work has been promoted to a global (ISO) level.

WORKING GROUPS IN CEN

The working group is the place where standards are made, but on what grounds? TCs are obliged to produce business plans, which are publicly accessible at http://www.cenorm.be/cenorm/businessdomains/technical committeesworkshops/centechnicalcommittees/index.asp. They are meant to put forth the strategies and implementation plans of the TCs, but these plans are of varying quality.

Business plans as used in the commercial domain contain budgets, management structures, targets, deadlines, and other subjects relating to the delivery of a product in good time, good shape, and within an agreed budget. The TCs' business plans lack important elements such as budgets and management structures because the work is voluntary, and both the TCs and the working groups have a very limited set of tools to enforce the plans.

The most concrete guidance for a WG is in the work item description. This is the description of what a new future standard is about, what it should be aiming at, and what it should contain. It may be more or less precise but should at least serve as a good starting point for the work to be done.

Any WG may work on one or a number of such work items at the same time. Work items are either derived from the TC business plan or they are suggested by a national member body or from within a WG. A new work item is then refined and accepted in a formal Europe-wide voting procedure.

The work item is thus a concrete order for a WG to undertake the design of a standard, and it contains a description of the product (the standard) to be designed. The timing is based on two elements:

1.   The WG's assessment of how long it will take to provide a draft proposal to be approved by the TC (within limits set by CEN)
2.   The formal rules on enquiry and balloting times

With the work item description and the timing, a working group has two important elements of a business plan for its work at hand. Other important issues are of course who does the work and under what management system.

MEMBERS DO THE WORK

CEN is based on the membership principle. The institutional members are the national European standardisation bodies, such as AFNOR for France and DIN for Germany. While the actual work is carried out in the international WG, the national bodies nominate delegates to the WG and monitor its work one way or the other. They often set up mirror groups for each or a number of WGs that also provide input from a national perspective. Anyone drawn to work in such a national body may be voted into the national mirror group, which is a mere formality.

Any member of a national (mirror) group wishing to participate in a European working group may ask his or her national group for promotion to become a member of the European WG. The European WG has a simple set of rules on how members can be accepted or rejected. In reality, as a leader of a CEN WG, I am normally happy to welcome new members as manpower is most of the time a scarce resource.

The delegation principle and the voluntary nature of the work makes it a fairly random affair whom you have in the group and who attends regularly. Thus, in principle a working group is a motley crowd, which is not to say that the members are not experts in their field, and generally speaking, the quality is very high. But there is little control over having the appropriate resources available at the right time.

Another problem is getting a reasonable representation of interests around the table. Depending on the standardisation subject, the true problem owners may or may not attend. Transport, for example, is a subject in which the problem owners are chronically absent, that is, those who organise transport or who carry it out. Technology providers and, for example, researchers define standards for the absent users. It can also happen that important industries or countries are not represented. Important administrations are also chronically absent.

The progress is slowed down if important members do not attend regularly, or if they are replaced during the design phase of a new standard, because the new or newly attending members have to be filled in and get (re)acquainted with the work.

Missing attendance from particular countries does not necessarily slow down the progress of the work. However, nonattending nations and relevant companies can and sometimes do cause a problem: Only when it comes to voting on a draft standard, such absent parties realize that the standard in its current shape is not in their interest. They then tend to block the standard or try to introduce changes, and in any case, they cause unnecessary delays.

Generally speaking, work progress depends more on the currently available workforce and everyone's goodwill than on precise planning.

MANAGEMENT

CEN's working groups are under the auspices of their respective TCs. The TCs in turn are under control of the Technical Board, which is working on a strategic level. High-level ad hoc committees and advisory bodies may be installed to identify new general directions, interact with the Europena Commission, modify work programmes, and so forth. This committee work regularly requires voluntary contributions from industry, associations, and so forth. The aims, depth, and outcome of such meetings and groups are thus not very predictable and may not recognise and reflect the needs of those doing the actual work in the working groups. As a matter of fact, the WGs are decoupled from the spheres of high-level decision making.

For the working group, only its dedicated TC is of relevance. A WG is led by a convenor who is an elected member of the working group. He or she reports to the TC. The group organises its own meeting schedule and is generally fairly autonomous. Overall, the work is governed by the work items assigned to the group.

A small set of documents for the convenor supports and guides a WG. The groups are advised to attempt consensus whenever possible. Convenors have very little constitutional power and thus primarily have to rely on their stamina, patience, management quality, and diplomacy. There is of course general goodwill in a working group to reach the goals as defined in the work items, but that does not automatically provide good working methods and time consciousness.

The management of a group is done on a voluntary basis, and it is not unlikely that a convenor is chosen for political reasons or because nobody else wants to do the job. In an ideal world, a leader would be chosen because of his or her management expertise; in reality, a working group may be happy to find someone who is expert enough and whose company comes up for the significant costs.

The quality of the convenorship of a large working group contributes to saving or wasting hundreds of thousands of Euro and a lot of time. But alas, there is no training available, there are often no or few management or secretarial resources available, there is no qualification required, and there are no institutional powers vested on a convenor.

TIMING

The timing is governed by CEN rules. Deadlines are given for the submission of drafts, and the way drafts are promoted to the different levels and balloting periods are fixed. However, there are no penalties for not meeting a deadline, and the WG working out a standard may apply for additional time. Only in cases in which the responsible TC has severe doubts about a working group to ever meet any deadline it will not grant an extension, and it will question and possibly withdraw the whole work item.

A working group will work as fast as it can, with the workload and the manpower defining the maximum speed. Often, though, members' or their companies' hidden agendas influence delays or threaten to produce less-than-optimum solutions.

Also, the work often incurs unintended delays, such as from the following:
- Nonoptimal management
- Lack of members' readiness to compromise
- Incompatible working methodology
- Cultural differences
- Insufficient support from the European Commission and the standardisation bodies
- Changes in briefings
- The necessity to translate standards in European languages

The time from the initial launch of a standardisation subject to its final approval can cost anything from 3 to 5 years. In IT terms, that is more than a product generation's life cycle and therefore it is often not acceptable. Various attempts have been made to speed up the process. One of them is the CEN ISSS workshop, which deserves qualified and comprehensive research in a separate paper.

Looking at regular standardisation, there is fairly little a working group and its management can do to speed up the process. The author codesigned a particular European standard in 1992. The draft was rejected, then redesigned and launched as a European prenorm (ENV). This time it succeeded and lived its first life cycle as an ENV until its regular revision in 2003. With new application requirements coming up, it took until the end of 2004 before the new version was finalized and launched for formal vote as a proper European standard (EN). Keep the fingers crossed that thirteen years after the idea of the standard was conveived it will be approved.

Industry often criticises standardisation to be too slow. That may be true for the balloting times and delays that are sometimes caused by CEN's bureaucracy. From inside a working group, it looks different, though: The same industry that complains about slow progress does not delegate sufficient manpower to WGs and sometimes slows down the work on purpose if it suits its specific interest.

DIRECT SUPPORT OF THE WORK

A usual working group meeting comprises of an intense exchange of opinions, of focused drafting and redrafting, and so forth. At the end of a meeting, the focus is then lost and the voluntary participants all return to their day-to-day work. To keep the work going in between meetings is very difficult. Homework and additional small team meetings are possible, but not very common and cannot be enforced.

To support the groups in particular between meetings, it would be useful to have a number of support measures at the disposal of the WGs.

- Document servers with exclusive access for WG members. These facilities may be available from some national CEN member bodies, but not from all, and CEN itself does not provide this facility.
- In the absence of simple document servers, it would obviously be too much to ask for facilities for cooperative work. Such facilities would be useful for the time between meetings.
- Working groups require good cross-referencing facilities for their work, such as a central database and retrieval system. CEN does not possess such a system.
- A relevant part of standardisation work is to shape standard drafts in the required format. This is editorial work for which the working group should receive support, but there is no funding for such work. There is not even a good system to align the terminology use.

All the above items make quality assurance and control a difficult matter. Only because many working group members are active in more than one group do they happen to know about neighbouring activities, and an effective informal network makes it possible to reduce double work.

HIGH-LEVEL SUPPORT

A working group interfaces on a working level with other international standardisation groups, and individual delegates interact with their mirror groups. On a management level, the WG deals with its technical committee only, but of course, the work is governed and influenced by a number of other factors. The role of he European Union is of particular importance.

The Union acknowledges the importance of standardisation and supports it in a number of ways. Its support can be very general, or it can be targeted on a high level, or it can be very specific and explicit. Examples for targeted support are the establishment of standardisation-related tasks in a research framework programme or the support for CEN's accelerated ISSS workshops. An example for specific support is the financing of project teams. Project teams are put on a specific normalisation project, for example, aimed at the production of a complete standard draft. Unfortunately, this type of comprehensive and effective support is now history.

Three European directorate generals (DGs) have relevant stakes in IT standardisation:

1. Information Society (DG INFSO)
2. Transport and Energy (DG TREN)
3. DG Enterprise

The DGs interact with CEN and other European bodies to understand both the requirements of the industry and those of the normalisation bodies, and also to influence the work, for example, in order to support European competitiveness or policies. While this is of principal relevance, working groups may be affected negatively. One example is the diversion of manpower because the EU launches new groups outside regular standardisation.

Another example are EU initiatives like eEUROPE, which have a tendency to draw members from standardisation groups because they are attracted by a possible fast track, possible funding, and other promises. After all, the number of experts available for any particular subject is limited.

While such initiatives may be useful as such, they do disturb the regular work not only because of the brain drain they cause, but also because they may complicate the WG's work. This is because new and additional interaction and interfacing is required. We are not promoting to fence off alternative activities. Indeed, they may bring in fresh and alternative ideas, but there is little reason to believe that the same people are able to solve the same problems more efficiently and effectively in a different setting with most of the principal problems being the same.

Another problem lies in the Union's tendency to come up with new themes every now and then and in particular for each new framework programme.

These themes sometimes have the quality of buzzwords that are echoed by framework project proposals and eventually find their way into standardisation. Due to the inertia of standardisation, this sort of fashionable activity does not frequently result in concrete standardisation work.

Unfortunately, neither the Commission nor CEN can overcome an important drawback of a number of standardisation subjects. The author, being primarily familiar with transport IT, for many years has been monitoring the absence of the users of transport-related standards. Standardisation work is carried out with very little engagement from true problem owners, which is an unhealthy situation. Promoting new standards from a high level without looking toward the other end, that is, the people available for their design, is not very healthy.

CONCLUSION

The ultima ratio of the requirements for today's standards has been nicely phrased by Charles H. Piersall (2003), chairman of ISO's TC8.

- Systems-based management
- Driven by the market
- Universally accepted as the "standard of choice"
- Timing and timeliness
- Developed concurrently with new technology
- Focused on interchangeability and performance
- Must define a product that can be manufactured, installed, used, maintained, and supported safely and efficiently at competitive cost

After reading this chapter, the reader may be tempted to ask why anyone still believes in official standardisation to achieve the above goals. There are indeed many odds against the successful, cost-effective, and timely implementation of standards. Only because the industry has many good reasons to promote and support standardisation does it cope with very significant deficiencies.

Does it have to? We believe there are a few fairly simple measures that would boost efficiency, save time, and take stress from the work. There are other more fundamental issues that are difficult to ease, for example, because they have to be seen in the context of global standardisation. We thus stick with the simple ones and conclude this chapter with a list of proposals.

Efficiency Boosters

CEN and/or the national standardisation bodies should provide working groups with effective support measures in the shape of personal and technical assistance.

Accept Reality

The European Commission should target its high-level activities in such a way that they do not interfere with the actual normalisation work. It should also focus more on how standardisation should be undertaken and supported rather than what should be normalised according to short-lived ideas.

Provide Funding

Voluntary work depends highly on industry interests and mercy, and lacks coherent planning, budgeted funding, and so forth. The costs of any funding would be easily offset against faster progress and better standards.

Improve Management

Standardisation work is generally successful because of the assembled know-how and dedication, and because of a good piece of luck. By supporting professional management and working standards, by reducing interferences from hidden agendas, and so forth, the element of randomness could be significantly reduced.

REFERENCES

Piersall, C. H. (2003). *Time to market.* Presented at the Second ISO Conference for Technical Committee and Subcommittee Chairs, Geneva, Switzerland.
Sherif, M. H. (2003). When is standardization slow? *Journal of IT Standards & Standardization Research, 1*(1), 19ff.

Section IV

IPR Problems

Chapter IX

Patent and Antitrust Problems Revisited in the Context of Wireless Networking

Gary Lea, University of New South Wales, Australia

ABSTRACT

The author seeks to illustrate some of the ongoing problems that patents present for those seeking to standardize in the ICT field. The chapter illustrates these problems by drawing on patent and international trade disputes surrounding the rollout of IEEE 802.11 family (colloquially, "WiFi") technologies during 2003 and 2004. It then presents several solutions including the introduction of a more systematic approach to dispute resolution by standards development organizations (SDOs) based around ADR procedures derived from the domain name Uniform Dispute Resolution Policy (UDRP), corresponding changes to dispute handling in international trade disputes and, in the long term, alternation to intellectual property laws to allow for appropriately-tailored standardization exceptions (at least at the level of interoperability).

INTRODUCTION

In a general sense, there is nothing new under the sun under the heading above. Since Morse tussled over telegraph patents in the 1850s, patents have been an issue for the ICT sector, and, since IBM's brush with the rules against the tying of the supply in the 1930s, the relationship of intellectual-property and antitrust-competition law in the ICT context has been a vexed one. However, what is noticeable is that the pace, scale, and complexity of litigation and associated regulatory activity relating to these issues began to increase dramatically during the 1990s (Soma & Davis, 2000); this suggests that a new approach needs to be taken to matter.

It is against this background of ever-rising patent litigation, ever-stiffer antitrust and competition scrutiny, and a generally deteriorating standardization situation (especially as outlined below for IEEE 802.11-family technologies) that the author puts forward two reformist propositions for consideration: First, there is a need to move away definitively from the traditionally preferred methods used for resolving patent disputes where standards are involved, and second, consideration needs to be given to making changes to patent laws themselves.

PATENT LITIGATION AND WI-LAN

It is a truth universally acknowledged that patent litigation is expensive and, consequently, engenders considerable fear, uncertainty, and doubt (FUD) on both sides of the dispute. The first point is eminently provable: For example, given that the costs of litigating a patent dispute in the United States as a patent owner is, at present, likely to be over $2 million for $25 million in damages, and more than $4 million if the damages claim is over that (Milo, 2004), it is necessary to consider whether to sue very closely and, bearing in mind costs of roughly the same magnitude in other jurisdictions, the same issue also arises there.

Therefore, in the present context, nobody ever undertakes patent litigation lightly, but one may still legitimately ask whether parties have undertaken it advisedly. Since the 1990s, commentators have been pointing out that alternative dispute resolution (ADR) makes for quicker, cheaper disposal of patent and other IP disputes (Terdiman, 1997), and, for reasons discussed below, the need to maintain a collaborative frame of mind in a standardization context directly boosts the arguments in favour of nonlitigious approaches. Put another way, one could begin to think of those patent holders who litigate where standardization is an issue as either misguided or worse: misguided in that, in many instances, patent holders' lawyers will not always have pushed the ADR alternatives to litigation when trying to protect legitimate IP interests, or worse in that, as will be seen below, in some instances, some companies will use litigation as an anticompetitive weapon, simply relying on having deeper pockets or a bigger legal team to force the competition off the road.

Consider all of the above against the following factually based example. On June 23, 2004, Wi-LAN Inc. announced that it was suing the U.S. network equipment giant Cisco Systems Inc. in the Canadian courts for breach of its Canadian patent 2,064,975 (*Wi-LAN Escalates Patent Enforcement Program*, June 23, 2004). The patent in question related to wideband orthogonal frequency division multiplexing (W-OFDM), which, put at its very simplest, is one of a number of techniques used for the concurrent encoding of data on multiple radio frequencies with a view to improving efficiency in bandwidth usage in mobile telecommunications systems. One particularly useful feature of W-OFDM is that it allows multiple low-power radio-frequency wireless local-area networks to be set up, each of which can operate with no significant interference to its neighbours.[1]

On June 30, 2004, Cisco had responded cautiously, saying only that Wi-LAN appeared to be asserting that its patents affected the WiFi industry as a whole and that it was going to review its own position. Wi-LAN's representatives, including its president and CEO (chief executive officer), Dr. Sayed-Amr El-Hamamsy, and vice president of corporate communications, Ken Wetherell, have made the following points in the press to date:

- Wi-LAN filed notice with the IEEE of its willingness to license on a RAND basis some time ago.
- Philips Semiconductors and Fujitsu Microelectronics took licenses in 1999 and 2003 respectively for a one-off lump sum plus continuing royalty payments.
- Most other WiFi companies have not, and by filing against a big name, Wi-LAN was, indeed, "serving notice" of its intentions on the WiFi industry as a whole.
- Wi-LAN was initially filing in Canada because of home-player advantage, but would consider filing against Cisco and others in the United States on the basis of related U.S. patents numbers 5,282,222 and 5,555,268.
- Wi-LAN was seeking a court order to force Cisco to take a license on terms and to pay damages, including punitive damages.
- Wi-LAN also drew attention to its recent acquisition of a portfolio of patents, which, it claimed, were necessary for the implementation of WiMAX, the next generation of wireless broadband technologies (Nobel, 2004; Smith, 2004).

In the patent litigation involving Wi-LAN, the possible pathways out from the initial position described above would be as for any other case, that is:

- a pretrial settlement, including at the door of the court,
- the plaintiff winning,
- settlement during litigation, or
- the plaintiff losing.

So, what were the parties' relative positions at the start of the process? Turning first to settlement, then, subject to what is said elsewhere, this might have been possible. In the present instance, Wi-LAN could attempt to exert leverage by making it clear that the further the litigation process goes on, the tougher the settlement terms that will be imposed;[2] this form of arm-twisting is a common tactic, though one that can backfire very badly if antitrust- and competition-law or tort-law claims are raised, either by way of defense or counterclaim.[3] Generally speaking, however, settlement during proceedings is very common in patent disputes anyway, mainly because, like other intellectual-property-infringement proceedings, they are inherently complex, uncertain, protracted, and greatly influenced by any interlocutory remedies granted.[4] These factors all combined both facilitate and encourage settlement to some extent. As a typical example in the mobile technology area, see the outline of the settlement made between Extended Systems and Intellisync in March 2004 (*Extended Systems and Intellisync Settle Lawsuit*, March 4, 2004).

If, hypothetically, any attempts to settle failed, the case were to be pursued to trial, and Wi-LAN won, Cisco would have to pay damages based on a notional royalty fee applicable to each unit of infringing product sold plus, if a trial judge saw fit, an undetermined sum by way of punitive damages. If, conversely, the case went to trial and Wi-LAN lost, then it would have to cover considerable legal costs, it would see its stock price plummet, and, bearing in mind costs relative to the company's size, it could even end up insolvent and/or bought up by a rival. Thus, even when holding the apparent trump card of one or more patents, the potential risks involved for a patent owner in suing for infringement are always very high (Gifford, 2003).

Just beyond Cisco's immediate response would lie a very interesting exercise: a careful evaluation of the scope and strength of Wi-LAN's patents by its patent attorneys. Like the rest of the WiFi industry, Cisco would be aware that if it transpired that Wi-LAN's assertions about the fundamental and broad nature of its patents were to be correct, the vast majority of WiFi equipment ever manufactured to IEEE 802.11a and 802.11g is infringing. Given this, consider-able efforts would be made to examine the width of the patent claims[5] and to find any weaknesses in any one or more of them in terms of lack of novelty or obviousness when compared to the appropriate prior art base. In essence, in relation to Wi-LAN's patent, the question that arises in relation to each claim made under it is whether that claim is:

- valid and infringed,
- valid but not infringed (i.e., the claimed infringement lies outside the scope of the claim), or
- invalid (i.e., for lack of novelty or for obviousness, etc.) and therefore inherently incapable of being infringed.

Wi-LAN, for its part, would undoubtedly hope that companies other than Cisco would see what might be happening to Cisco, change their attitudes, and come quickly into its licensing scheme. This is because of the cost of patent suits, the loss of senior-management time, and the potential risk of losing. Taking the first issue, if Wi-LAN's assertions are correct, the company could be in line to sue multiple defendants in multiple jurisdictions; whilst this might result in a large payout, the legal costs of mounting such a challenge could make significant inroads into any award made. In the second instance, the management time issue is not just a quantitative issue but a qualitative one: If management's attention is being diverted from growing the business into lawsuits, experience shows that both strategic work and immediate decision making can suffer badly.

Finally, on the risk of a patentee losing, one must remember that no matter how strong a patent seems, there is never any cast-iron guarantee that a patent owner will win a suit: The defendant may pull a prior art rabbit out of the hat, thereby invalidating the patent in suit, or, in a marginal case, the judgment might just simply go against it. For the latter, whilst appeals are possible, the time and cost implications are often too much for all but the largest companies to bear.

PATENT LITIGATION RECONSIDERED

In summary, patent litigation, like any other kind of litigation, has a flavour of the game of Chicken, where the two contestants drive cars at each other until one or the other veers off the road. The problem is that there can be massive inequalities of nerve and driving prowess amongst the players of this particular variant, which translates into issues of legal funding, legal representation, the solidity of the IP base, and patent-litigation tactical play.

Moreover, whilst the consequences for any plaintiff that loses are often far more serious than most other branches of the civil law, for any defendant that loses, there are not only the considerable legal cost implications, but also the potential risk of having court orders made to recall and destroy products, thereby destroying business goodwill, leading to possible litigation from customers and having to pay compensation to them as well. The spectre of the $873,000,000 payable in the (in)famous *Polaroid vs. Eastman Kodak* case in damages and a court-controlled customer refund scheme still haunts many patent attorneys.

Beyond this, though, what is worrying here are the implications of this litigation, not just for the parties, but for the ICT industry as a whole. First of all, although this author has argued elsewhere (Lea, 2000) that patent blocks are seldom absolutely insoluble (i.e., because of the patent holder's overriding need to deal to secure some form of return on R & D [research and development], whether monetary or in kind, and to justify continuing to pay the not-inconsiderable patent renewal fees), it is a matter of concern that such massive uncertainty, wasted time, wasted commercial opportunity, and lost cost can happen for everybody involved before such a resolution actually occurs.

Second, it also becomes clear that even in those other situations where some form of legally prohibited sharp practice has occurred, the theoretical ability to apply antitrust- or tort-law rules to correct a problem, again discussed elsewhere by this author (Lea & Hall, 2004), may often be practically irrelevant because virtually the same time, cost, and uncertainty considerations will crop up as in an innocent-behaviour situation. This is not to say that antitrust- and competition- or tort-law claims, whether by way of defense or counterclaim, should never be considered, but merely that their application is, by their very nature, confined to a relatively narrowly defined set of bad or recalcitrant behaviours and will, therefore, only be efficacious in the most extreme instances.[6]

THE CHANGING BALANCE BETWEEN PATENTS AND STANDARDS

Traditionally, if it were possible for an ICT standards-development organisation (SDO) or its membership to resolve a particular patent or related issue in house (i.e., without involving lawyers), then this was the preferred way of doing things. However, increasing litigation against either SDOs or their members in that capacity under the two headings described above is rendering this approach increasingly difficult to sustain (Verbruggen & Lorincz, 2002). Moreover, one situation in which SDOs, including the IEEE, have always been very unwilling to get involved is where patent litigation breaks out, partly for fear of attracting some kind of secondary or derivative lawsuit, especially an antitrust or competition suit. The point to bear in mind here is that this is also likely to happen if an SDO does not intervene in some way.

So, then, the industry players fight it out in the courts and, in doing so, make obvious part of the reason for the increase in patent litigation: There are simply more patents to trip over. Whether because of changes in examination standards, growth in international access to national patent systems, expansion of patentable subject matter, or simply more applications being filed, it does not matter. The fact remains that, albeit with traceable annual dips and spikes, there are significantly more patents in force globally than at the start of the 1990s. However, much more important than even the raw numbers is the change in patent litigation culture, which has swept the ICT industry over the same time frame. Tansey (2005) has gone so far as to posit that, for some companies, patent litigation has become a do-or-die issue that is changing the rules of engagement within industries.

This change of attitude also applies in the more general context of simply being participants in standards-development processes. One phenomenon previously observed but rapidly increasing is "standards jumping," where companies second-guess what draft of a standard will finally be ratified, as what happened right at the introduction of IEEE 802.11g itself (Kewney, 2004). This can, in its turn, open up additional possibilities for patent misuse (Mueller, 2002). Furthermore, even leaving aside the obvious increases in competition between different SDOs, it is also noteworthy that there is rapidly increasing competition between standards-development committees within the same organisation as what may be termed "draft standard creep" comes to the fore. Again, although previously observed, its increasing prevalence within IEEE wireless networking standardization is again notable ("IEEE Groups Fight for Control of Key Standards," 2004).

One tried and tested solution to the problem of IPR disputes (being litigation settlement as part of some wider "pooling"[7] or one-stop-shop licensing scheme) was suggested and initial moves were made in relation to at least some of the IEEE 802.11 family of technologies. On April 14, 2004, Via Licensing, a subsidiary of Dolby Laboratories, brought together a number of industry players in Tokyo with the hope of creating arrangements similar to those dealing with MPEG 2 and MPEG 4 issues. However, what is interesting is the slow and difficult progress that has been made since that time toward the avowed goal of "cutting down on destructive lawsuits" ("Wireless Industry Intellectually Challenged," 2004). With so many players and such entrenched positions already arising from multiple patent disputes, it is clear that pooling or one-stop shops are best established ahead of time rather than after the event. This "act sooner, not later" argument also applies, a fortiori, to using ADR itself.

IPR POLICY PROBLEMS

Moreover, even where a relevant one-stop shop does operate, there would be another problem arising from the limitations of IPR policies in terms of encouraging or even forcing relevant IPR disclosure; this continues to be a major issue in the wake of the infamous *Rambus vs. Infineon* litigation (Alban, 2004). The Federal Circuit Court of Appeals acting in that case also picked up on what it described as "the staggering lack of details" in the relevant JEDEC IPR policy, indicating that, in legal terms at least, the document operating at the time of Rambus' membership and departure had more holes than a Swiss cheese.[8] In the wake of this withering rebuke, Jakobsen (2004) has recently suggested not just fine-tuning the IPR policies in the way this author has previously advocated, but introducing such radical measures as a fully lawyer-tested set of model policies for SDOs by sector, and a full professional ethics screening of SDO committee members. Similarly, there are increasing legal calls for a massive overhaul of rules relating to membership acquisition and participation (Webb, 2004).

This lack of clarity is evident almost everywhere. One key difficulty that still bedevils IPR policies is the attempt to develop a consistent and comprehensive definition of what actually constitutes an essential IPR. Leaving aside the issue of what IP is actually covered, ETSI deals with this issue by reference to the impossibility of substitution:

...on technical (but not commercial) grounds, taking into account normal technical practice and the state of the art generally available at the time of standardization, to make, sell, lease, otherwise dispose of, repair, use or operate equipment or methods which comply with a standard without infringing [the IPR in question].[9]

However, although this is far from being the worst definition publicly available, even this has been criticised, and more than one affected company has sought to challenge it (Godby, 2003).

Another serious problem stems from the lack of willingness or, in some cases, the simple inability of many SDOs to define what constitutes fair or reasonable in the context of (F)RAND licensing and the increasing amount of antitrust and competition litigation arising from that failure to define (Curran, 2003). In a recent law and economics analysis, Patterson (2002) makes the interesting suggestion that, in order to achieve maximum efficiency, it would be desirable to limit and define license fees by separating out those things attributable to exploitation of the patent from those things attributable to the deployment of the standard, for example, a fee portion calculated against cost savings achieved by allowing compliance with a standard, a fee portion for increasing the desirability of a product or service as a result of the particular technical feature the patent introduces, and a fee portion for direct impact on performance, interoperability, or other essential criteria.

Furthermore, Patterson (2002) argues, economic efficiency would also seem to dictate throwing away the long-held tenet that SDOs should not get involved in license negotiation: If truly representative of membership, that should be their prime function since they would possess greater bargaining power. Interestingly, there are also some suggestions that, in relation to the more specific economic conception of social efficiency, SDOs also tend to make inefficient decisions on whether to adopt a technology affected by a particular patent or set of patents (Teece & Sherry, 2003); this, however, is still a matter of some contention in terms of the strength of the effect when compared to that of the actions of the individual patentee (Carrier, 2003; Patterson, 2003).

RELATED COMPETITION AND ANTITRUST ISSUES

Most significantly, a problem with settling disputes in a way that results in subsequent collaboration (a virtual sine qua non for standardization) is that, particularly since the 1990s, there have inevitably been suspicions of such settlements sliding into or constituting anticompetitive collusion.[10] Worryingly, in a recent economic analysis, Willig and Bigelow (2004) have concluded that a key and regularly intervening regulator, the U.S. Federal Trade Commission (FTC), often examines many of the wrong indicators (e.g., the type and scope of payments made) and tends to rely on too-inflexible per se rules in its determination, suggesting that it would be better off merely determining competitiveness on a rule-of-reason basis arising from comparative likely vs. actual dates of entry of competition.[11] The message coming from that is simple: It is better to design processes and policies that are sufficiently honest in terms of transparency, openness, and accountability, and that are legally bulletproof so as to avoid protracted scrutiny in the first place.

Like its U.S. counterparts, the European Commission (EC) has also been developing a system of regulation for standardization activities that are increasingly sophisticated and comprehensive: Dolmans (2002) notes that there are now mechanisms under competition provisions of the EC treaty to deal with such issues as membership and access to process, spillover restriction effects, the depth of standardization activity, the selection of technology, access to information and essential IPRs, IPR policy disclosure issues, and compulsory licensing. Therefore, again, in order to enjoy any kind of protection from antitrust- and competition-law claims in Europe, then SDO operating procedures, including IPR policies, will have to be fundamentally altered to guarantee that they are regarded as fair, accountable, and transparent bodies serving a public interest. Part of this process should also involve moving toward a more mature stance on IPR litigation.

Finally, another familiar but rising phenomenon is the use of antitrust- and competition-law-based defenses in IP infringement cases, especially in the ICT standards context. For example, in addition to the (now relatively limited) patent-misuse doctrine, under U.S. law, any failure to disclose material information relating to the patentability of an invention that is the subject of a current application may constitute a form of fraud on the USPTO, and, consequently, any attempt to enforce a patent thus obtained would be a violation of the Sherman Act; this would also be true for sham patent litigation designed to disrupt a competitor's business relations with third parties (Abbatt, 2004).

Changing IPR Dispute Resolution

Given all the matters listed above, including the costs and FUD factor surrounding patent litigation, it is not surprising that interest in ADR has noticeably grown since the 1990s (Blackmand & McNeill, 1998). ADR, including for present purposes arbitration, conciliation, and mediation, can be viewed as an ideal substitute for IP litigation and, as has already been suggested, in order to preserve the collaborative environment of standards-setting work, should be considered as a matter of course and not as some legally exotic afterthought (Lemley, 2004). Indeed, it is noticeable that even U.S. courts themselves are beginning to use ADR as a prefilter for patent disputes more generally (Anonymous, 2005).

So, what should be done in the present context? In the short term, SDOs of all shapes and sizes will have to devise a scheme to resolve patent blocking and licensing problems more rapidly and systematically among what might be termed persons of goodwill.[12] The author proposes that a central, standing arbitration body be either created or appointed,[13] and then take up one or more mandates from one or more groups of SDOs under specific, uniform terms of reference to deal with IPR and standards disputes as defined.

Crucially, willingness to submit to such arbitration should be made a prerequisite for continuing membership of an SDO in that (a) membership should be made contractually binding and (b) the arbitration requirement should be tied into that, either within a revised IPR policy or as a freestanding obligation. In the case of nonmembers, the arbitration system should be made contractually binding wherever possible (e.g., offering some form of paid associate membership status, including the arbitration obligation, in return for standards information or other privileges) and should be strongly encouraged; members should make coordinated best endeavours to bring the nonmember to the table.

Although the author himself has been sceptical of many operational aspects of ADR in the past, it should now be possible to create a scheme of the sort required, particularly bearing in mind the very ingenious solutions deployed in response to other IP problems over the last few years, especially in the field of domain names and trademarks. Ideally, therefore, the arbitration scheme should take the best elements of the UDRP system developed from 1999 onward by ICANN in conjunction with WIPO (World Intellectual Property Organization) arbitration and other arbitration providers;[14] that is to say, it should be quick, specifically tailored, and low cost. In such an ideal world, an arbitration panel comprising a senior patent attorney in the chair and two assessors drawn from the relevant technical field should be able, within a matter of no more than 3 months by e-mail, to do the following:

- Determine and state publicly whether the patent in question is essential, possibly by reference to a direct legal and technical assessment of the possibility of infringement if use is made of a particular process in accordance with a given standard
- Make binding awards of equitable remuneration against a published formula and methodology for applying the same, both for grant of a license in the future and past unlicensed uses where a patent is deemed essential

The question of determining patent validity is a much more sensitive one: Bearing in mind the unlikelihood of any scheme being accepted if patent owners feel that it places the continued existence of their patents directly at risk, the author believes that the panel should neither make a decision on this nor air any opinion on it. However, it would not be unreasonable for an arbitration panel to be able to receive and subsequently publish (unedited) observations on validity from interested third parties as part of its arbitration award report.

Finally, the author suggests that consideration should also be given to applying the same system, mutatis mutandis, to other types of IPR cropping up in standards disputes (e.g., copyright, sui generis database rights, etc.). To this end, it is noteworthy that the International Telecommunications Union (ITU) has already undertaken such a study and has made some adjustments to its IPR policies, such that IPR should no longer just mean patents and allied rights.

Changing Intellectual Property Law

In the longer term, the question of some form of alteration to patent legislation to interface it properly with standards development and deployment will have to be canvassed by SDOs. This will fill the gap between what is achievable via the contracting-in scheme above and the very specific applications of tort and antitrust law already briefly noted. Along these lines, it is interesting to note that the European Parliament has recently been considering inserting a specific provision into the planned harmonisation directive on the patentability of computer-implemented inventions, whereby:

Member States shall ensure that wherever the use of a patented technique is needed for the sole purpose of ensuring conversion of the conventions used in two different computer systems or network[s] so as to allow communication and exchange of data content between them, such use is not considered to be a patent infringement. (Committee on Legal Affairs and the Internal Market, 2003)

While this particular provision is unlikely to survive into the final text because of opposition from the European Union (EU) Council of Ministers, it serves as a useful baseline for future work. As a starting position in the present context, the author proposes that certain clearly defined forms of exploitation of patented processes or products in the course of developing a standard, manufacturing in accordance with a standard, or operating in accordance with a standard should be deemed to be noninfringing provided that the patent holder receives equitable remuneration; the latter will be agreed between the parties or, in default of such agreement, by an approved arbitration body.[15] Again, suitable similar changes, mutatis mutandis, should also be considered for other types of IP laws and other changes sought. In this context, there have also been several interesting suggestions as to limitations of remedies, including damages and injunctive relief, in patent disputes where standardization issues are involved (Merges, 1994).

It is clear that the legal changes referred to above should also be tested against the limitations imposed by international IP treaties; if and where there is a conflict, the treaties themselves should be altered to ensure that the necessary changes in domestic law can be legally effective. However, provided the end provisions are suitably drafted, there should be no conflict anyway. Indeed, in relation to patents, the changes even as they stand in their basic form arguably do not fall foul of the restrictions applicable to compulsory licenses of patents under Article 5(4) of the Paris Convention;[16] this is because they do not relate either to failure to work or insufficiency of working the patent per se. In short, even if the proposed mechanism is classifiable as a compulsory license, which is itself an arguable proposition, it would instead fall under the general powers to impose such licenses in Article 5(2) of the Paris Convention.

Similarly, turning to Article 30 of the Trade-Related Aspects of Intellectual Property Rights (TRIPS) Agreement of 1994, the generality of the proposed scheme would already appear to meet the "triple hurdle" criteria of being

a.   a limited exception,
b.   not unreasonably in conflict with normal exploitation of the patent, and
c.   not unreasonably prejudiced toward the legitimate interests of the patent owner, taking account of the legitimate interests of third parties.

INTERNATIONAL TRADE AND STANDARDIZATION DISPUTE RESOLUTION

The need for the changes outlined above is made even more abundantly clear by events in the international arena. With the progressive reduction of tariffs and quotas in world trade achieved by multilateral agreements administered by such bodies as the WTO (World Trade Organization) or made bilaterally between states, technical standards and related issues of conformity testing remain, almost by default, one of the biggest barriers to international trade. In principle, this should not happen because of the Technical Barriers to Trade (TBT) Agreement of 1994. Born out of the same round of agreements as TRIPS, TBT was supposed to ensure transparency and fair play; in practice, this does not always happen on the ground because of the way that standardization processes are steered by particular interest groups or because of a simple refusal to accept a result on other grounds.[17]

The increasing risk here is that the drive for globalization in standards will founder where states and regions have the capacity to act independently and break away from standardization processes. In the WiFi context, we have had a potentially disastrous near miss at the end of 2003, chronicled in detail by

Updegrove (n.d.). For some time, the Ministry of Information Industries in People's Republic of China (PRC) has voiced concerns about the security element of WiFi products as defined by the Wired Equivalent Privacy (WEP) component of IEEE 802 family standards.

In November 2003, the Standardization Administration of China (SAC) declared that only WiFi products containing its preferred choice, the domestically developed WLAN (wireless local-area network) Authentication and Privacy Infrastructure (WAPI) component, would be allowed to be sold in the PRC; as Updegrove (n.d.) makes clear, this effectively meant that products from the United States, EU, and other global trade partners could only be sold in the PRC if localised in conjunction with one of 24 local entities licensed to use WAPI technology. The deadline for compliance with the SAC order was set for December 1, 2003, then deferred until June 1, 2004, but, fortunately, in the meantime, after 11th-hour intervention by the then U.S. Secretary of State, Colin Powell, an agreement was reached between the United States and PRC that WAPI compliance should be postponed indefinitely whilst both countries promoted efforts within SDOs at all levels to move toward a global 3G (third-generation) WLAN standard. There are also a number of analogous strategic issues that have arisen in connection with the introduction of 3G mobile telephony technologies into the PRC (McGinty & Da Bona, 2004).

The author suggests that in light of this experience and the likely increase in such incidents in the future, the WTO, ITU, and WIPO will have to undertake major reviews of their individual and collective responsibilities in relation to standardization. Again, the time has come for a centralised standards dispute-resolution procedure between countries or regional blocks, one that is quick, specifically tailored, and low cost. Existing systems within the ITU and WTO simply do not meet these criteria, especially in relation to speed, which, in the case of standardization, is often of the essence today, bearing in mind the ferocious rates of change in the relevant technologies. At the very least, a new memorandum of understanding is needed between the three bodies concerned, plus internal reforms to create and empower the envisaged joint Standardization Dispute Resolution Committee: Suitable links will also have to be established with ISO/IEC and other suitable SDOs.

CONCLUSION

Whether looking at dispute resolution in relation to patents, antitrust, or international trade issues at all levels, the author is reminded of the old phrase, "This is no way to run a railroad." Whilst bad or recalcitrant behaviour cannot ever be totally eliminated at any level, the time has come for a more systematic, transparent, and legally binding approach to dispute handling in those areas where there is still the possibility of common ground between the parties; indeed, as has been seen, the use of ADR may actually contribute to preserving that goodwill in the first instance. Ultimately, the simple motivation for moving away from litigation en masse is that relying on commercial common sense alone is no longer enough to resolve some of the kinds of disputes that are erupting inside the standards framework, if only because that latter quality seems to be ever diminishing.

REFERENCES

Abbatt, W. (2004). Bringing out the big gun: Invoking the Sherman Act in patent litigation. *IP & Technology Law Journal, 16*(12), 5-7.

Alban, D. (2004). Rambus v Infineon: Patent disclosures in standards-setting organizations. *Berkeley Technology Law Journal, 19*, 309-331.

Anonymous. (2005). Courts act to promote settlements. *Managing Intellectual Property, 146*, 87-88.

Blackmand, S., & McNeill, R. (1998). Alternative dispute resolution in commercial intellectual property. *American University Law Review, 47*, 1709-1734.

Carrier, M. (2003). Why antitrust should defer to the intellectual property rules of standards-setting organizations: A comment on Teece & Sherry. *Minnesota Law Review, 87*, 2019-2037.

Committee on Legal Affairs and the Internal Market. (2003). *Report on the proposal for a directive of the European Parliament and of the Council on the patentability of computer-implemented inventions* (Doc. No. A5-0238/2003). Retrieved from http://www.europa.eu.int

Curran, P. (2003). Standards-setting organizations: Patents, price-fixing and per se legality. *University of Chicago Law Review, 70*, 983-1009.

Dolmans, M. (2002). Standards for standards. *Fordham International Law Journal, 26*, 163-208.

Extended Systems and Intellisync settle lawsuit. (n.d.). Retrieved from http://www.intellisync.com/press_releases/030404.html

Gifford, D. (2003). Developing models for a coherent treatment of standards-setting issues under the patent, copyright and antitrust laws. *IDEA, 43*, 331-384.

Godby, G. (2003, June). Essential intellectual property rights: What, why and how? *IT Law Today, 7*.

IEEE groups fight for control of key standards. (2004, July 6). *Wireless Watch.* Retrieved from http://www.theregister.co.uk/2004/07/16/ieee_groups_slug_it_out/

Jakobsen, K. (2004). Revisiting standards-setting organizations' patent policies. *Northwestern Journal of Technology & Intellectual Property, 3*, 43-60.

Kewney, G. (2004). *WiFi alliance warns chip makers over 802.11n claims.* Retrieved from http://www.theregister.co.uk/2004/10/14/wifi_alliance_threatens_members/

Lea, G. (2000). Ever decreasing circles? The crisis in standards setting in the wake of US v Microsoft. *Yearbook of Copyright & Media Law, 5,* 166-186.

Lea, G., & Hall, P. (2004). Standards and intellectual property rights: An economic and legal perspective. *Journal of Information Economics & Policy, 16*(1), 67-79.

Lemley, K. (2004). I'll make him an offer he can't refuse: A proposed model for alternative dispute resolution in intellectual property rights disputes. *Akron Law Review, 37,* 287-327.

McGinty, A., & Da Bona, D. (2004). 3G licensing in China: A waiting game. *Computer Law & Security Report, 20*(6), 480-486.

Merges, R. (1994). Intellectual property rights and bargaining breakdown: The case of blocking patents. *Tennessee Law Review, 62,* 75-106.

Milo, P. (2004). Avoiding patent litigation: Editorial. *Evaluation Engineering, 46*(6), 6-7.

Mueller, J. (2002). Patent misuse through the capture of industry standards. *Berkeley Technology Law Journal, 17,* 623-684.

Nobel, C. (2004). *Wi-LAN sues Cisco over data transfer patents.* Retrieved from http://www.eweek.com/article2/0,1759,1617076,00.asp

Patterson, M. (2002). Inventions, industry standards, and intellectual property. *Berkeley Technology Law Journal, 17,* 1043-1083.

Patterson, M. (2003). Antitrust and the costs of standard-setting: A comment on Teece & Sherry. *Minnesota Law Review, 87,* 1995-2017.

Smith, T. (2004). *Cisco sued in Wi-Fi patent clash.* Retrieved from http://www.theregister.co.uk/2004/06/24/cisco_wifi_lawsuit

Soma, J., & Davis, K. (2000). Network effects in technology markets: Applying the lessons of Intel and Microsoft to future clashes between antitrust and intellectual property. *Journal of Intellectual Property Law, 8,* 1-51.

Tansey, R. (2005). Get rich or die trying: Lessons from Rambus' high-risk predatory litigation in the semiconductor industry. *Industry & Innovation, 12*(1), 93-97.

Teece, J., & Sherry, E. (2003). Standards setting and antitrust. *Minnesota Law Review, 87,* 1913-1994.

Terdiman, D. (1997). ADR may promise a new era for patent litigation. *New Jersey Law Journal, 149,* 9-11.

Updegrove, A. (n.d.). *Breaking down barriers: Avoiding the China syndrome.* Retrieved from http://www.consortiuminfo.org/bulletins/pdf/may04/trends.pdf

Verbruggen, J., & Lorincz, A. (2002). Patents and technical standards. *IIC, 33*, 125-146.

Webb, R. (2004). There is a better way: It's time to overhaul the model for participation in private standard-setting. *Journal of Intellectual Property, 12*, 163-227.

Wi-LAN escalates patent enforcement program. (n.d.). Retrieved from http://www.wilan.com/news/20040623.htm

Willig, R., & Bigelow, J. (2004). Antitrust policy towards agreements that settle patent litigation. *The Antitrust Bulletin, 71*, 655-673.

Wireless industry intellectually challenged. (2004, March 8). *Wireless Watch.* Retrieved from http://www.theregister.co.uk/2004/03/08/wireless_industry_intellectually_challenged/

ENDNOTES

[1]   For a basic primer explaining OFDM and comparing it to other multiplexing schemes, see E. Lawrey's "OFDM as a Modulation Technique for Wireless Communication" (1997, student thesis) at http://www.skydsp.com/publications4thyrthesis/index.htm

[2]   Primarily, financial consequences (e.g. higher settlement fee, higher license fees) plus pressure for formal admissions of liability and, occasionally, public apologies.

[3]   See comments made on this point in *Rambus Inc v infineon Technologies AG* 318 f.3d. 1081 (Fed. Cir., 2003).

[4]   That is, injunctions or other court orders made prior to full trial.

[5]   That is, do Cisco's Aironet and linksys WiFi products fall inside them?

[6]   That is, where, on an economically rational assessment, the massive cost/time/uncertainty burdens of litigation are outweighed by the dire consequences of letting the behaviour(s) in question stand.

[7]   For an EU competition law perspective on this practice, see R. Franzinger's (2003) "Latent Dangers in a Patent Pool: the European Commission's Approval of the 3G Wireless Technology licensing Agreements" in the *California Law Review, 91*, 1693-1727.

[8]   For suggested reforms, see N. Tsilas's (2004) "Towards Greater Clarity and Consistency in Patent Disclosure Policies in a Post-Rambus World" in the *Harvard Journal of Law & Technology, 17*, 475-531.

[9]   ETSI Rules of Procedure, Art. 15.6, ETSI IPR Policy.

[10]   For a balanced view of the issues as they stood at that time, see S. Gates's (1998) "Standards, Innovation, and Antitrust: Integrating Innovation Concerns into the Analysis of Collaborative Standards Setting" in the *Emory Law Journal, 47*, 583-657.

11   Willig, R. & Bigelow, J. (2004). Antitrust policy towards agreements that settle patent litigation. *The Antitrust Bulletin, 71,* 655.

12   That is, here, those companies that are members of the SDO in question, otherwise adhere to its IPR Policies or are, more generally, prepared to deal fairly with an SDO or its members.

13   For example, the World Intellectual Property Arbitration and Mediation Centre.

14   See http://www.icann.org/udrp and http://arbiter.wipo.int/domains/index.html, respectively.

15   That is, prefereably the same one or ones dealing with arbitration under the SDO membership and related persons scheme outlined.

16   That is, the Paris Convention on the Protection of Industrial Property 1883.

17   For example, national security, planned boosts to local industry/technology, etc.

Chapter X

Intellectual Property Protection and Standardization

Knut Blind,
Fraunhofer Institute for Systems and Innovation Research, Germany

Nikolaus Thumm,
Swiss Federal Institute of Intellectual Property, Switzerland

ABSTRACT

This chapter presents the first attempt at analyzing the relationship between strategies to protect intellectual property rights and their impact on the likelihood of joining formal standardization processes, based on a small sample of European companies. On the one hand, theory suggests that the stronger the protection of one's own technological know-how, the higher the likelihood to join formal standardization processes in order to leverage the value of the technological portfolio. On the other hand, companies at the leading edge are often in such a strong position that they do not need the support of standards to market their products successfully. The results of the statistical analysis show that the higher the patent intensities of companies, the lower their tendency to join standardization processes, supporting the latter theoretical hypothesis.

INTRODUCTION

Over the last decade both the number of patent applications submitted to national and international patent offices and the number of standards claimed at standardization bodies have risen tremendously. In patenting, a 'pro-patent era' began in the mid-1980s. At the European level, it accompanied the establishment of a coherent legal European framework, introducing new national and European legislation for different technological fields. Standardization processes, measured by their output (i.e., the number of formal standards) also increased, especially in Europe (Blind, 2002a). One indication of this trend was the creation of new standardization bodies such as the ETSI, the European Telecommunication Standards Institute. Both phenomena have already been the subject of scientific analysis.[1]

The ambivalence of intellectual property rights and *de facto* industry standards, or *de jure* standards for technological development, is triggered by two different economic mechanisms. Intellectual property rights (IPRs) provide knowledge producers with the temporary right of exclusive exploitation of the benefits deriving from the new knowledge. In this way, IPR provides knowledge producers with the publicly desirable incentive to invest in R&D. They provide holders with a temporary monopoly position, but IPR limits the free diffusion of technological knowledge. Potential users can either not get access to required knowledge or have to pay for it (licensing). Some IPRs, like patents, include at least a positive element of diffusion by the publication of the protected specifications.

In contrast to intellectual property rights, standards released by standards development organizations are decisive for the diffusion of new technologies. They make information about new technologies available to everyone for a small fee and come near to being a classical public good. Innovation researchers until now have concentrated primarily on the analysis of mechanisms that foster the generation of new technological knowledge. However, only the broad diffusion of technology triggered by standards and technical rules can foster economic growth.

Intellectual property rights and standardization are important social institutions that play active roles in technical innovation. They share certain similarities as institutions: for example, both patenting and standardization essentially serve to codify technical information into non-dubious, replicable language. At the same time, their roles are essentially different. A patent describes the parameters of a technology (product or process) over which the patentee owns limited rights, while standard specifications are elaborated by diverse interest groups in order to provide common ground for the future development of new technologies. This common ground consists of not only standards to reduce the variety of possible technological trajectories to a minimum, but also of compatibility standards that allow the exploitation of network externalities and of quality standards for increasing consumer acceptance.[2]

The traditional point of conflict between IPR and standardization occurs when the implementation of a standard, by its essence, necessitates the application of proprietary technology. Both processes bring together two (p. 11) seemingly contradictory processes: the creation of variety and its successive reduction through selection. Effective long-term adaptation requires that these two processes be kept in balance (p. 11) (Carlsson & Stankiewicz, 1994).

Since involvement in standardization processes is accompanied by the danger that the other participants could use the disclosed and unprotected technological knowledge for their own purposes, R&D-intensive companies whose knowledge is either insufficiently protected by IPR or extremely valuable may be reluctant to join standardization processes. Although a variety of protection instruments exists, it is difficult to measure their effectiveness. One formal means of legal protection is patenting. Since applying for a patent entails significant costs (such as fees and costs for legal advice), the expected economic value must be higher than the actual expenses. Thus, a patent application will be made only if the value of the technological know-how reaches a certain level. In addition, the know-how to be protected must also be of value to competitors and not only the company itself. In other words, the number of patent applications indicates not only the intensity with which knowledge protection instruments are used, but also the dimensions of the expected economic value of the company's technological knowledge base. Since IPR tends to concentrate in the areas of greater technical complexity, it becomes virtually impossible to adopt a standard without incorporating proprietary material.

The interaction of standardization and intellectual property rights has not yet been the subject of in-depth analysis. The only research to date is a study, "Interaction of Standardization and Intellectual Property Rights," funded by the Directorate General Research of the European Commission (EC Contract No G6MA-CT-2000-01001) (see Blind et al., 2002). The study was carried out by the Fraunhofer Institute for Systems and Innovation Research (FhG-ISI) and the Institute for Prospective Technological Studies (IPTS) on a small sample of 149 European companies, more than half of which were from the United Kingdom, Germany, and France, covering a wide range of industries; however, with a focus on research and development and manufacturing of chemicals. Except for a few individual case studies, this survey connected the IPR strategies of companies with their standardization activities for the first time. We have used the survey data derived from this study as the basis for the empirical results presented in this article.

Based on the previously outlined differences between the strategy of participating in standardization and that of using intellectual property rights, we propose several hypotheses that are investigated in the following analysis:

Summarizing the findings of our statistical analyses, we have to find an explanation for the much higher negative ratio in the case of patent intensity in comparison to the slightly negative ratio of the R&D. Regarding this aspect, one has to consider that most standard development organizations require that patent holders are willing to license their rights at reasonable and non-discriminatory conditions. Consequently, some companies with broad and valuable patent portfolios are more reluctant to join formal standardization processes, whereas for the R&D intensity itself, the insignificance can be explained by the ambivalence of the complementary and substitutive relationship. One further dimension has to be mentioned. Besides formal standardization processes, companies can also choose to join informal standardization processes, which are more flexible regarding patent rights of the involved parties. Consequently, companies with strong patent portfolios have alternatives with lower opportunity costs to promote the diffusion and marketing of their technologies.

CONCLUSION

This chapter presents the first attempt at analyzing the relationship between strategies to protect intellectual property rights and their impact on the likelihood of joining formal standardization processes, based on a small sample of European companies. On the one hand, theory suggests that the stronger the protection of the company's own technological know-how, the higher the likelihood to join formal standardization processes in order to leverage the value of its own technological portfolio. Behind this hypothesis is the assumption that companies should try to achieve both a combination and strong link of standards and IPR. This would provide them with an even stronger market position. A strong IPR portfolio is an essential weapon within a standards war, when companies try to achieve a de facto standard position in the market (Shapiro & Varian, 1999, p. 295). On the other hand, companies at the leading edge are often in such a strong position that they do not need the support of standards to market their products successfully. The statistical results to explain the likelihood to join standardization processes support the latter theoretical hypothesis, because the higher the R&D and patent intensities of companies, the lower the tendency to join standardization processes.

The final question that has to be addressed is: What follows from this result for policy makers? If companies with a strong technological base stay away from standardization processes, the quality of the standards released may fall behind the technological leading edge and their broad diffusion may be inhibited because of this. Consequently, the positive economic impacts of standards will not be full exploited. In order to solve this shortcoming, solutions have to be found that create additional incentives for technologically strong companies to join standardization processes, and reduce simultaneously the threat that they suffer from unintentional spillovers of technological know-how to other participants of the standardization processes.

ACKNOWLEDGMENTS

The authors thank three anonymous reviewers and the participants of the SIIT2003 in Delft and the EURAS 2003 workshop in Aachen for their comments. Furthermore, we are grateful for the financial support of DG Research of the European Commission for financial support (EC Contract No G6MA-CT-2000-02001).

REFERENCES

Antonelli, C. (1994). Localized technological change and the evolution of standards as economic institutions. *Information Economics and Policy, 6,* 195-216.

Arundel, A. (2001). The relative effectiveness of patents and secrecy for appropriation. *Research Policy, 30,* 611-624.

Arundel, A., van de Paal, G., & Soete, L. (1995). *Innovation strategies of Europe's largest industrial firms: Results of the PACE survey for information sources, public research, protection of innovations and government programmes.* Directorate General XIII, European Commission, EIMS Publication.

Bekkers, R., Duysters, G., & Verspagen, B. (2002). Intellectual property rights, strategic technology agreements and market structure; the case of GSM. *Research Policy, 31*(7), 1141-1161.

Blind, K. (2001). Standardisation, R&D and export activities: Empirical evidence at firm level. *Proceedings of the Third Interdisciplinary Workshop on Standardization Research* (pp. 165-186), University of the German Federal Armed Forces, Hamburg: University der Bundeswehr.

Blind, K. (2002a). *Normen als Indikatoren für die Diffusion neuer Technologien, Endbericht für das Bundesministerium für Bildung und Forschung im Rahmen der Untersuchung "zur Technologischen Leistungsfähigkeit Deutschlands" zum Schwerpunkt "methodische Erweiterungen des Indikatorensystems."* Karlsruhe: ISI.

Blind, K. (2002b). Driving forces for standardisation at standardisation development organisations. *Applied Economics, 34*(16), 1985-1998.

Blind, K. (2004). *The economics of standards: Theory, evidence, policy.* Cheltenham: Edward Elgar.

Blind, K., Bierhals, R., Iversen, E., Hossain, K., Rixius, B., Thumm, N., & van Reekum, R. (2002). *Study on the interaction between standardisation and intellectual property rights.* Final Report for DG Research of the

European Commission (EC Contract No G6MA-CT-2000-02001). Karlsruhe: ISI.

Blind, K., & Thumm, N. (2003). Interdependencies between intellectual property protection and standardisation strategies. *Proceedings of EURAS 2002* (pp. 88-106), Aachener Beiträge zur Informatik, Band 33, Wissenschaftsverlag Mainz, Aachen.

Drahos, P. (1996). *A philosophy of intellectual property.* Brookfield, Singapore, Sydney: Aldershot.

Farrell, J., & Saloner, G. (1985). Standardization, compatibility, and innovation. *RAND Journal of Economics, 16,* 70-83.

Geroski, P. (1995). Markets for technology: Knowledge, innovation and appropriability. In P. Stoneman (Ed.), *Handbook of the economics of innovation and technological change.* Blackwell.

Grandstrand, O. (1999). *The economics and management of intellectual property.* Cheltenham: Edward Elgar.

Hall, B. H., & Ziedonis, R. H. (2001). The patent paradox revisited: An empirical analysis of patenting in the U.S. semiconductor industry, 1979-1995. *Rand Journal of Economics, 32,* 101-128.

Harabi, N. (1995). Appropriability of technical innovations: An empirical analysis. *Research Policy, 24,* 981-992.

Heller, M., & Eisenberg, R. (1998). Can patents deter innovation? The anticommons in biomedical research. *Science, 280,* 698-701.

Jungmittag, A., Blind, K., & Grupp, H. (1999). Innovation, standardization and the long-term production function: A co-integration approach for Germany, 1960-1996. *Zeitschrift für Wirtschafts und Sozialwissenschaften, 119,* 205-222.

Kortum, S., & Lerner, J. (1999). What is behind the recent surge in patenting? *Research Policy, 28*(1), 1-22.

Lemley, M. A. (2002). *Intellectual property rights and standard setting organizations.* Contribution to the public hearing on competition and intellectual property law and policy in the knowledge-based economy in 2001 and 2002. Available online at: http://www.ftc.gov/opp/intellect/index.htm.

Mazzoleni, R., & Nelson, R. (1998). The benefits and costs of strong patent protection: A contribution to the current debate. *Research Policy, 27*(3), 273-84.

Meeus, M. T. H., Faber, J., & Oerlemans, L. A. G. (2002). *Why do firms participate in standardization? An empirical exploration of the relation between isomorphism and institutional dynamics in standardization.* Working Paper, Department of Innovation Studies, University of Utrecht.

Merges, R. P. (1999). *Institutions for intellectual property transactions: The case of patent pools.* Working Paper, Revision 1999, University of California at Berkeley.

Ordover, J.A. (1991). A patent system for both diffusion and exclusion. *Journal of Economic Perspectives, 5*(1), 43-60.

Rammer, C. (2002). *Patente und Marken als Schutzmechanismen für Innovationen.* Studien zum deutschen Innovationssystem Nr. 11-2003, Zentrum für Europäische Wirtschaftsforschung (ZEW), Mannheim.

Rapp, R. T., & Stiroh, L. J. (2002). *Standard setting and market power.* Contribution to the Public Hearing on Competition and Intellectual Property Law and Policy in the Knowledge-Based Economy in 2001 and 2002. Available online at: http://www.ftc.gov/opp/intellect/index.htm.

Salop, S. C., & Scheffman, D.T. (1983). Raising rivals' costs. *American Economic Review, 73*(2), 267-271.

Salop, S.C., & Scheffman, D.T. (1987). Cost-raising strategies. *Journal of Industrial Economics, 36*(1), 19-34.

Scherer, F. M., & Ross, D. (Eds.). (1990). *Industrial market structure and economic performance.* Dallas, Geneva: 3.A.

Shapiro, C. (2001). Navigating the patent thicket: Cross licenses, patent pools, and standard setting. In A. Jaffe, J, Lerner, & S. Stern (Ed.), *Innovation policy and the economy* (vol. 1, pp. 119-150). Boston, MA: MIT Press.

Shapiro, C., & Varian, H. (1999). *Information rules. A strategic guide to the network economy.* Boston, MA: Harvard Business School Press.

Swann, P. (2000). *The economics of standardization.* Final Report for Standards and Technical Regulations, Directorate Department of Trade and Industry, University of Manchester.

Thumm, N. (2001). Management of intellectual property rights in European biotechnology firms. *Technological Forecasting and Social Change, 67*(July), 259-272.

Thumm, N. (2002). Europe's construction of a patent system for biotechnological inventions: An assessment of industry views. *Technological Forecasting and Social Change, 69*(December), 917-928.

Wakelin, K. (1998). Innovation and export behaviour at the firm level. *Research Policy, 26,* 829-841.

ENDNOTES

[1]   Especially patents: For a comprehensive analysis of the recent surge in patenting in the U.S. (see Kortun and Lerner, 1999).

[2]   See Swann (2000) for a recent overview of the different effects of standards.

[3]   Compare Scherer (1990), p. 623.

4   This is empirically underlined by the very low usage of patents as a source for innovation; cf. Eurostat (2000), p. 30.
5   Geroski, P. (1995). p. 97. Analogously, Ordover (1991) prefers a strong patent regime, which facilitates a broad diffusion of knowledge, in coordination with an efficient licensing system.
6   A number of motives for the biotechnology industry are explained in Thumm (2001).
7   Various surveys demonstrate that manufacturing firms estimate secrecy higher than appropriation methods (Arundel, van de Paal, & Soete, 1995; Arundel, 2001; Harabi, 1995).
8   So-called patent portfolio races, see, e.g., Hall and Ziedonis (2001) for the semiconductor industry and Heller and Eisenberg (1998) for the biotechnology industry.
9   For a systematic overview on various patenting strategies, see Grandstrand (1999), p. 232.
10   In addition, these companies depend even more than big firms on the regulatory framework (see Thumm, 2002).
11   In the survey among U.S. companies, Cohen et al. (2000) find that patents are very important in the chemical sector, whereas in the manufacturing industry, in electrical machinery, and in medical instruments, secrecy is the most important protection measure.
12   Cf. Arundel (2001), p. 622. Secrecy in general is more important for process innovation, where its level of importance turns out to be independent from the firm size. Larger companies tend to be more familiar with formal appropriation methodologies, and consequently, for product innovations the importance of secrecy decreases with the firm size.
13   A previous and extended version can be found in Blind and Thumm (2003). The applied regression approach is developed in Blind (2004).
14   In the case of privately owned de facto standards caused by network externalities, the R&D decision will change towards a socially ineffective speed up of R&D. Cf. Kristiansen (1998).
15   Therefore, Antonelli (1994) goes even further and characterizes standards as non-pure private goods.
16   See above, since secrecy is the most important protection strategy, whereas patenting is only of secondary importance.
17   However, there are examples where companies tried to influence the formal standardisation process into specification which seemed to be very good in the technological sense or not very promising concerning market acceptance in order to increase the market shares of their technologies or products.

This chapter was previously published in the Journal of IT Standards & Standardization Research, 2(2), 61-75, July-December, Copyright © 2004.

Section V

Applications

Chapter XI

Standards for
Telecommunication Services

M. H. Sherif, AT&T, USA

ABSTRACT

Management of standardization must be part of an overall strategy for knowledge management. This principle is illustrated with the help of a planning tool for standardization in telecommunication services. The tool integrates knowledge gained from studies on the management of innovation to understand the role that external and internal standards play in the development and operation of telecommunication services. We show how the scope of standardization should differ according to the timing of the standard within the life cycle of the technology as well as the type of interfaces to be standardization. This has implications on the role of standard bodies particularly because the product cycles of equipment manufacturers and service providers are not always synchronized.

INTRODUCTION

Standards are a means to impart specialized technical know-how to a wider audience and promote its application in commercial systems. The production of standards is influenced by technological factors, business needs, and market expectations. This explains the increased interest in telecommunication standards following the deregulation of the industry and current attempts to converge service offers irrespective of the access and transmission technologies.

This chapter presents a methodology to plan for standardization in telecommunication services. In particular, we show how success depends on the integration of several streams of information such as the nature of the innovation, the life cycle of the technology, and the level of details needed in the standards. We take advantage of studies on the management of innovation to gain insight into the role that external and internal standards play in the development and operation of telecommunication services. First, we give an overview of the different types of innovations in telecommunication services. Next, we differentiate between the standardization of equipment and of services. We then show how the scope of standardization should differ according to the timing of the standard within the life cycle of the technology as well as the type of interfaces to be standardized. The implications on the role of standard bodies are discussed. Finally, all these elements are integrated to address standardization within an overall strategy for knowledge management.

INNOVATIONS IN TELECOMMUNICATION SERVICES

Public telecommunication services are available to subscribers that share a common infrastructure managed by a service provider. Depending on the degree of changes they introduce in the technology or in the existing value network,[1] innovations can be grouped into four categories as shown in Figure 1: incremental, architectural, platform, and radical (Abernathy & Clark, 1985; Betz, 1993; Sherif, 2003a, 2003b).

Incremental (or process or modular) *innovations* build upon well-known technological capabilities to enhance an existing technology through improved performance, enhanced security, better quality, and reduced cost within the established value network. It is estimated that half of the economic benefit of a new technology comes from process improvements after the technology has been commercially established (Christensen, 1997). As a result, process innovations are more readily integrated within the technological and financial plans, which explains why they tend to fit the dominant industrial structure (Betz, 1993).

Architectural innovations (sometimes called systems innovations) provide new functional capabilities based on rearrangements of existing technology

Figure 1. Classification of innovation in terms of the value chain and the technological competencies

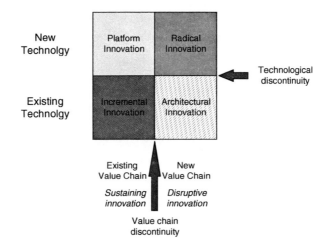

to satisfy unmet needs (simplicity, cost, reliability, efficiency, convenience, etc.; Betz, 1993). Clearly, a market pull is the main stimulus for this category of innovation, which tends to modify the supply chain and to reorganize the market segments, ultimately forming a new value network (Christensen, 1997).

Platform innovations correspond to a quantum leap in performance following a technology transition within the boundaries of an existing value chain (Betz, 1993; Christensen, 1997). They usually demand the integration of sophisticated resources and the exploitation of expertise gained along vertical lines that are beyond reach of small- or medium-sized companies (Christensen). These innovations are complex programs that require large capital investments and as a consequence, favor the dominant players that have the necessary financial wherewithal and can assemble the necessary technical and managerial expertise.

Radical innovations provide a totally new set of functional capabilities that are discontinuous with the existing technological capabilities or value networks. Once a radical innovation becomes the dominant design, improvements continue as platform or incremental innovations.

Services offered on public telecommunication networks rely on network elements used for switching, routing, and transmission in addition to information-processing systems to support the operations. These operations-support systems (OSSs) monitor the performance; detect troubles; localize faults; manage inventories, capacity, and security; and poll accounting records to prepare bills. In addition, each operator devises its own methods and procedures (M&Ps) to meet its goals with respect to quality, service availability, reliability, and so forth.

Telecommunication services offered to the public on a subscription basis differ from free services or from connectivity services within enterprise networks in that they depend on the accuracy of billing processes and systems as well as other support systems that ensure quality. The complexity of the operator's pricing models and billing processes is one aspect of the problem. In many cases, operators must make legacy systems as well as systems that they have inherited through successive business restructurings (e.g., mergers and acquisitions, spin-offs, etc.) work with newer systems that support new services.

Innovations in public telecommunication services can affect one or more of the network elements, the OSSs, as well as the operating procedures. Radical innovations disrupt the existing order because they introduce new networking technologies that require new OSSs, require new M&Ps, and are often associated with new user applications. Platform innovations consist of improved networking technologies, improved OSSs, and improved applications. Incremental innovations build on mature network technologies with enhancements to the M&Ps and/or the applications. Finally, architectural innovations consist of sustained networking technologies, improved OSSs, and new applications. All these innovations—with the possible exception of those that are incremental—require some degree of unlearning of habits and practices. Figure 2 summarizes the various categories of innovations in public telecommunication services (Sherif, 2003b).

Incremental innovations increase profitability by reducing costs and improving efficiencies. The telephone answering machine is one such innovation (Vercoulen & van Wegberg, 1999); it did not perturb the structure of the public telephone network while it increased the operators' revenues because callers

Figure 2. Innovation types in telecommunications services

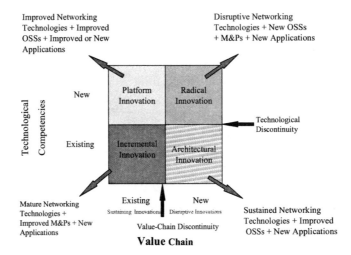

left a message to the absent party so that the operators could charge for the connection.

Architectural innovations integrate existing components to offer new tele-communication services. The reverse charging (800) service changed the way telephone calls are paid for, that is, by the called party instead of the calling party. Bluetooth (and the IEEE 802.11x standards) is a marriage of local-area networks (LANs) and wireless communications. Another example is the i-mode service whose father, Kei-ichi Enoki, explained that it is a new concept made with existing technologies (Nakamoto, 2001). This service combines information processing with wireless access to the public switched telephone network (PSTN), a new embodiment of what the minitel demonstrated back in the 1980s with wireline access. The international callback service was an architectural innovation to take advantage of the differential of pricing following the deregu-lation of telecommunications to allow cheaper overseas calls.

According to the classification in Figure 2, frame relay and the asynchro-nous transfer mode (ATM) are platform innovations of packet switching along the connection-oriented paradigm. Gigabit Ethernet can be viewed either as a platform innovation (from a local-area-network viewpoint) or an architectural innovation (from a wide-area-network viewpoint).

The introduction of modems in the late 1950s was a radical innovation because it allowed the transmission of digital data among computers over the analog telephone networks. Another radical innovation is packet switching, which branched into two approaches. The connection-oriented approach contin-ued along the lines used in telephony so as to maintain the overall call quality. The connectionless approach of IP (Internet protocol), however, requires a major overhaul of the OSSs, new rules for traffic management, and retraining of human resources.

The more radical the technical innovation, the less likely that available market knowledge and existing customers will be able to provide guidance (Betz, 1993; Christensen, 1997). For example, the ubiquitousness of wireless access modifies the order of the criteria used for ranking subjective speech communi-cation (Johannesson, 1997). Radical innovations incorporated in the main stream of the industry lead to process and incremental innovations. For example, the replacement of copper with fiber for cables—a radical innovation—opened the way for several generations of transmission equipment each operating at a higher speed than its predecessor.

End-to-end telecommunication services rely on the smooth integration of many components. This is why innovations in telecommunication services encompass organizational and business aspects in addition to technology. Because of the evolution of systems, organizations and technologies are not necessarily synchronized; for any given technology, the market dynamics for network equipment may be considered a leading indicator for the evolution of

Figure 3. Relationship between the market penetration of a technology in equipment and in services for telecommunications

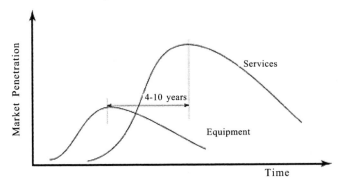

network services. Similarly, the peak of the equipment revenues lags the peak of service sales by anywhere from 4 to 10 years (McCalla & Whitt, 2002; Sherif, 2003a). This trend is sketched in Figure 3.

The reason for the time lag is the considerable amount of effort for a new service to be generally available on a public network. Network elements need to be tested and deployed. This process includes vendor selection and a thorough laboratory evaluation that can last anywhere from 12 to 18 months. New network engineering rules have to be conceived and verified while the OSSs are modified to accommodate the new services. Maintenance personnel need to be trained and sales people prepared to explain the offers to potential customers. The magnitude of the up-front investment explains why once the service is up and running, operators prefer to exploit it as long as they can. Furthermore, unless there are substantial savings (e.g., 10 times) in terms of cost or quality, customers prefer to avoid the disruptions that are associated with service migration. The persistence of technologies in networks explains the long tail of service-market penetration and revenues even after equipment sales have tapered off.

We now shift out attention to the consequences of the previous observations on the standardization in telecommunication services.

SERVICES VS. EQUIPMENT STANDARDIZATION

Public telecommunications require specific agreements among the parties involved (equipment manufacturers, network operators, service providers, and end users). The diminishing role of governments in mandating the rules for telecommunications means that these agreements will be based on voluntary standards, that is, on the business strategies of the various firms involved either through the technologies of communication or of information processing or both.

The traditional focus of standardization in telecommunications has been on the network elements and subcomponents because the outlay for standardization is easier to justify in this case. Another incentive for equipment manufacturers to cooperate is the increased value of a communication network with the number of its users, the so-called network externalities, up to a point. The absence of a common interface standard is a burden on all parties. For example, having a dual digital transmission standard (the m-law at 1544 kbits/s in North America and Japan with the A-law at 2048 kbits/s for the rest of the world) adds a step to all digital transmission equipment. Similarly, the fragmentation of the market for cellular telephony in the United States into islands of competing digital standards (IS-95, IS-136, GSM [Groupe Spécial Mobile], as well as Nextel and Motorola's proprietary systems) increases the cost of interconnection and prevents the consolidation of the industry.

Standards protect service providers against supply interruptions or the monopolistic power of a sole source. Finally, they provide a framework for bilateral and multilateral negotiations among carriers.

It is well known that a manufacturer "will accept and use standards only if it believes that it cannot influence the market directly and that standards can" (Cargill, 1989, p. 42). Even though information technologies are embedded in a wide variety of telecommunication products and services, the standardization for OSS interfaces is lagging behind that of network-equipment interfaces. Attempts at defining a common management interface across different administrative domains, the so-called TMN-X interface, have not addressed the information flow needed for service management, with the possible exception of the exchange of trouble reports for private lines (Covaci, Marchisio, & Milham, 1998). As a result, OSSs remain stand-alone systems that operate more or less independently of other systems, and automated end-to-end service creation, provisioning, and maintenance are not possible (Chen & Kong, 2000). Thus, services that cross several administrative domains (e.g., on international links) are managed by phone, fax, and e-mail. Obviously, this is an error-prone process that can increase the total cost of operation by about 10-20%. Trouble detection and isolation is even more problematic as it requires coordinated manual activities. Another consequence of this lack of standardization is the expense needed to update or develop new OSSs for each new service without being able to reuse what is existing for other service offers. As a result, a typical large operator would end up operating and maintaining few hundreds of such systems for all the various services and networks that it operates and maintains.

Today, these problems are compounded by the increase in the number of interfaces through deregulation, the entry of new players, the introduction of new technologies, and the requests for new services. Virtual private networks, for example, appeal to enterprises because they give customers more control of their part of the network so that they can track their traffic or change the parameters affecting their service. This of course raises many questions on how the network engineering rules will be designed and how the network provider will arbitrate among conflicting requests. Perhaps one of lessons from the failure of so many joint ventures or mergers among telecommunication operators (Curwen, 1999) is the difficulty of consolidating distinct OSSs or at least making them communicate with each other.

The Timing of Standardization in Telecommunications

In the past, a long gestation and a process of gradual evolution and refinement preceded many telecommunications standards. Today, the majority of standards are adopted during, and sometimes before, product design.

The timing of the standards can be considered on the basis of the marketing needs of a specific product (time to market, time to scale, and time to profitability). Another way is to look at the technology itself to base the specifications on the inherent capabilities of the technology at a given phase of its life (Sherif, 2001).

Marketing View of Standards

With the introduction of a product or a service in mind, standards can be anticipatory, enabling (participatory), or responsive as shown in Figure 4. *Anticipatory standards* are essential for the widespread acceptance of the product or service. *Enabling* (participatory) *standards* proceed in lockstep with the introduction of the product or service. *Responsive standards* codify a product or service present in the market or define the expected quality of a service and performance level of the technology. In this discussion, the cycle

Figure 4. Standardization within the product or service life cycle

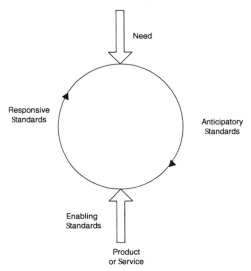

starts when the need for a product or a service arises irrespective of whether the anticipatory standards are ready or may have to be developed.

While the model gives us an appreciation of the a posteriori relationship of the standard to the market development, it does not take into account the intrinsic capabilities of the technology. As a consequence, it does not provide a tool for planning the content and the speed of standardization at a given juncture of the technology life cycle. This is the main purpose of the discussion in the next section.

Technological View of Standards

To consider the technological aspects of standardization, we need to consider the five stages of a technology life cycle, that is, innovation, improvement, maturity, substitution, and obsolescence (Betz, 1993; Khalil, 2000). These stages are shown in Figure 5.

Taking in to account the improvement of the performance of the technology that the end user perceives (Betz, 1993), we can posit as optimal the relationship depicted in Figure 6 between the innovation type and standardization.

Anticipatory Standards

Anticipatory standards are crucial for the successful interoperation of communication systems. Their specification runs parallel to the production of prototypes, to pilot services, and to field trials. They also provide a way of sharing ideas through a systematic way of distilling investigations and experimental data into useful engineering knowledge. This is useful when the risks are high because collaboration with other competitors working on the sets of problems can increase the chances of success. The incentives to standardize in this phase are less when one organization is so far ahead of the pack that other organizations

Figure 5. Technology life cycle

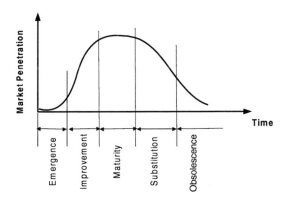

Figure 6. Relation of the category of innovation and standardization with the technology life cycle

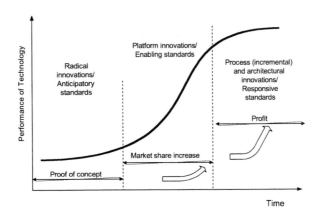

will have little to offer. However, in the case of radical innovations, standardization may also help legitimize the new technology.

The wireless access protocol (WAP) is a specification for Internet access through mobile terminals defined by a consortium of companies (i.e., it is an anticipatory standard). This was an architectural innovation to support a large number of available and planned wireless protocols and a variety of portable terminals (palmtop, phone, computer, personal digital assistant) with a variant of HTML (hypertext markup language), wireless markup languages (WMLs), suitable for the display data on mobile terminals. Its commercial failure shows the importance of a correct reading of the market pull.

Some earlier examples of anticipatory standards are the X.25 packet interface, ISDN (integrated services digital network), TCP/IP (transmission-control protocol/Internet protocol), the secure socket layer (SSL) for secure end-to-end transactions, Bluetooth, IEEE 802.11, and so forth. The UMTS (Universal Mobile Telecommunications Service) is an anticipatory standard for new services for voice and data (144 to 344 kbits/s outdoors, 2 Mbits/s indoors). Thus, the transition to third-generation wireless may be viewed as a discontinuity in both the value chain and in the technology competencies. It is still not clear what services would attract interest and what configuration of customers, competitors, and suppliers will emerge.

Anticipatory standards are susceptible to wrong specifications whenever field experience is lacking and when the market requirements are unclear or ill defined. For example, the open system interconnection (OSI) transport and management protocols became hopelessly entangled in attempts to satisfy the requirements of all parties involved. Also, sticking to wrong market assumptions may lead to a dead end, such as in the case of Group 4 facsimile (facsimile on ISDN). Ideally then, anticipatory standards should offer a minimum set of features to stimulate the market for services and, with feedback from pilot studies, define the production environment of the new technology. Such a restricted scope reduces the chance of overspecifications that could lead to onerous implementations. It avoids premature commercial conflicts that can stall the standardization provided that it is clear what aspects are expected to evolve and when they will evolve.

Enabling (Participatory) Standards

The definition of enabling standards proceeds in parallel with market growth and enhancements to the technology and the products. Thus, enabling standards describe refinements in the production system, in the product systems that embody the technology, as well as in the application systems.

The advantage of enabling standards is that they diffuse technical knowledge and prevent market fragmentation. This comes at a cost for the manufacturers whose technical members are leading the effort to specify the standards. In addition to developing a product, they may have to release technical information earlier than anticipated to conduct the necessary tests. As further changes and refinements are made to the draft standards, the manufacturers and service providers must review their products periodically to bring them into conformity with the new standards. Finally, enabling standards signal that product differentiation is shifting to areas not covered by the standards (cost, quality of implementation, service support, etc.).

One pitfall of a long wait before working on an enabling standard is the risk of incompatible approaches that would fragment the market. V.90 is a good example of enabling (participatory) standards where such risks were avoided. To expand the total market, the developers of the precursor proprietary modems agreed to collaborate within the ITU (International Telecommunications Union) to come up with a common specification.

The GSM specifications constitute a standard for digital mobile telephony because they defined a platform for future growth both for service operators and for manufacturers. Once this platform was established, network operators grew through acquisitions[2] that called for architecture innovations to integrate the different operations-support systems of individual operators. The speech- and video-coding algorithms, some of which are discussed in the appendix, give another example of enabling (participatory) standards. Typically, the standardization of these algorithms occurs in two steps. In the competitive phase, all candidates are evaluated with respect to the agreements defined in the terms of reference. If no candidate emerges as the overall winner, the best proposals from a short list are combined in a solution that it is superior to each of the initial candidates.

Responsive Standards

A responsive standard relates to the manifestation of the technology in a service system (Betz, 1993). They codify knowledge already established in practice through precursor products or services. For service providers, GPRS and SMS are responsive standards in the sense that the market was already established and the performance of the GSM system had reached a saturation level. In contrast, manufacturers would tend to see these standards as anticipatory because they are needed for the design and deployment of the equipment. Clearly, there is some tension between manufacturers and service providers in terms of the urgency of coming up with a specification. Other examples of responsive standards are the V.90 modem, the various methods for the evaluation of voice quality through objective and subjective means, the measures for the overall quality of services, and so forth. In the IT world, the traditional role of ISO TC97 was to improve existing specifications and turn them into international standards. SGML (standardized general markup language) was built on the generalized markup language (GML). The root of the high-level data link protocol (HDLC) was the synchronous data link control (SDLC) that IBM had developed for its System Network Architecture (SNA).

Lack of Standardization

When a stand-alone product is well entrenched (e.g., Microsoft Windows or Oracle's database), there may be no incentive to standardize. Another reason for not standardizing is that the producer may be seeking a niche market, the product may be ephemeral, or both. These two conditions are clearly satisfied in the case of networked computer games. Conversely, pressures to standardize suggest a quest for some stability. For example, a modem manufacturer with a large market share worked on the formalization of the modem AT command set to form network externalities.

There is also the temptation to avoid standardization and rely on proprietary protocols in architectural innovations. This is illustrated by the cases of the minitel and i-mode as well as the smart-card industry. More recent examples include Apple's iPod and Bloomberg's terminal for real-time financial information. The major reason is that architectural innovations are combinations of existing components, therefore switching costs of users and not technology is the real barrier to competition. However, the benefits of network effects may encourage manufacturers to standardize to increase the market size as shown in the case of Bluetooth and WAP.

As previously mentioned, standardization in telecommunication services has traditionally focused on network elements and has rarely considered support systems. Many network operators are now realizing that OSS standardization facilitates back-office integration and reduces the cost and time to roll out any new technology in the network (Cochran & Dunne, 2003). This explains the various "zero provisioning" initiatives to increase the efficiency of provisioning through the automation of the relevant processes. To facilitate the standardization of OSS interfaces, IT providers, given their traditional reluctance to open interfaces, have to be convinced that this is in their best interest. OSS standardization can be then presented as an opportunity to provide operators with efficient solutions for operations and maintenance once they have built their networks and reduced their equipment purchases. The main selling point would be that most of the benefits from platform innovations take place later in the technology life cycle.

DETAILS OF THE STANDARDS AND THE INTERFACE LAYER

Figure 7 depicts an arrangement of the interfaces to be standardized arranged in a layered format. They include standards for reference, for similarity, and for flexibility. In addition, standards for performance and quality can be present at all layers. At each layer of the OSI reference model, standards can define a reference interface, ensure similarity or compatibility, or aim at a smooth evolution to new services.

- **Layer 1—Reference Standards:** Reference standards provide measures to describe general entities in terms of reference units. The classical example is the unit standards for measurable physical qualities, for example, ohm, volt, watt, dBm, and so forth. Other examples in the information and telecommunications field include the ASCII (American Standard Code for Information Interchange) character set, the various standardized high-level computer languages, the OSI model, the E.164 numbering plan for international telephone service, Internet addresses, and so forth.
- **Layer 2—Similarity Standards:** Similarity standards define aspects that have to be identical on both sides of the communicating link as well as the allowed variations or tolerances, if any. In the case of physical goods, the

Figure 7. A layered architecture for technical standards

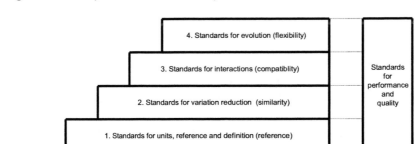

purpose of similarity or uniformity is to achieve a certain economy of scale. In the telecommunications and information applications, similarity is essential to establish the linkage successfully. Examples include the nominal values of signal levels, the shapes (or masks) of current pulses, source-coding algorithms for interactive speech and video, algorithms for line coding on transmission links, encryption algorithms, computer operating systems, and so forth. The appendix illustrates this discussion with a brief history of some speech-coding algorithms.

- **Layer 3—Compatibility Standards:** Compatibility has several aspects such as multivendor compatibility, backward compatibility (multivintage compatibility), product-line compatibility, and so forth (Gable, 1987). Many protocols today consist of a core and many options, which could be a source of interoperability problems. Compatibility standards may be referred to as profiles, functional standards, interface templates, user agreements, or implementation agreements. Profiles, functional standards, or interface templates designate a fixed set of options to perform a given service. User agreements or implementation agreements define the various profiles that a given customer can use. Although compatibility can be verified if implementations are available for testing before the standard is approved, maintaining compatibility as the technology evolves requires the ability to accommodate extensions not yet defined. This will be the scope of flexibility standards. The appendix gives an example in which a similarity standard can cause backward compatibility problems.

- **Layer 4—Flexibility Standards:** Flexibility standards focus on *compatible heterogeneity*, that is, the capability of a single platform to interoperate with different systems and its upward and downward compatibility. For example, the negotiation procedures used in the modem recommendations of the ITU-T (ITU Telecommunications Standardization Sector), such as Recommendation V.34, enable the exchange of user data with legacy modems through common ground rules. A hypothetical smart-card flexibility standard may work with contact and contactless readers, and may contain several applications (bank card, electronic purse, wallet, etc.).

- **Standards for Performance and Quality:** The quality of a telecommunication service results from the collective effect of the performance at several layers. Assessment of the performance involves a mixture of subjective and objective parameters to assess the quality of transmission in the presence of impairments. It covers the user interface to the service offers, the network performance from the aspects of switching and transmission, as well as the operational aspects and the service-support functions (Comité Consultatif International Télégraphique et Téléphonique [CCITT], 1993). Quality and performance standards operate at all layers: Some examples are the transmission-planning guidelines given in ITU-T

Recommendations G.113, G. 114, and G.131; the various methods for the evaluation of voice quality using multilingual tests through objective and subjective means (e.g., ITU-T Recommendations E.510, G.120, P.84); the service quality metrics for Group 3 facsimile (E.451, E.452, E.453, E.456); the measures for transmission quality or switching quality of service for each service (e.g., E.800, I.356, X.140); and so forth.

In a competitive environment, the work on performance telecommunications standards raises several issues regarding what proprietary information can be released during the work on standardization and how to carry the performance evaluation. Some of the concerns relate to the representativeness of the test conditions, the objectivity of the evaluation, and the availability of testing resources. In the precompetitive days, organizations could volunteer to host the testing and carry out the centralized processing of all the data collected by the other laboratories. Today, it is highly unlikely that an organization would offer its resources without adequate compensation or would be able to reschedule its other activities to perform the necessary tests.

ROLE OF VARIOUS STANDARDIZATION BODIES

There are two main categories of standardization bodies: formal standard organizations and consortia. We consider three types of consortia: proof-of-concepts consortia, implementation consortia, and application consortia. Proof-of-concepts consortia are associated with new ideas that are being investigated to understand their potential applications and limitations. An example is the Object Management Group (OMG), which developed the Common Object Request Broker Architecture (CORBA). The focus of implementation consortia, such as the Optical Internetworking Forum (OIF), is the interworking of network elements. Application consortia deal with end-to-end applications such as Digital Imaging and Communication in Medicine (DICOM) for the exchange of digital images for biomedical, diagnostic, and therapeutic applications; the VoiceXML (Voice Extensible Markup Language) Forum that created and promoted VoiceXML for speech-enabled applications; the Foundation for Intelligent Physical Agents (FIPA) that produces specifications for heterogeneous and interacting agents and agent-based systems; and the World Wide Web Consortium (W3C) for Web-based applications.

The Internet Engineering Task Force (IETF) started as a proof-of-concept organization, then expanded its scope to ensure the interoperability of implementations. It then added to its activities some applications such as e-mail or electronic commerce. Today, it is the main place for writing specifications of

Figure 8. Decision matrix for the selection of a standard-development organization

	Anticipatory	Enabling (Participatory)	Responsive
Flexibity	ISO/ITU IETF	Application Consortia	ISO/ITU
Compatibility	ISO/ITU/IETF Proof of Concept Consortia	IETF Implementation Consortia	ISO/ITU
Similarity	ISO/ITU	ISO/ITU	ISO/ITU
Reference	Proof of Concept Consortia	National Regional Organizations	ISO/ITU

Type of interface standard

Timing of standard

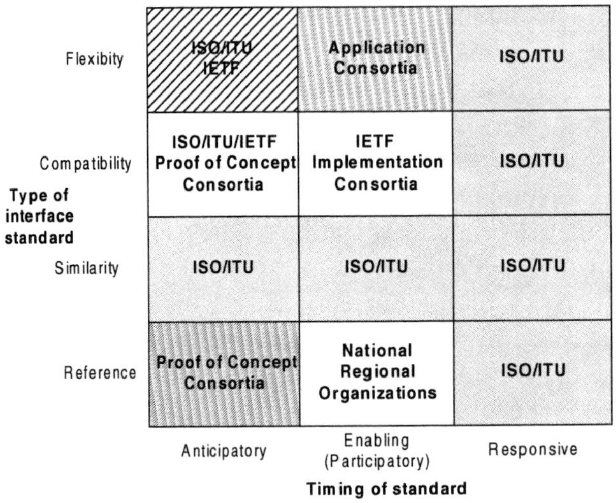

network elements that use the IP protocol. However, it still does not directly address service-related issues such as network management and maintenance and operations-support systems.

The classification system shown in Figure 8 may help organize the partition of work among formal standard development organizations (SDO) and consortia. It takes into account the layer to be standardized, the timing of the standard within the technology life cycle, and the success of each type of organization in the past.

Responsive standards are associated with process and incremental innovations to improve the telecommunication services or to support new services on the existing infrastructures. In all these instances there is a need for considerable deliberation and consultation to avoid disruptions to existing services and ensure end-to-end service interoperability across platforms and networks. These are the areas in which traditional SDOs have excelled. In extreme cases, the standardization happens when the technology is known, the product is already commercialized, and the outcome of the market competition has been settled. For example, ANSI C was defined much later than the introduction and widespread use of the language.

In the case of enabling (participatory) standards, reference and similarity interfaces need to be more stable than other types of interfaces. This is why the former are more suitable to formal SDOs while the consortia handle the latter better. Collaboration takes place also among national, regional, and international

Figure 9. Evolution of GML to SGML, HTML, XML

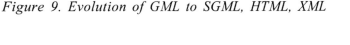
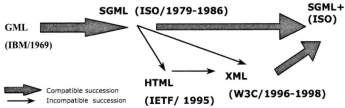

organizations because end-to-end telecommunication services do not know geographic and political boundaries. Based on past experience, it is also possible to state that when the technology is new and the product is new, small-group discussions unencumbered by procedural rules would be more appropriate. In this case, proof-of-concept consortia offer a suitable forum. This was demonstrated by the success of the TCP/IP standardization in the early phases of the Internet (late 1970s and early 1980s) in an environment akin to that of proof-of-concept consortia. However, as the technology proves itself, more formality is needed, either by handing over the work to formal SDOs or by an evolution of the original organization structure for standardization, as in the case of the IETF.

Many consortia base their work on existing formal standards. The interplay between application consortia and SDOs is illustrated by the following example. SGML was adopted by ISO in 1986 as an incremental evolution of the GML that IBM had developed to manage electronic documents. HTML and XML descend from SGML. HTML is an architectural innovation based on SGML because its field of application is document retrieval over the Internet, which is in a different value chain than the original application of database management. Finally, XML is an incremental innovation of HTML from the W3C. This is summarized in Figure 9 (Egyedi & Loeffen, 2001).

There is also collaboration in the reverse direction. For example, the OMG collaborated with T1M1, the ANSI-accredited North American standard group working on the Telecommunications Management Network (TMN), to develop a framework to manage the objects of the TMN using the CORBA specifications.

STANDARDS AND
KNOWLEDGE MANAGEMENT

Contribution toward a standard implies a policy of knowledge management, that is, that of generating, keeping, and/or releasing information. Without such a discipline, organized and consistent participation in the standardization process may be difficult or inefficient. Firms approach standardization in several ways that can be listed from the least responsive to the more responsive as follows.

1. Do nothing and wait until a trend appears.
2. Passive participation in the standards bodies to collect the necessary documents and understand the background of the standard
3. Active participation in an existing standard group to collect the necessary information and understand the significance of the specifications, and in influencing the direction of the process
4. Building or joining alliances, for example, to be able to influence the course of standardization through participants in the development of the specifications
5. Starting a new consortium and attracting other interested parties to approve or promote a given technology

It is possible to sketch some guidelines on the course of action that a service provider may take based on the information presented in this chapter.

As was stated earlier, architectural innovations integrate many technological capabilities to meet the needs of new markets. Even if the service interfaces are kept proprietary (e.g., i-mode or minitel), standardized components reduce the cost of operations and maintenance. This is a case where building alliances provides the service provider with indirect influences over the standardization of the various components instead of active participation in each organization.

In the case of radical innovations, the typical decision is to wait until some prototype network element with the new technology is available. This approach uses the least amount of resources provided that two major risks are avoided: (a) waiting too long until the chance of improving the standard development is lost, and (b) not having a good understanding of the technology leading to erroneous strategic decisions.

Typically, service providers alternate between passive and active participation, particularly in application and implementation consortia. They may also start specific consortia to deal with the applications related to the operational and business applications such as the TeleManagement Forum.

CONCLUSION

This chapter organized empirical observations to provide a guide to the development of standardization strategies in an environment of constant change. The standardization strategies of equipment manufacturers and service providers are not always synchronized. Furthermore, the details that are needed in each standard depend on the nature of the interface to be defined. These two factors affect the role of standard organizations as well as the optimum pace of activities. Other important factors include the nature of the interface to be specified, the phase in the technology life cycle, as well as the market goals of the various firms. It is hoped that future studies will refine the arguments presented here to enrich the tool kit that firms as well as standard organizations use to plan their activities.

REFERENCES

Abernathy, W., & Clark, K. N. (1985). Innovation: Mapping the winds of creative destruction. *Research Policy, 14*, 3-22.

Banerjee, S., Feder, H., Kilm, T., & Sparrell, D. (1985). 32-Kbit/s ADPCM algorithm and line format standard for US networks. *Proceedings of GLOBECOM'85, 3*, 1143-1147.

Baskin, E., Krechmer, K., & Sherif, M. H. (1998). The six dimensions of standards: Contribution towards a theory of standardization. In L. A. Lefebvre, R. M. Mason, & T. Khalil (Eds.), *Management of technology, sustainable development and eco-efficiency* (pp. 53-62). Amsterdam: Elsevier.

Benvenuto, N., & Bertocci, G. (1987). Prevention of quantizer overloading in the 32 kb/s ADPCM algorithm. *Proceedings of the IEEE International Conference on Communications (ICC 87), 3*, 1483-1486.

Benvenuto, N., Bertocci, G., Daumer, W. R., & Sparrell, D. K. (1986). The 32-kb/s ADPCM coding standard. *AT&T Technical Journal, 65*(5), 12-22.

Betz, F. (1993). *Strategic technology management.* New York: McGraw-Hill.

Cargill, C. F. (1989). *Information technology standardization: Theory, process and organizations.* Bedford, MA: Digital Press.

Chen, G., & Kong, Q. (2000). *Integrated telecommunications management solutions.* New York: IEEE Press.

Christensen, C. M. (1997). *The innovator's dilemma: When new technologies cause great firms to fail.* Boston: Harvard Business School Press.

Cochran, R., & Dunne, E. (2003). *Executive briefing to the ATM Forum, the MPLS/Frame Relay Forum and the BCD Forum.* Westwood, MA: Vertical Systems Group.

Comité Consultatif International Télégraphique et Téléphonique (CCITT). (1984). *Annex 1 to report of Working Party XVIII/2: Report of the work of the ad hoc group on 32 kbit/s ADPCM* (COMM XVIII-R 28-E). International Telecommunication Union, Geneva.

Comité Consultatif International Télégraphique et Téléphonique (CCITT). (1987). *Ad hoc group on DCME-based solution for voiceband data transparency: Report to Working Party XVIII/8* (Temporary Document No. 9, XVIII/8). Hamburg, Germany.

Comité Consultatif International Télégraphique et Téléphonique (CCITT). (1993). *Handbook on quality of service and network performance.* Geneva, Switzerland.

Covaci, S., Marchisio, L., & Milham, D. J. (1998). Trouble ticketing X Interfaces for international private leased data circuits and international Freephone service. *IEEE Network Operations and Management Symposium (NOMS'98), 2,* 342-353.

Cox, R. V. (1997). Three new speech coders from the ITU cover a range of applications. *IEEE Communications Magazine, 35*(9), 40-47.

Curwen, P. (1999). Survival of the fittest: Formation and development of international alliances in telecommunications. *Info, 1*(2), 141-158.

Dimolitsas, S., Sherif, M., Colin, S., & Rosenberger, J. R. (1993). The CCITT 16 kbit/s speech coding recommendation G.728: Editorial. *Speech Communication, 12,* 97-100.

Egyedi, T., & Loeffen, A. G. A. J. (2001). Succession in standardization: Grafting XML onto SGML. *Proceedings of the Second IEEE Conference on Standardization and Innovation in Information Technology (SIIT 2001)* (pp. 38-49).

Gabel, H. L. (1987). Open standards in computers: The case of X/OPEN. In H. L. Gabel (Ed.), *Product standardization and competitive strategy* (pp. 91-123). Amsterdam: North Holland.

Johannesson, N.-O. (1997). The ETSI computation model: A tool for transmission planning of telephone networks. *IEEE Communications Magazine, 35*(1), 70-79.

Khalil, T. (2000). *Management of technology: The key to competitiveness and wealth creation.* Boston: McGraw-Hill.

McCalla, C., & Whitt, W. (2002). A time-dependent queueing network model to describe the life-cycle dynamics of private-line telecommunication services. *Telecommunications Systems, 19,* 9-38.

Nakamoto, M. (2001, December 17). NTT DoCoMo. *Financial Times IT Review,* IV.

Schmidt, W. G. (1986, July). Can the world agree on the next digital-speech coding algorithm? *Data Communications,* 157-166.

Schröder, G., & Sherif, M. H. (1997). The road to G.729. *IEEE Communications Magazine, 35*(9), 48-54.

Sherif, M. H. (1997). Study Group 16 and the interface between the Internet and public networks. *Proceedings of Telecom Interactiv'97,* Geneva, Switzerland.

Sherif, M. H. (2001). A framework for standardization in telecommunications and information technology. *IEEE Communications Magazine, 39*(4), 84-100.

Sherif, M. H. (2003a). Technology substitution and standardization in telecommunications services. *Proceedings of the Third IEEE Conference on Standardisation and Innovation in Information Technology (SIIT 2003)* (pp. 241-252).

Sherif, M. H. (2003b). When is standardization slow? *International Journal of IT Standards & Standardization Research, 1*, 19-32.

Sherif, M. H. (2004). Characteristics of projects in telecommunications services. *Proceedings of the 13th International Conference on Management of Technology,* Washington, DC.

Sherif, M. H., & Sparrell, D. K. (1992). Standards and innovation in telecommunications. *IEEE Communications Magazine, 30*(7), 22-29.

United Kingdom. (1987). *Some networking aspects of 16 kbit/s coding applied to Digital Mobile Radio (DMR)* (Contribution D. 1259, CCITT SG XVIII). Hamburg, Germany.

Vercoulen, F., & van Wegberg, M. (1999). Standards selection modes in dynamic complex industries: Creating hybrids between market selection and negotiated selection of standards. *SIIT'99 Proceedings* (pp. 1-11). Piscataway, NJ: IEEE.

ENDNOTES

[1] A value network is defined by the attributes used to rank products, services, or technologies and determine their cost structures (Christensen, 1997, p. 32, 39-41). Changes in the attributes or their ranking provoke discontinuities in the value chain that can alter the industrial structure and offer opportunities to new entrants.

[2] Vodaphone, for example, acquired Airtouch in the United States in 1999 and Mannesman in Germany in 2000.

APPENDIX

Standardization of Speech Coding at 32 kbits/s, 16 kbits/s, and around 8 kbits/s

This appendix relates to a few similarity standards in the area of speech coding to illustrate the cases of anticipatory, enabling, and responsive standardization.

32 kbits/s Algorithm

The standard 32 kbits/s speech-coding algorithm reduces by half the bandwidth needed to transmit speech without reducing voice quality. It is uses the adaptive differential pulse code modulation (ADPCM) in which speech samples are represented in a binary format by the quantized difference between the actual sample and its predicted value before transmission.

To select an ADPCM algorithm, the CCITT organized a contest and combined the best features from the various contenders. The final algorithm was evaluated for voice and voiceband data (Banerjee, Feder, Kilm, & Sparrell, 1985; CCITT, 1984). After its initial deployment in the United States, however, it was discovered that it caused some low-speed frequency-shift keying modems to malfunction. A modification was quickly found and forwarded to the CCITT so that a worldwide standard was promptly attained (Benvenuto, Bertocci, Daumer, & Sparrell, 1986; Benvenuto & Bertocci, 1987). The unexpected increase in data and facsimile traffic on international links pushed for extending the algorithm to support modems operating at 9.6 kbits/s. However, because some algorithms had already been implemented, an agreement on a single algorithm could not be reached, and the final decision was left to higher levels of the CCITT (1987). After extensive debate and long negotiations, a solution was reached through a consensus (not unanimity) by allowing dissenting organizations to save face in the "spirit of international cooperation" (Schmidt, 1986; Sherif & Sparrell, 1992).

While the initial development was an anticipatory standard, the standard to support 9.6 kbits/s modems was responsive. This case illustrates the difficulties in reaching agreement in the case of responsive standards.

16 kbits/s and 8 kbits/s Speech-Coding Algorithms

The CCITT/ITU-T Recommendation G.728 on low-delay code-excited linear prediction (LD-CELP) was approved in May 1992 for the telephone quality of speech at 16 kbits/s. This recommendation culminated efforts over a 4-year period that required the collaboration of a multinational group of experts on speech coding, network transmission, and software instrumentation. The goal of such activity was to define a single coding algorithm that could be used worldwide for all potential applications and be transparent to the network (Dimolitsas, Sherif, Colin, & Rosenberger, 1993).

In the late 1980s, various methods available on a regional level were proposed for the coding of speech at 16 kbits/s. For example, the European speech-coding experts had selected a regular pulse-excitation linear predictive coder with a long-term predictor for GSM. Inmarsat adopted an adaptive, predictive coding algorithm for its Standard-B maritime satellite communication system. In Japan, NTT developed a 16 kbits/s speech code using adaptive, predictive coding with an adaptive bit allocation for use in digital leased-line networks.

Not only was the proliferation of regional and incompatible algorithms detrimental to global interworking, but the modest quality of these algorithms threatened to impose undue constraints on the overall network transmission planning. For example, the integration of digital mobile radio systems into the fixed network was understood to impose strict performance requirements on the mobile ratio systems (United Kingdom, 1987). Accordingly, a formal set of performance requirements was agreed to in Geneva in June 1998 even though it was not clear whether any candidate could meet the strict condition of a one-way coding delay of 5 ms.

The original list of possible applications contained a reference to the presence of the algorithm in the core of the PSTN. This term was removed late because of two main reasons. Its presence would have caused substantial revision of echo-control procedures because it introduced additional delay and nonlinearity in the speech path. Also, because the 16 kbits/s algorithm could not transport voiceband data at higher bit rates, its presence in the PSTN between end offices would have required the introduction of automatic methods to separate voiceband data from speech signals so as to route the former on a different route.

Because this was a responsive standard, its commercial diffusion was not as wide as the 32 kbits/s speech-coding algorithm. In contrast, the 8 kbits/s algorithm of G.729/G.729A as well as the dual-rate algorithm of G.723.1 at 5 1/3 kbits/s and 6.4 kbits/s can be considered as competing anticipatory standards that made them readily available when the need arose (Cox, 1997; Schröder & Sherif, 1997).

Chapter XII

Tightrope Walking:
Standardization Meets Local Work-Practice in a Hospital

Gunnar Ellingsen, University of Tromso, Norway

ABSTRACT

Traditionally, uniform and standardized IT-solutions in health care are considered mechanisms for increased control, efficiency and quality. Unfortunately, in spite of existing studies of the actual experiences of standardization, such as how they come into being, and how they are intertwined with local practice, unreasonable belief in standardization seems to prevail. Acknowledging the origin of standardization and its local character, however, does not mean that standardization is futile or should be avoided. It rather means that standardization efforts should balance the management level's need for increased coordination and control, and the local level's need for flexibility. The aim of this chapter is to strike this balance as it elaborates the implications and the "costs" for local practice in order to make a standard work. Empirically, the chapter draws on a standardization effort of discharge letter production in the University Hospital of Northern Norway.

INTRODUCTION

As information systems increasingly are employed in hospitals and primary care, they are dependent on standardization. From a technical point of view, standardization enables integration of information systems based on various infrastructures, developed with different tools and running at different locations. Related to use of the information systems (the social aspect), standardization serves as a means for collaboration, shared meaning and far-reaching coordination among different health care professionals.

Standardization efforts are, however, often promoted in a top-down and uniform manner with weak local influence. This is unfortunate as standards are not merely a technical or neutral device ready to be put into use. Rather they are socially constructed, achieved as results of negotiation processes (Bowker & Star, 1999; Lachmund, 1999; Rolland & Monteiro, 2002; Hanseth & Monteiro, 1997). Failing to acknowledge how standardization comes into being often results in lack of adoption, resistance in use or only temporary validity (Bowker & Star, 1999, p. 293). As a standard is intertwined with local practice, it both shapes local practice and is being shaped by it. Consequently work is required to reach agreement about a standard and, subsequently, maintenance-work is required to keep it "alive."

Acknowledging the local and partly unpredictable character of standardization does not mean that standardization is futile. It rather means that standardization efforts must be targeted to a level that is acceptable for those involved. In this chapter, I define this to be striking the balance between the management level's need for increased coordination and control, and the local level's need for flexibility. I underscore that this balance is not just out there, ready to be revealed by scientists. Rather I choose to construct it as a way to emphasize the different interests and the negotiations around the two different perspectives outlined above. More specifically, the chapter will elaborate on the costs in a standardization effort, not as an argument for discarding standardization, but as an argument in the process of defining a balanced solution between the management and the different local contexts. This chapter argues that the costs involve both additional work for some actors, restructuring of work and implications for quality. I also elaborate the implications for the management of standardization efforts

Empirically, the case draws on the work of physicians at the University Hospital of Northern Norway, with special focus on the production of discharge letters. The discharge letters are summaries of patients' stays and play several roles. Firstly, they inform general practitioners and local hospitals what has happened during the stay, current status and prognosis. Secondly, they distribute responsibilities for follow-ups between the hospital and the general practitioner. And thirdly, the hospital physicians themselves frequently use these letters whenever the patients return to the hospital, or when they for other reasons need to reconstruct the case.

The management at the hospital aimed at both increased efficiency and improved quality through standardization of the discharge letters. The motivation behind this was that sometimes the discharge letters could be delayed several weeks in the university hospital. This became a reiterating problem for the local hospitals and the general practitioners who needed the letters as soon as possible as a part of their ongoing work with the patients. There was also expressed concern about the lack of readability, and there were even complaints about discharge letters that actually lacked important information. As a part of this effort, it was necessary to instruct the physicians to work in a routine way. This turned out to be difficult because work practice in a large university hospital is extremely heterogeneous. Heavy resistance surfaced among the physicians who felt that the interests of the management were not aligned with their own. As a result, the initial strive for a completely standardized solution was translated into a recommendation.

The remainder of this chapter is organized as follows. The next section elaborates more thoroughly the theoretical foundation and is followed by a reflection on the research design. After that the background and the status of the discharge letter project is provided. The next section presents three case vignettes, which contain illustrations of physicians' work in three different contexts (two departments at the hospital and a general practitioner's office). The analysis and the conclusion appear in the following two sections. The conclusion also provides some guidelines for design of standardization efforts.

THEORY

Traditionally, uniform and standardized solutions are considered key mechanisms for increased control efficiency, simplicity, quality and collaboration. Unfortunately, this is often promoted in a simplistic manner. An example is when the former chairman of CEN/ TC 251 (De Moor, 1993) in a dogmatic way points to how efficiency in health care is closely connected to standardization:

...to make sure that unsuitable circumstances (e.g., proliferation of incomplete solutions) are not allowed to take root...[so] standardization must be started as soon as possible in order to set the development in the right track. (De Moor, 1993, p. 4)

Accordingly, it is important to eliminate standards that have evolved out of local contexts. In a similar way, many authors (especially in management literature) consider standardization to be well-suited instruments of control.

Brytting (1986, p. 145) argues that "Diversity almost by definition means trouble for managers" as it results in uncertainty and lack of overview. Standards is accordingly considered a solution as they are assumed to "generate a strong element of global order in the modern world" (Brunsson & Jacobsson, 2000).

The problem with these perspectives is that standardization seems to be defined as an "overall good." Firstly, it is argued that standardization must be carried out before it is "too late," and before non-optimal solutions and standards get a foothold. Secondly, it appears to be high belief in a deterministic relationship between standardization and control.

This clearly overrules local work practices and how standards actually are established. It is by no means certain that the best standard wins, which was clearly illustrated by the failed efforts to replace today's keyboard layout 'QWERTY' with the assumed "best one" (David, 1985). Real standards are rather constructed as a result of negotiations and compromises. It is an extremely complex socio-technical task where the standards must be adapted to different local contexts (Hanseth & Monteiro, 1997).

Like standardization of technology, the routinization of work is regarded as a foundation for increased managerial control. In her book about extremely routinized service-work in McDonalds franchising companies, Robin Leidner (1993) describes how management maximizes its control of the work process in order to predetermine how to conduct those tasks and thus promote increased efficiency and simplicity.

Leidner (1993) also elaborates the relationship between standardization of work and quality. Standardization of work and services in McDonalds franchising companies are considered a key tool for offering a proper level of quality to the customers. A precondition, however, is that the customers accept standardization of themselves as customers and play their role defined by McDonalds (p. 25). This implies that customers parting from the predefined script and instead wanting a more flexible service might easily experience that they 'equate quality with greater customization of service interactions than the companies will allow' (p. 34). Consequently when both efficiency and quality are major goals, quality becomes subordinated (p. 173).

In spite of my affinity with the work of Leidner (1993), there are a couple of aspects in my study that deviate from hers. Firstly, in my case, the customers (the discharge letter receivers) actually wanted more standardization. They considered it a strategy for quick feedback and good readability. Secondly, physicians represent an autonomous profession, which is far from the experience of the service workers in Leidner (1993). This implies that imposed routines are doomed to fail if they are not consistent with the physicians' own interests.

Studies from CSCW literature, partly in a similar way, warn that collaborative aspects of introducing new technology are often overlooked. Despite careful planning, many of these efforts fail as a result of unforeseen 'group' aspects.

What comes into play is, according to Grudin (1989):

Designers' main focus has been aiming at support for the group as a whole and less on the individual user.

Rejecting unrealistic expectations of standardization does not imply that standardization is superfluous. On the contrary standardization is a condition for making things work together over distance (Bowker & Star, 1999, p. 14) as well as it promotes both local and shared meaning (p. 293). However, standardization efforts must be approached with soberness. It means acknowledging the work involved in establishing information infrastructures, about how standards come into being; about who becomes visible and invisible and why. It also means acknowledging that standards have no single universal meaning.

Several researchers have elaborated how standardization both shapes and is shaped by local practice. For instance, Timmermans and Berg (1997) in their study of the development of clinical protocols, Lachmund (1999) in his study of auscultation sound, Winthereik and Berg (2001) in their study of the use of diagnostic codes in primary care, and Rolland and Monteiro (2002) in their study of an infrastructural information system in a global maritime classification company. Lachmund (1999, p. 440) argues, for instance:

The universalization of medical knowledge and practice and the forging of vernacular particularity went hand in hand as part of one single process.

This implies that it is not possible to achieve standardization without some tinkering with the imposed structure. Timmermans and Berg (1997, p. 291) argue:

this tinkering with the protocol, however, is not an empirical fact showing the limits of standardization in practice, but rather a condition for the functioning of the structure in the first place.

The key question, then, becomes how to balance local use and heterogeneity against uniform solutions, and at the same time take into account how these extremes transform and influence each other. This question is at the core of the work on information infrastructure (Bowker & Star, 1999; Hanseth & Monteiro, 1997). Bowker and Star's (1999) study of how the World Health Organization's global efforts to capture the variety of death certificates in a single form, serves as an exemplary expression of this kind of work. The involved classification scheme, International Classification of Diseases (ICD), constitutes an impressive attempt to coordinate information and resources about morality and morbidity globally (Bowker & Star, 1999, p. 21). Accordingly, the ICD is recognized as an important infrastructural component of medical and epidemiological software, as it tries to collect global information across several unique contexts. The historical, political and international span has complicated administration of the ICD, and cooperation has been hampered by different ways of recording and reporting. Different cultures, for instance, place different emphasis on causes of death, which influence the way coding has been conducted, and different national schools of medicine may disagree about issues such as simultaneous causes of death (Bowker & Star, 1994). The achieved level of standardization has been influenced by, and balanced against, the interests of various stakeholders. For instance: the conflict of interests between government on the global level, and the individuals in each community on the other.

REFLECTIONS ABOUT METHODOLOGY

This study belongs to an interpretative approach to the development and use of information systems (Klein & Myers, 1999) relying on four types of data: participative observations, interviews, informal discussions and documents. The observations took place from January to March 2001 in two hospital departments as well as observation in one general practitioner's office. The major observation was conducted in the Department of Cardiac and Thoracic Surgery. In total, 42 hours were spent observing work. In addition, from March to February 2000, I conducted 19 hours of observation in three other departments, which were used as background material.

I was allowed to move freely around in the wards. I participated in several morning meetings with physicians and nurses and participated in coffee- and lunch breaks; I joined groups of physicians discussing in corridors, on-duty rooms, examination rooms and pause rooms. In one department, I also observed several patient examinations. Sometimes I "shadowed" physicians in their work. In these situations I could pose questions in order to clarify and elaborate observations. The extent and format of these questions varied with what was possible without intruding too much with ongoing work. Questions were postponed when the work was recognized as hectic, during formal group meetings or in front of patients.

In sum I conducted 27 semi-unstructured interviews during the periods mentioned above. Each interview lasted at least two hours. Sometimes it was difficult to make appointments for interviews because of the hectic everyday work. I solved it by agreeing that I should be in the surrounding area, for example waiting and observing in the ward until there was some time available. Even during some interviews, interruptions occurred, caused by incoming phone calls or colleagues that needed to talk. One interview with a surgeon was also interrupted as a patient was in immediate need for surgery.

SETTING THE STAGE

Local hospitals and general practitioners regularly receive discharge letters from the University Hospital of Northern Norway. However, it has been a recurring problem for the discharge letter receivers that these letters appear to be delayed in the hospital departments. There has also been complaints about bad readability and a lack of important information in the discharge letters.

The problem was also apparent at the university hospital, which had defined a quality standard for discharge letter production time to be a maximum seven days. Unfortunately, two completed time studies[1], in 1999 and 2000 respectively, revealed that about 50% of the discharge letters were sent after this limit. The bottleneck was identified to be delays in physicians' dictating, proofreading and signing. Lack of formalized practice was also promoted as a key problem:

The departments have very different practice when they produce discharge letters. The extent of the letters fluctuates between one and six pages and some departments do not use templates. Accordingly some discharge letters are badly adapted to the receivers.

In the autumn of 2000, the management initiated the discharge letter project as a means to improve the situation. Three hospital departments participated in the project. Two of them are referred to in this chapter, the Department of Cardiac and Thoracic Surgery and Department of Medicine. Four general practitioners from two health centres participated. The project also asked the 11 other hospitals in Northern Norway, as receivers of discharge letters, to respond to some of the project's suggestions. The project had two objectives:
1. Increase efficiency in discharge letter production.
2. Improve the quality of the content of the discharge letter (in order to make them more useful and readable for the general practitioners).

To deal with the problem, the project management suggested deploying a standardized discharge letter template for the whole hospital (a strategy that is also suggested on the national level (KITH, 2001)). The text in the discharge letter should then be classified in accordance with the template and should be a foundation for routinized reuse of information from the electronic patient record:

A patient stay will accumulate several documents, like an admittance report, several notes, surgery report and lastly a discharge note. The discharge letters can be produced by reusing [categorized] information (project documentation).

Figure 1. The suggested strategy of producing discharge letters. The shaded area represents predefined extracts or summaries from existing documents.

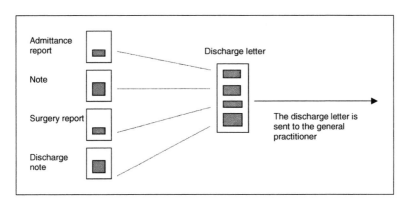

The idea is illustrated in Figure 1.

The commitment towards routinized reuse of existing information was presented to the physicians as a means for considerable efficiency gains. One of the physicians remarked:

They presented some examples where they had saved 70-80-90% time in the production of discharge letters by "cut-and-paste" and reuse of information. All you had to do was to just flow through this template with this kind of built-in structure.

It soon turned out that the aim of a standardized template for the whole hospital failed. One physician argued, "It is not possible because the departments are so different." As a result, the strategy of a standardized template for the whole hospital was abandoned and the focus on routinized reuse was de-emphasized. Instead, each of the involved departments established their own template, which was put into use in their respective departments around the turn of the year 2000/2001. The primary aim was now to increase readability for the discharge letter receivers.

THE CASES

Medical practice varies enormously—within different domains, departments, hospitals and countries (Atkinson, 1995; Berg, 1998). I have no ambition of justifying this variation in any systematic or comprehensive manner. Rather, I merely aim at motivating appreciation of this variation through the sampling of two wards at the University Hospital of Northern Norway and the work context for a general practitioner. The observations are targeted at the process of producing and receiving discharge letters. Characteristic features of these work-situations in the different contexts are:

1. **Department of Cardiac and Thoracic Surgery:** A hectic, highly specialized department with relatively narrow problems of concern, where most of the patients have had a full examination in another department. This is reflected in the discharge letters, which are short, precise and usually based on free-text.

2. **Section of Nephrology, Department of Medicine:** A section with a lot of patients with chronic diseases, like kidney failure. These patients need regular follow-up. The associated discharge letters are extensive and sometimes require a high degree of structure.

3. **The general practitioners' offices:** The general practitioners are the receivers of discharge letters in the health centres. They often read extracts of the discharge letters prior to, and during the patient consultation.

Department of Cardiac and Thoracic Surgery: Highly Specialized

The Department of Cardiac and Thoracic Surgery is responsible for cardiac surgery for adults in the Northern Health Region of Norway, as well as regionally responsible for general thoracic surgery. Most of the patients have already received a full examination by another department or (local) hospital. The patients normally stay for about a week. After their surgery, they are transferred back to another department within the hospital or to a local hospital. The discharge letter is the key vehicle for communicating to the recipients the relevant insights gained and further follow-up. The following vignette illustrates the work in the department:

Olivera is seated with a pile of patient records in front of her in one of the patient examination rooms. She has been a junior doctor in this ward for only two weeks. There is a computer on the desk with the electronic patient record system, DocuLive EPR. The new discharge letter template is nailed to the wall right behind the computer. It has been in use in the department for two months. The necessity of being brief is strongly emphasized as the template only contains eight sections, and for several sections, there are constraints on the number of lines. For instance:

- Anamnesis (maximum 4 lines)
- Examinations (maximum 2 lines)
- Treatment (maximum 3 lines)

This is followed up at the bottom of the template where it is written in big and bold types, "the discharge letter must not exceed one page."

Olivera uses the template actively when she dictates, looking up at the wall and following the sections downwards. She dictates pretty slowly. She stops at each new section and looks at the template in order to be guided to the next section. She says aloud: "personalia," "receivers," "diagnoses," "treatment codes," etc. She takes her time and spends almost half an hour in dictating this report. In the middle of the dictating she also stops and leaves the room in order to get advice from one of her senior colleagues.

Being explicit about the use of the template and its structure, she has also instructed the secretary to follow it. Figure 2 shows the discharge letter after it has been written by the secretary. The fonts in bold represent the section headlines in the template. Olivera has dictated precisely in accordance with the template.

Steinar, a head physician, rushes into the on-duty room. He sits down by the desk and starts to dictate. He is very fast; there is no pause in his dictation. He knows exactly where to look for information and does not use the template for

Figure 2. The discharge letter written by the new junior doctor, Olivera

---

Background:<years> year old man, unmarried, no children. Polio in <year>. Subsequent paralysis in the left lower extremity. Low-degree angina pectoris, not examined. In May <year> femorodistal bypass. The patient has had a postoperative fistula, which has not dried up through conservative treatment, and he had now been admitted for operative treatment.

Examination: Angiography performed the <date> indicated that the applied graft had clogged up. It was decided that this graft be removed without risking the circulation in the lower extremities.

Treatment: Surgery <date> with revision of the infected vessel prosthesis in the left lower extremity (see enclosed surgery description).

Progress, complications: The patient was sent to the infections ward because of infection in all wound cavities.

Supplementing examination: Microbiological examination of the wound secretion showed yellow staphylococci. Recommended anti-bacterial therapy Diclocil and Claforan.

Medication at discharge: Ismo 50 mg 1 tabl. daily, Fragmin 2500 IN x 1, subcutant, Diclocil 500 mg x 3, Keflex 500 mg x 3, Paracet 1x4 tabl. daily.

Status at discharge: The patient is released in relatively good general condition without signs of infection around the edge of the wound. The patient requires further antibiotics treatment and also physical therapy in order to better mobilise and especially because of paralysis in the left foot caused by polio. The patient is sent to (...) nursing home.

Follow-up: The patient has been called for post-surgery examination in 3 months. If complications should occur within 30 days, we ask to be notified.

---

support. This also implies that he does not instruct the secretary to reuse text from the admittance report. The dictating is accomplished within 3 minutes and Steinar asks the secretary to write it immediately. One of the nearby junior doctors comments that Steinar's discharge letters are "state-of-the-art:" they are very surgical, they are short, precise and do not contain anything superfluous. The nurse responsible for the discharge of the patient asks for the discharge letter since the ambulance plane is waiting. The secretary has just finished writing the discharge letter into the electronic patient record and she calls Steinar and requests him to sign the letter. This is done electronically. She says that Steinar and the rest of the head physicians normally do not need to proofread the discharge letter because they are so experienced. She adds: "Did you notice that I did not use the template? The reason is that I knew that Steinar did not use the template; still it follows the structure in the template except from that he has exchanged "blood results" and "medications." The literal part of the discharge letter looks like Figure 3.

Figure 3. The discharge letter dictated by the head physician

The patient is married and lives with his wife. Some coronary disease in the family.

Angina pectoris from <year> Inguinal hernia. Possibly infarkt in -<year>. In the beginning of January 2001, he has had unstable angina pectoris. Examination has indicated <disease-1> and <disease-2> as well as EF in the lower normal area. Pseudo aneurism in the groin.

He received surgery in <date> with a coronary bypass and surgery to a pseudo aneurism in the groin.

Postoperatively, this has proceeded well. He has had a short period of self-terminating auricular fibrillation. He may according to plan be moved to <local hospital> hospital today <date>90.

Medication:
Zocor 10 mg vespere. Albyl-E 160 mg x 1. Mycomust *tablets* x 4. Sotacor tablets 40 mg x 2.

Hb 11.6. Whites 10.7. CRP 217. Na 141. K 3.8. Temperature 38.0°.

The patient has had reaction with CRP and some temperature, but we are unable to give a good explanation why. He is in good shape. We have not administered antibiotics, but leave that, if necessary, to the colleagues at <hospital>

No appointment for control here. Will be followed up by cardiologists locally in the regular manner.

If any complications should arise after the initial 30 days, we would like to be notified.

Section of Nephrology, Department of Medicine: Structure for Chronic Patients

The Section of Nephrology is a part of the Department of Medicine. The section has many patients related to chronic diseases (such as kidney failure) who come for periodic controls. In addition, as a part of Department of Medicine, the section has to respond to emergency patients with undiagnosed problems. Excerpts from the work of Samuel, an experienced physician, are presented below.

The physician has just managed to break away from the daily buzz of patient-related work to produce discharge letters. The patients involved have been discharged from the hospital a couple of days ago, but this is his first opportunity to finish off work related to their departure. He is able to find an unoccupied office at the ward where he brings his pile of paper-based patient records. Obviously, the others at the ward know where he is as they pop into the room to make inquiries. He also has to respond to his beeper, but this does not force him to leave the room.

"RETURN DIALYSE" is written in large letters on the front covers of several of the paper records. This means that these paper records are stored in the Peritoneal Dialysis (PD)-section in a special archive. They belong to a special type of patients who come in regularly. As a result, only the secretaries in the PD-section write these reports in order to ensure that everything is done right.

Samuel points to two different discharge letter templates and says that they use both of them. One is general for the department while the other is special for PD-patients.

The physician starts by looking up laboratory results from the laboratory system, then dictates the results while reading them from the screen. After that he turns to the x-ray-system, reads the x-ray description and makes a summary of it on the fly as he dictates. He follows the structure in the general discharge letter template when he dictates and he marks it by explicitly naming each section headline. Afterwards he says, "Sometimes I can cut and paste parts of the x-ray description, it depends on how much is important—the short ones I dictate."

The following discharge letter is based on an emergency admission. This time he dictates partly the same information that existed in the admittance report. He says that the reason why he did not instruct the secretary to reuse the first sections (the summary) was incorrect information in the admittance report. He had to correct this information based on conversations with the patient and his wife.

For the next patient, on the other hand, the physician instructs the secretary to copy from X to Y in the admittance report. Now the secretary will reuse this text. Afterwards the physician says that he knew what the documentation contained because he had dictated the admittance report himself.

The physician does not know the last patient, nor does he know who has discharged her. There is no handwritten discharge form either (a form that is written by hand and functions as a preliminary discharge letter). This complicates the dictation. After quite a while reading, he starts dictating. He dictates the first part of the discharge letter all over again, that is, he decides not to use part of the existing text because, as he says, "I made a summary of it because there was so much there. There is no point reusing all that once more." He dictates in the category 'admittance findings,' after which he stops and says, "no ... this is a comment and must be put into another category." He explains later that it is important to distinguish between facts and assessments. He also studies the chart and dictates current medications. He works for about half an hour with this letter.

The final patient is a chronic Peritoneal Dialysis patient. In addition to the dictation, he retrieves the patient's Peritoneal Dialysis form (see Figure 4) from the computer. He copies it and pastes it into the discharge letter. It contains a lot of important measurements related to the patient's condition. It acts as a working 'memory' as: "This is a patient that regularly returns to the section and he needs clear-cut rules for who is responsible for what. As is possible to see here [pointing], PET analysis is not performed during this stay, but down here [pointing to the bottom of the form] you can see that it has been decided that it will be carried out during the next stay." When Peritoneal Dialysis patients are

Figure 4. Reused documentation from the PD-section

Date	10.01.01
Estimated dry weight	87-88 kg
weight	87-88 kg
Q antity of urine	2000 ml
Q antity of dialyse solution	12730 ml
Ultra filtrate	1000 – 10 0 ml
Blood pressure	172/ 106
KT/V ew n	1.16
-total	3.09
K eatin-clearance	74,6 l/ week
PET	Not performed at this stay
Prealbumin/a lbumin	35
HB	12.6
iron	17
TIBC	62
ferritin	205
Ca / ion. Calcium	2.36/ 1.22
Phosphate	1.92
PTH	20.4
Bag strength	CAPD bag strength Locolys 2.3 % 4 x 2 liter at day time, 2.5 liter Extrane al at night time
Exit-site	Good
Next control	1. At Medical policlinic. With MRU, 2 months. With measuring of rest function. 2. New PD-control with PET in May-02
Signature	NN

hospitalized, it is standard procedure to check the most recent discharge letter in order to see whether any special tests are planned.

The General Practitioner's Office

Centrum Health Station is located in the middle of the city, a couple of kilometers from the hospital. The health station consists of a co-localization of 9-10 general practitioners that offer first level services to the public. This means that the general practitioners serve as the first-line gatekeepers to the hospitals. Although the major part of the discharge letters is paper-based, the health station increasingly receives discharge letters electronically.

We are in the office of one of the general practitioners. The room is sterile, white and equipped with a lot of medical devices. Two examination benches are placed beside two of the walls. In one part of the room, there is also a desk containing a computer. As the desk lacks huge piles of papers and patient records, the work context reflects tidiness and order. On one side of the desk, there is an empty chair, signalling that this is the patient chair. The general practitioner is in-between two patient consultations and just has time to read a discharge letter from the hospital concerning one of his patients. He sits behind the desk and reads it on the computer screen. He does not wear the usual physician's white coat, just ordinary clothes, signalling a certain level of informality.

The discharge letter is relatively extensive and the content of the letter is divided into sections, but the headlines are not in boldfaced types and the letter lacks underlining. The general practitioner reads the discharge letter thoroughly, following the sections in the letter. After a while, he makes a halt and says that the structure in these letters is important, because structure makes extensive letters more readable and, as he says, "then you can just jump to the section you are most interested in, but if they are very unstructured and extensive, you might lose important information as a result of the limited amount of time available."

He also underscores that "what is also very important and which is often missing in the discharge letters, is an understandable conclusion for how to follow up things, with a clear distribution of responsibility between the hospital and the general practitioner." Such a conclusion or summary encompasses information about what has happened during the hospitalization, treatment given, prognosis, further plans and treatment strategy, sick notes and what kind of information is given to the patient about his condition. The summaries are also useful as they very quickly can reconstruct the patient cases before and during short patient consultations (typically 15 minutes). The general practitioners put these summaries into their own electronic patient record where they can be read directly from the computer screen.

The general practitioner rounds off by reading the conclusion (summary) aloud from the current discharge letter, "status ...ok ...follow-up...in fact, here

I miss something about the loss of blood and the follow-up, as this is a condition where I understand that this patient must have lost a lot of blood."

Incomplete summaries such as the current one force the general practitioners to compose the summaries themselves based on bits and pieces from the discharge letter. In relation to the current discharge letter, the general practitioner says, "the fact that they [hospital physicians] haven't made a summary where the treatment is included forces me to more advanced cut-and-paste. I have to cut and paste the CT-description, the treatment, medications by departure and the follow-up." Correspondingly, if the discharge letters are based on paper, then the physicians use a yellow marking pen to mark the sections and sentences they want included in the electronic patient record, whereupon the secretary will do the actual writing.

ANALYSIS

In this section I take the perspective of the different key actors and analyze how the degree of standardization influences their comprehension of quality. The point is to illustrate that quality is not globally given, but measured in accordance with local practice and in accordance with whoever is "strong" enough to make his viewpoint valid. This also means that when quality for some actors decreases, this will amount to the "cost" by introducing the standardized solution. In a similar way, standardization and classification processes also become a question of promoting and hampering different perspectives. Promoting a certain perspective implies de-emphasizing others. Promoting a perspective too strongly then, also becomes a question of how useful it is, for whom and whether the "cost" is acceptable. Estimating the total "cost," however, is difficult since both standards and work practice are interwoven and have the potential to transform the mode of work as well as reflect back, and possibly shaping the form of the introduced standard. As we also know, introducing a new standard for producing and reusing information feeds directly into the core of physicians' documentary work as summarizing and thus reusing previous information is an integral part of medical work (Berg, 1998, p. 298).

Quality—For Whom?

An important motivation for the project is to increase the quality of the discharge letters. Superficially, quality is a term that is easy to agree upon. Focusing on the different key actors, however, highlights the heterogeneity of comprehension of quality. There are some partly different, some partly aligned and some completely different opinions of quality. From a managerial point of view, quality is tightly connected to efficiency:

We could save considerable time for the patients and the general practitioners if the departments use templates together with more cut and paste [automatically reuse predefined information].

The hospital management tries to hook up the general practitioners' interests with their own interests. On the surface there seems to be some sort of alignment of interests as both pursue structure in the discharge letter. The general practitioners prefer structured discharge letters as a means for more surveyable reading than discharge letters based on free text. The hospital management, on the other hand, considers a common structure an opportunity for routinized reuse of information and as a means to reduce discharge letter production time. The alignment of hospital management and general practitioners, however, is only valid up to a certain point as a general practitioner complains: "Predefined reuse is possible, but very easily it becomes repetition of things and inclusion of information that is irrelevant." On the contrary, what is considered a quality discharge letter from the general practitioner's point of view is:

What is very important and which is often missing in the discharge letters is an understandable conclusion for how to follow things up and with a clear distribution of responsibility between the hospital and the general practitioner.

When turning to the Department of Cardiac and Thoracic Surgery and their mode of work, quality discharge letters are considered "short, precise without containing anything superfluous" and as you may recall, produced very quickly. Usually they are also based on free text. The head physicians of course represent the concrete expression of this kind of quality. One of them puts it like this:

We are an expedition department. We don't give a total assessment of a patient, and the discharge letters are shaped accordingly. They become relevant for the current condition, where the status of departure and follow-up for the patient is important.

This illustrates that good quality is evaluated in close relationship with the role of the department. The template that has recently been introduced in this department illustrates very well both how the template and the work influence each other and how this relationship is transformed. Even if this template contains structure, it reflects the mode of quality work in this department. The number of categories is low, for several sections it is recommended to be "just a few sentences", and finally at the bottom of the template page, it is emphasized that the discharge letter must not exceed one page. On the other hand, the actual use of the template also influences work for the physicians. In spite of how preciseness and brevity is underscored in this template, adhering to it shapes the mode of the work, as the head physician puts it:

The discharge letters become twice as long when I use the template. I can express myself much better in three sentences about how the patient stay has been, rather than using a template that induces a whole page.

For the junior doctors, however, standardization and a certain degree of routinization serve as a means to ensure that the quality of their work is compatible with the standard of the department. The junior doctor, Olivera, is a typical example when she lets herself be completely guided by the template. Similarly, the junior doctor, Pasi, who has some experience, uses the template more superficially. Consequently the standard may be seen both as giving the junior doctors skills (Leidner, 1993, p. 175) and as a carrier of prescriptions for 'good medical practice' (Timmermans & Berg, 1997, p. 296) as:

The junior doctors that come to the department from for instance the Department of Oncology are used to very long discharge letters, generally 2-3 pages. Therefore they include things that are not that relevant. They must learn to dictate the discharge letters once more, and when they use the template, their discharge letters become shorter.

In this way the template not only prescribes quality measures, but also influences and transforms the way the junior doctors conduct their work

Turning to the Department of Medicine illustrates very well how the quality of the discharge letter is linked to everyday work. The physician underscores how difficult it is in emergency cases (50% of the department's patients) to structure information in accordance with predefined rules:

It is a very big difference from a surgical point of view, with clear-cut cases and fixed surgical procedures, to the internal medical point of view, with a lot of emergency cases [and the corresponding uncertainty] (physician, Department of Medicine).

Restructuring the Work Chain

In medicine, an important part of the work is regularly summarizing previous (and thus reusing) information (Berg, 1998, p. 298). A typical situation is when the physician in the Department of Medicine "on the fly" makes a brief summary of the x-ray description, which a physician from the Department of Radiology has produced. He selects the information he considers relevant and uses it as a part of the discharge letter. Another place in the work chain, but nevertheless a similar

Figure 5. An instance of a usual work chain in the hospital

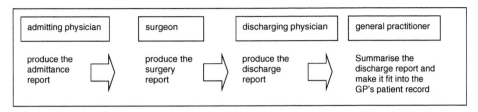

example, is when a general practitioner composes a summary by using a yellow marker pen on the discharge letter as a way to instruct the secretary what to put into their electronic patient record. An illustration of the work chain is presented in Figure 5.

This underscores that the physician who produces the summary is different from the physician who has produced the actual text. The suggested new way of producing summaries is that the physician that originally produces the whole text also produces a well-written summary that can be automatically reused by the "next physician in line." In this way, discharge letter production time is expected to decrease. Correspondingly, with summaries of discharge letters–if they are well-produced by the hospital physician, they can automatically be put into the general practitioner's electronic patient record. A typical instance of such a work chain is illustrated in Figure 6.

The major point here is that the responsibility for a piece of work is handed over from the consumer of the information to the producer. This implies a major restructuring of the work chain because responsibility for work along the whole chain will change. Physicians who write a document regarding a patient must now do some additional work that is not directly beneficial to them. Rather, they do it for the physician responsible for the next activity, which is supposed to improve efficiency in the whole work chain. Suggesting this way of doing things is however not without costs:

Figure 6. The suggested restructured work chain

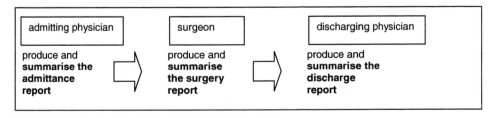

The physicians use a lot of time when they dictate the summary. (head physician)

It becomes even harder when the discharge letters cover complicated cases, as a junior doctor puts it:

I would have difficulty in saying in 3 sentences what I have used over 2 pages to express.

Suggesting that the hospital physician should summarize the discharge letters for the general practitioner, is not unproblematic, as a general practitioner puts it:

Why should the hospital physicians do the work for us, they have probably enough other things on their mind. It also implicates extra work for them to express the discharge letter in a few sentences.

This is perfectly aligned with Grudin's main point, when he argues that groupware might fail, as "it requires that some people do additional work, while those people are not the ones who perceive a direct benefit from the use of the application" (Grudin, 1989, p. 248). This additional work would not be immediately visible to 'outsiders' or decision-makers because of little or no sociological illumination of most of the back regions of medical institutions (Atkinson, 1995, p. 34). Emphasizing too strongly that the hospital physicians should produce a discharge letter summary for the general practitioners, and the general practitioner's use of just extracts of the discharge letters, may have some unexpected side effects, as a general practitioner puts it:

It is not uncomplicated. It becomes a question of how we manage this information, i.e., that we simultaneously expect to achieve comprehensive discharge letters from the hospital and that we only choose to use a tiny part of the information.

Marginalizing the Rest Categories

The structure in the discharge letter templates represents in itself selections. Somebody has chosen to visualize exactly these categories based on a certain set of motivations. An example of this is when a head physician from a local hospital for efficiency reasons suggests that:

In order to save space, I would have collapsed [the two template headings] family *and* social *into one,* family and social. (Head physician)

The motivation behind this suggestion is obviously to obtain a more readable discharge letter on the road to a general discharge letter template. However, the health personnel in the Section of Nephrology are depending on that exactly these categories are kept apart in their follow-up of the small amount of chronic PD patients:

In our section we have split up family *and* social *in order to be conscious towards the social and medical initiatives a chronic patient needs. One loses that consciousness if everything is squeezed together in one category that maybe contains a sentence that informs about marital status and whether they have children.* (Physician)

This illustrates what appears natural, eloquent, and homogeneous in one context may appear forced and heterogeneous in another (Bowker & Star, 1999, p. 131). Pushing too hard on a common discharge letter template implies marginalizing the needs of certain patient groups and especially those groups that have insufficient power to be influential. What is left implicit becomes doubly invisible: "It is the residue left over when other sorts of invisible work have been made visible" (p. 247). Consider also the Peritoneal Dialysis form (Figure 4) that is an integral part of the discharge letter for the PD patients. The physician in the Department of Medicine underscores:

We have dialysis patients who come regularly for inspection. Every time certain things must be carried out (...), in part some extremely important computations. (...) Those computations are extremely important because they indicate if it is necessary to change modus and whether the medication is sufficient.

The physician argues that discharge letters from this section need a certain structure because the detailed computations in this form are a presumption for the quality assurance in the follow-up. It also allows an increase in the overall activities through taking coordinating tasks out of the hands of the medical staff (Timmermans & Berg, 1997, p. 296).

The point I am trying to make is that this takes us away from a standardized discharge letter as an "overall" good. An "over all" good discharge letter does not exist in an ultimate sense but is situated, highly dependent on the actual content of both work-practices and standards. It follows that the question of "how much standardization is needed on a general level" becomes meaningless as it can only be addressed in the sense of more or less standardization (an approach advocated in De Moor, 1993; McDonald, 1993; Brunsson & Jacobsson, 2000; Brytting, 1986). A concrete example is when De Moor (1993, p. 6) emphasizes one of CEN TC 251's statements of principles: "Do not 'over'-

standardize." The clear limitation of this is that the actual content of both work-practices and standards are left untouched.

CONCLUSION AND SOME IMPLICATIONS FOR DESIGN

This chapter has illustrated how an ambitious standardization effort has been transformed into separate recommendations for the hospital departments. Comparing the results with the original visions of increased efficiency, the project must be regarded a failure. However, a positive side effect occurred. Establishing local templates in each of the involved departments has made explicit the actual work-practice within each of the involved departments. Especially for junior doctors, this might speed up their learning curve in adhering to the quality standard of the department.

It would, however, be frustrating for practitioners and disappointing for IS researchers to deny the possibility of transformations, of changes to the contents and organization of knowledge work. Socially informed accounts of design and use of information systems should not, as Berg (1999) compellingly argues, be misconstrued as an (implicit!) argument that existing practise is a contingent, delicate balance too fragile to be touched. In the following, I outline the implications for standardization processes in general.

First, *the different stakeholders must be educated in the various perspectives*. In spite of failure, the described project has spelled out the general practitioners' experiences as receivers of discharge letters. This has made the hospital physicians more aware of what to include, and how to present the content, when they produce discharge letters. This promotes what Boland and Tenkasi (1995) call "perspective taking," the ability or capability to take the knowledge of other communities of knowledge into account. This suggests that physicians at the hospitals, general practitioners and managers need to learn more about and take into account the various arguments in standardization efforts. For managers it is surely in their own interests to pay more attention to the complexity of the underlying situations with which they are dealing. For merely by developing a deeper understanding of the complexities of the different local contexts, by appreciating the intricate nature of problems which may arise, and also by maintaining a scepticism about any management techniques which purport to provide simple solutions, managers will be less liable to walk recklessly into standardization efforts that otherwise are doomed to fail.

Secondly, *standardization efforts must emerge out of practice*. What is underscored in this chapter is that a "good" discharge letter does not exist in itself. As we know, the discharge letter plays several roles in several contexts. A good discharge letter depends on the content of the work, the purpose with the discharge letter and who the reader is. The implication is that the order of appearance in standardization efforts is crucial. Such projects must found their efforts on the local practices and not the other way around where the starting point is standardization for its own sake. Therefore is the failed discharge letter project not an illustration of physicians who are not willing to see the big picture? On the contrary, as one of the physicians emphasized, "If it is meaningful, we can always do additional work", thus emphasizing the importance of the content of the work. The fact that the physicians are a powerful profession and rejected a standardized discharge letter for the whole hospital exactly made this point explicit. The physicians very easily were able to reject any standardization suggestions that they felt were useless.

Thirdly, *the establishment of several related standards is better than striving for the ultimate best one.* There will always be hospital physicians, general practitioners and patients that have very special needs. An example is the health personnel in the Section of Nephrology who are completely dependent on their PD-template, which serves a pinpointed purpose. Therefore it is unreasonable to impose a hospital-wide standard for discharge letters. Accordingly allowing several standards to emerge that serve a well-defined purpose must be the strategy.

Fourthly, *it is difficult to actually assess the "cost" from a co-construction perspective.* Everyday practice may change as a result of standardization, signalling that the standard itself must be prepared for changes. Accordingly defining the "costs" in standardization projects is notoriously difficult. Unfortunately this tends often to be overlooked as some authors very easily seem to write about this issue, as if they have the ability to assess who is marginalized, and who's not, whereas this is inherently problematic when looking at changing configurations.

REFERENCES

Atkinson, P. (1995). *Medical talk and medical work.* Sage Publications.

Berg, M. (1998). Medical work and the computer based patient record: a sociological perspective. *Methods of Information in Medicine, 38,* 294-301.

Berg, M. (1999). Accumulating and coordinating: Occasions for information technologies in medical work. *Computer supported cooperative work, 8,* 373-401.

Boland, Jr., R. J., & Tenkasi, R. V. (1995). Perspective Making and Perspective Taking in Communities of Knowing. *Organization Science, 6*(4), 350-372.

Bowker, G., & Star, S. L. (1994). Knowledge and infrastructure in international management: Problems of classification and coding. In L. Bud-Frierman (Ed.), *Information acumen* (pp. 187-213).

Bowker, G., & Star, S. L. (1999). *Sorting things out: Classification and its consequences*. MIT Press.

Brunsson, N., & Jacobsson, B. (2000). *A world of standards*. Oxford: Oxford University Press.

Brytting, T. (1986). The management of distance in antiquity. *Scandinavian Journal of Management Studies*, November.

David, P. A. (1985). Clio and the economics of QWERTY. *American Economic Review, 75*(2), 332-37.

De Moor, G. (1993). Standardisation in health care informatics and telematics in Europe: CEN 251 activities. In De Moore, C. McDonald, & J. Noothoven van Goor (Eds.), *Progress in standardization in health care informatics*. Amsterdam: IOS Press.

Grudin, J. (1989). Why groupware applications fail: Problems in design and evaluation. *Office: Technology and People, 4*(3), 245-264.

Hanseth, O., & Monteiro, E. (1997).Inscribing Behaviour in Information Infrastructure Standards, *Accounting, Management & Information Technology, 7*(4), 183-211.

KITH (2001). *The Medical content in discharge letters—"The good discharge letter."* Report from the Centre for IT in healthcare (KITH) on commission from the Ministry of Health and Social Services.

Klein, H. & Myers, M. (1999). A set of principles for conducting and evaluating interpretive field studies in information systems. *MIS Quarterly, 23*(1), 67-94.

Lachmund, J. (1999). Making sense of sound: Auscultation and Lung sound codification in nineteenth-century French and German medicine. *Science, Technology & Human Values, 24*(4), 419-450.

Leidner, R. (1993). *Fast food, fast talk—Service work and the routinization of everyday life*. Berkeley: University of California Press.

McDonald, C.J. (1993). ANSI's Health Informatics Planning Panel (HISPP)—The purpose and progress. In De Moore, C. McDonald, & J. Noothoven van Goor (Eds.), *Progress in standardization in health care informatics*. Amsterdam: IOS Press.

Rolland, K. H., & Monteiro, E. (2002). Balancing the local and the global in infrastructural information systems. *The Information Society, 18*(2).

Timmermans, S., & Berg, M. (1997). Standardization in action: Achieving local universality through medical protocols. *Social Studies of Science, 27*, 273-305.

Winthereik, B. R., & Berg, M. (2001). "We fill in our working understanding": On achieving localisation in an information system for primary care. *Proceedings of the 24th Information Systems Research Seminar in Scandinavia*.

ENDNOTE

[1]   Both in 1999 and 2000, The Centre for Research, Quality and Development (FoKUs) at the hospital completed a time study in order to estimate discharge letter production time. The number of involved departments was 7 and 16 respectively, and the number of discharge letters was 565 and 1,368 respectively.

This chapter was previously published in the Journal of IT Standards & Standardization Research, 2(1), 1-22, January-June, Copyright © 2004.

Chapter XIII

Unified Citation Management and Visualization Using Open Standards:
The Open Citation System

Mark Ginsburg, University of Arizona, USA

ABSTRACT

Scientific research is hindered when there are artificial barriers preventing the efficient and straightforward sharing of bibliographic information. In today's computing world, the barriers take the form of incompatible bibliographic formats and constraining operating-system and vendor dependencies. These incompatible platforms isolate the respective camps. In this chapter, we demonstrate and discuss a new approach to unify citation management: the Open Citation System (OCS). OCS uses open XML standards and Java-component technologies. By providing converter tools to migrate citations to a centralized hub in BiblioML format (an XML tag set based on the UniMARC standard), we then make use of XML topic maps to provide an extensible framework for visualization. We take as an example the ACM classification code and show how the OCS system displays citations in a convenient focus and context hyperbolic tree interface. We conclude by discussing future directions planned to extend the OCS system and how open citation management can supply an important piece in our inexorable march toward a worldwide digital library.

INTRODUCTION

The Internet infrastructure enables many large- and small-scale citation databases that researchers enjoy today. Some of the citations are directly available from online digital libraries (DLs). Others are parsed from free-standing Web documents and placed into an online DL as the CiteSeer project accomplishes with its autonomous citation indexing (Lawrence & Giles, 1999a). The DLs, coupled with traditional journals, conference proceedings, and stand-alone citation collections, place voluminous citations online.

It would seem we are well on the way to the goal, proposed by Cameron (1995), of having a "universal, bibliographic and citation database linking every scholarly work ever written." Yet, researchers struggle with several incompatible "spoke" formats based partially on historical reasons separating the physical and social sciences. Unified citation management is the goal, but at present, it is a fractured endeavor with ad hoc bridging attempts littering the landscape but no cohesive themes or design principles underlying the efforts.

What should a unified citation-management system accomplish? At the very least, it should provide a convenient interface to convert citations to and from various spokes: a set of spoke-to-spoke converters. It should also be able to group citations into a database to support fielded queries in order to retrieve specific subsets for the task at hand. Later, we will review some of the software systems that accomplish these basic goals. Thus, researchers can move back and forth between formats using simple Web forms and construct simple fielded queries on databases. The systems generally input and output serial lists of citations in the requested format. It is difficult to know which topics a given set of citations covers, though. Researchers must do fielded queries, for example, by author, title, abstract, or keyword to retrieve citation subsets. Ad hoc visualization efforts local to various spoke formats, while possible, have not been widely adopted in practice and do not address the fundamental problem of the citation information islands.

This chapter seeks to approach the problem of citation-topic exploration with more design cohesion and stability by using an integrated set of open XML (extensible markup language) standards and Java components. As Krechmer (2002) points out, open standards are "a codification that a society (collection of users and implementers) wishes to maintain unchanged for a span of time," which directly yields the desired stability. Our citation-management system prototype, the Open Citation System (OCS), is introduced later in the chapter. The OCS implements converter classes, written in Java and implemented as a Web-based servlet, to transform spoke formats into a special XML tag set: the bibliographic markup language (BiblioML; Cover, 2001) "hub." After discussing the advantages of the choice of XML as the hub format, we present additional XML tools, such as XML topic maps (XTMs; Auillans, 2002; *ISO/IEC 13250 topic maps*, 2000), that work in conjunction with a Java servlet and a hyperbolic tree applet to provide the researcher with a key additional functionality: topic visualization. The ability of the OCS to make use of many open-source components speaks to the strength of the open-source-development community and the ease of integration. Currently, the OCS prototype is online to demonstrate both the conversion (http://louvain.bpa.arizona.edu/ocs/biblio.html) and visualization (http://louvain.bpa.arizona.edu/ocs/tree.html) functions.

Thus, the OCS is an open-standard initiative to bridge the citation-format schism by layering visualization on top of the more mundane format-conversion function. After presenting the OCS architecture and a system walk-through with illustrative prototype screen shots, we conclude with a description of current and future research plans.

CITATION MANAGEMENT AT PRESENT: THE "UNIX vs. WINDOWS" BATTLE

Journals and professional societies offer explicit standards and electronic templates and tools on various operating-system platforms (Unix, Mac, Windows) to help a researcher print a reference section (IEEE style, Kluwer style, APA [American Psychological Association] style, and so on). Yet, while citation formatting is fairly standard across platforms, there is no unified manner to store and manage bibliographic citations.

To date, there persists a sharp schism between two bibliographic-management approaches. Physical scientists and engineers, who are familiar with a variety of Unix operating systems, often prefer BibTeX (Goossens, Mittelbach, & Samarin, 1993). BibTeX is the bibliographic engine provided with the LaTeX macro collection, which in turn makes use of the TeX typesetting engine (Lamport, 1986). The combination of LaTeX and BibTeX has been popular for several decades in various Unix environments.[1] This may reflect the scientist's willingness to work at a low level with mathematical formulae, tabular environments, and other detailed typesetting tasks. The combination of the Gnu public-licensed product Emacs[2] (a powerful editor), BibTeX for bibliographic management, the X-Window environment, and the AUC TeX3 enhanced Emacs environment for LaTeX/BibTeX provides a powerful graphical environment for document preparation. BibTeX files may be culled according to specific interests or personal involvement and made available for sharing amongst peers. Some examples are the following: Iamnitchi (2001) makes personal computer-science publications available on the Web, and also on the Web, the LIDOS Project (Herzog & Huwig, 1997) makes a searchable artificial-intelligence BibTeX archive available.

Social scientists often prefer Microsoft (MS) Word as a visual document software package and resort to vendor packages, such as EndNote and ProCite (http://www.endnote.com) to provide citation management in the familiar MS

Word visual interface. These tools provide a proprietary storage structure and MS Word toolbar integration, as well as live links to various Internet bibliographic resources. Efforts to avoid proprietary citation management in MS Windows exist, such as visual LaTeX/BibTeX implementations for MS Windows and tools to manage BibTeX entries from inside MS Word (Canós, 2001), but they show very little adoption compared to the MS Word and EndNote or ProCite user community.

By way of empirical evidence, we consider the ISWorld Web site (http://www.isworld.org), the worldwide home page for MIS (management information systems) researchers. This organization dedicates a special section to EndNote bibliographic resources (http://www.endnote.auckland.ac.nz/) and another to ProCite resources (http://www.sistm.unsw.edu.au/research/isworld/). The latter site explains how to convert EndNote files into ProCite format, but the conversion is only one way: As of this writing, EndNote cannot handle ProCite-formatted citations. Some of the contributors to this section, recognizing that not all their peer researchers have a compatible version of EndNote to view their submissions, also make their EndNote files available in BibTex and/or HTML (hypertext markup language) format. Citations stored in the EndNote and ProCite vendor packages have a proprietary binary format and it is necessary to export the data to a readable Refer format for subsequent postprocessing and conversion to BibTeX. Conversely, a BibTex user would need to postprocess his or her .bib file into Refer format for import into EndNote or Procite.

The multiple incompatible formats require converters to share citations across the camps. These converters are economically inefficient (Farrell & Saloner, 1993) but a necessity at present since no unified format is apparent. The next section details some efforts to make citation research easier in the face of the splintered environment.

RELATED WORK

Many digital libraries, to retain flexibility and not commit to one camp or another, prefer a flexible middle approach: to store their citations in an internal database and export, on request, citations to a specific output format. This is the approach of the arXiv technical-report repository (http://www.arxiv.org). The contributing authors enter the citation metadata and then upload their articles to the arXiv repository.

One of the output formats supported by arXiv is the Dublin Core (DC) XML format (Sugimoto, Baker, Nagamori, Sakaguchi, & Tabata, 2001). To fetch the citation metadata from arXiv, an open protocol is used: the Open Archives Initiative Metadata Protocol Handler (OAI-MPH; Breeding, 2002; Cole, Habing, Mischo, Prom, & Sandore, 2002), which is very convenient. The OAI was designed to be a catalyst to the establishing of a low-entry and well-defined interoperability framework applicable across domains (Lagoze & Sompel, 2001). Because XML is metadata, or data-describing data, it is well suited to be used in the OAI framework and more specifically by the OAI-MPH protocol. OAI-MPH allows users to make date-based queries on the archive. Users can request the date of creation, date of deletion, or the latest date of modification (Lagoze & Sompel). Another option is a set-based request, which is an optional construct for grouping items in the OAI repository for the purpose of selectively harvesting records. Using XML as the platform-independent standard makes sense as it separates the semantics (the field names, such as title, author, and so on) from the presentation (output style; Khare & Rifkin, 1997) that the client might elect.

Another approach is taken by the CiteSeer project (http://citeseer.nj.nec.com/cs), hosted by the NEC Corporation, which has a very large repository of scientific articles. Unlike the arXiv project, CiteSeer does not require authors to supply metadata up front. It has parsing routines to recognize common citation formats and place them into an internal database. The users benefit from automated citation analysis and graphical displays (Lawrence & Giles, 1999a). It allows authors to submit their own work and it also has Web crawlers that go out and proactively search for previously unentered articles (Lawrence, 2001a, 2001b; Lawrence, Bollacker, & Giles, 1999; Lawrence & Giles, 1999a, 1999b). This two-pronged approach has enjoyed widespread adoption and is part of the inexorable march of the scientific community toward a global digital library (Fox & Marchionini, 1998).

The Georgia State University ISBib project (Chua, Cao, Cousins, Mohan Straub, & Vaishnavi, 2002) is another example of a DL that collates specific articles for a professional audience. In this case, numerous MIS journal data and metadata are collected and stored in an internal MS Access database. However, although the schema allows for citation-topic classification, in practice that field is not populated since the manual effort to classify the collected articles was too great (C. E. H. Chua, personal communication, 2002). It allows for fielded query and subsequent export to Refer (EndNote) or BibTeX format.

Stanford's Interbib (Chua et al., 2002; Paepcke, Chang, Garcia-Molina, & Winograd, 1998) project, part of the Stanford Digital Library Project (SIDL), also offers a straightforward spoke-to-spoke conversion service from BibTeX and Refer to HTML and MIF (Framemaker) using Python routines written by Andreas Paepcke. In addition, the conversions write data to a database for subsequent fielded search (Paepcke, 2003).

Also, there are several new initiatives exploring the attractive hub-and-spokes architecture that we adopt in the OCS system. For example, the American Association of Physics Teachers and the University of Oklahoma's Library and Information Science department are currently considering a Web service to output a requested spoke format (e.g., BibTex, Refer, EndNote, etc.) in a requested citation style (e.g., APA, MLA, etc.). The well-motivated plan is to give digital libraries the value-added ability to return citations in a format customized to the library users (L. Barbato, personal communication, March 9, 2005). Similarly, the Swiss Federal Institute of Technology is hosting the bibconvert service (http://dret.net/bibconvert/), which uses a shared internal data model as the hub and provides conversions between various spoke formats (E. Wilde, personal communication, January 5, 2005).

The following section introduces the Open Citation System and demonstrates how its use of XML and Java-component technologies gives it a distinctive approach to citation management.

THE OCS

Initial Decisions

The OCS project started by "scratching" the author's "personal itch" (Raymond, 1998) of struggling with incompatible citation formats and incomplete topic mappings across multiple digital libraries. At the beginning of the OCS project, we decided upon XML as a citation storage layer. Not only does XML separate presentation and semantics, as mentioned previously, but it also has several standards pertinent to citation management that the OCS can leverage.

First, we chose a subset of BiblioML as our basic citation XML tag set. BiblioML (Brown, 2002; Cover, 2001) is a specification based on UniMARC (Anonymous, 1994) records. Its purpose is "the interchange of UniMARC bibliographic and authority records between applications" (Sévigny & Bottin, 2000). However, UniMARC is very complex and so likewise is BiblioML. In the OCS, we reduced the BiblioML fields to those a researcher would commonly find in a journal or book citation. An alternative, which would work just as well, is the Dublin Core (Dekkers & Weibel, 2002) metadata initiative.

The second convenient XML standard is extensible stylesheet language transformation (XSLT). This allows users to customize their own XSLTs into specialized formats easily and quickly, thus promoting the XML document set's reuse. We use XSLT to convert DC-format (Sugimoto et al., 2001) citations that we fetch, for example, from the arXiv repository, and convert it to the BiblioML hub format.

Other useful XML standards allow us to find and extract citation data from XML data stores. The Simple API for XML or SAX allows programmers to access and manipulate XML documents. The document object model (DOM) also allows programmers to access and manipulate XML documents, and it is better suited for the processing of smaller XML documents because it reads the whole XML document into memory before processing. The following discussion of specific OCS features will introduce additional XML-component technologies leveraged by the OCS.

Given the initial design decision to store citations internally in the BiblioML hub format, the first step was to code the converter classes to go from spoke formats, such as BibTeX or Refer, to the BiblioML hub and vice versa: our conversion classes are bidirectional.

Bibliographic-Record Converter

The core of the converter application consists of the conversion module. The conversion module is responsible for converting bibliographic records in a specific format into a different format. For the purposes of this project, it was necessary to be able to convert records between three different formats: EndNote/Refer, BibTeX, and BiblioML. An efficient and extensible design was chosen to provide conversion between the specified formats. The design consists of a central hub, which is a format-independent representation of a bibliographic record, and a number of spoke converters. Each spoke converter is responsible for converting between a specific format and the generalized hub format. The hub format must be able to encapsulate all of the data necessary to completely represent every spoke format without losing any information. Using this design, any format can be converted to any other format by simply writing one converter for each format (hub-and-spokes considerations in the realm of business documents are discussed in Wuestner, Hotzel, & Buxmann, 2002). In the case of OCS, three converters were implemented and one hub format was defined. The conversion module was designed using an object-oriented approach, so each converter and the hub format were implemented as Java classes. In OCS, a specific converter is referred to as a BiblioConverter, and the hub format is referred to as a BiblioRecord. The BiblioRecord class consists of a set of fields that hold all of the information necessary to fully encapsulate one bibliographic record, such as title, authors, bibliographic type, publisher, and so on. Each BiblioConverter used the same Java interface, which primarily consisted of two methods responsible for converting between the specific format and BiblioRecord objects. By using interfaces, the converter objects could be accessed in a common manner without knowing which specific converter was being called. The BibTeX converter was implemented using standardized documentation on the BibTeX file format. EndNote could not be implemented so easily. The EndNote library-file format is a proprietary, binary format that is not documented; thus, it was decided to use the EndNote/Refer plain-text file format. This format is well documented in the vendor user manual and can be easily parsed.

The BiblioML converter was implemented using the XML Xerces parser to parse and build BiblioML documents since BiblioML is an XML-based format. In order to convert from one format to another, two BiblioConverters are chained together by using the output of one as the input of the other. For example, in order to convert between an EndNote-formatted file and a BibTeX-formatted file; an

EndNote BiblioConverter would be instantiated. The EndNote file would be passed to the EndNote converter and a set of BiblioRecord objects would be returned. These BiblioRecord objects would then be passed to a newly instantiated BibTeX converter, which would return a BibTeX-formatted character stream. This stream could then be written to a file or a response stream if the converter is running on a servlet.

The actual bibliographic-record converter application consists of two servlets: ConvertServlet and ResultsServlet, which provide a simple user interface for the core converter module. ConvertServlet receives requests to convert a set of records from one format to another, and returns an HTML representation of the converted records and the option to download the converted records. The servlet generates the HTML representation by converting the records to BiblioML format and then transforming the XML to HTML using XSLT. The servlet is initialized by parsing and caching the selected XSL stylesheet that is responsible for transforming the records. The user can choose a set of records to convert by either uploading the file from the client machine or by downloading the file from an Internet source, as shown in Figure 1.

The ConvertServlet determines which of the two options has been selected and either uploads the local file or downloads the networked file. The appropriate converters are then instantiated and the records are converted to the user-selected format and to BiblioML. The resulting converted records are placed in the user's HTTP (hypertext transfer protocol) session, and the BiblioML records are transformed and written to the response stream. The ResultsServlet receives requests to download the converted records and returns the results in an appropriate manner. If the resulting records consist of multiple files, the files are first zipped up before being written to the response stream. The servlet also sets

Figure 1. Setting up a conversion request

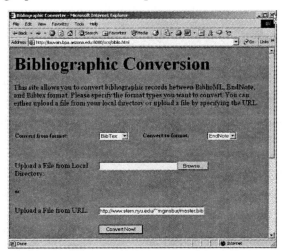

Figure 2. BibTeX records are converted into Refer format, and BiblioML hub objects are displayed to the user.

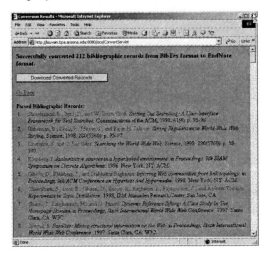

the content type of the result in the response header so that the user's browser can handle the response appropriately.

Figure 2 shows the situation after a conversion request has been processed; XML records have been transformed with XSLT into HTML in the user's Web browser.

Here XSL is used to colorize the resulting middle layer. The lettering of the authors' names are in red, the titles of the resources are in blue and are italicized, and all other information is in black print.

The architecture helps simplify programming by minimizing the number of converters needed to be written. Also, by designing the converters in this arrangement, additional converters could be written in the future with minimal changes to the system. All one needs to do is write a converter for the given format and add it on to the system like another spoke added to the hub.

Topic Visualization

The choice of the BiblioML format is fortuitous: There are unused tags in the specification that can serve as placeholders to link citations to a topic or set of topics. Another XML specification is the XML topic map (Auillans, 2002; *ISO/IEC 13250 topic maps*, 2000). XTM is an official ISO/IEC standard that allows the representation of parent and child nodes and was developed to facilitate locating and managing information through topic organization and relationship. Within an XTM, topics are linked to form a semantic network of information. This organized view of related topics lends itself well to visualization (Auillans). Of course, topic organization will differ across reader communities, and the maintenance of a set of ontologies will be an ongoing effort as older terms expire or change meaning over time and new terms are added (Pejtersen, 1998; Petrie, 1998). In this chapter, we consider only a static topic set that is known a priori.

We decided to make use of a well-known classification tree for the OCS visualization prototype: the Association for Computing Machinery (ACM) digital-library classifier (Coulter et al., 1998). Most ACM DL articles are coded with this classification scheme, which has undergone several revisions since its inception. Our plan of attack was to encode our BiblioML hub citation records with an ACM topic and tie the classification to a master ACM XTM document. A Java hyperbolic tree applet can display the original ACM tree to the researcher, and XML-component technologies, in conjunction with a Java servlet, can display results depending on user navigation in the applet. The following discussion details how it was done. See Appendix A.

The purpose of the bibliographic tree visualizer is to graphically represent a hierarchical categorization of citation records. The tree visualizer reads in a set of BiblioML records and an XML topic map and builds a graphical tree, which the user can browse. The tree visualizer consists of two major components: the BiblioServlet and the BiblioTreeApplet. The BiblioServlet is responsible for returning an HTML representation of all citations in a specific category. The BiblioServlet is initialized by parsing and caching the XSL stylesheet and loading all BiblioML records in the specified directory on the server. When the BiblioML records are loaded, they are parsed and cached, and the category that they reside in is extracted from the Note element of type SubjectAccess. This element can be used to store a reference to information concerning the record's subject. The BiblioServlet extracts the record's category by reading an XLink URL (uniform resource locator) to the topic that the record is a member of in a topic map from a Note element. The BiblioTreeApplet is initialized by retrieving the specified topic-map file, parsing it, and building a graphical tree representation.

So that the BiblioTreeApplet is not restricted to only retrieving topic-map files from the server from which it was downloaded, the HttpProxyServlet is provided, which acts as a proxy through which the applet can retrieve a topic-map file from any Internet source. In our case, the ACM topic map is parsed using the MinML2 parser and a tree data structure is built. The FloatingTree component uses this tree data structure to build a graphical representation that is displayed to the user. When the user clicks a node in the tree, an event is fired to notify the applet to retrieve all citations in the selected topic. The applet sends a request to the BiblioServlet containing the selected topic in the topic map. The BiblioServlet searches through all cached BiblioML records for all citations that are either in the selected topic or in a subtopic of the selected topic. Each BiblioML record that is found is transformed to HTML and written to the response stream, which is returned to the user's browser. Figures 3 and 4 show subsets of citations retrieved in the bottom frame as the user navigates the tree in the upper frame.

See Table 1 for a complete listing of the XML technologies used in the OCS prototype. Many of the components are under the aegis of the open-source Apache project or the World Wide Web Consortium. The high availability and rapid development of these standards means the OCS modular system is able to take advantage of advances in any of its underlying components.

Figure 3. Exploring the ACM classification hierarchy and expanding the root node into the three main branches

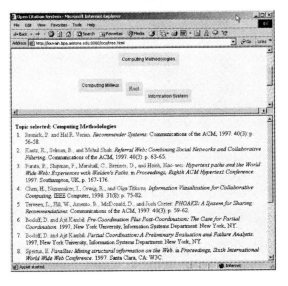

Figure 4. Exploring Computing Milieux. The Computers and Education child node is highlighted, and the results are retrieved in the bottom frame.

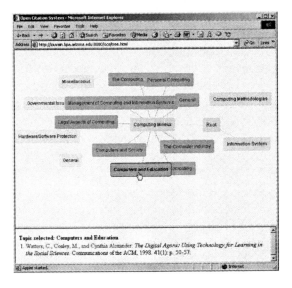

Table 1. XML components used in the OCS

XML Component	What Does it Do?	Use in the OCS Prototype
MinML2	A minimal XML parser that speeds the performance of applets	Downloaded with the hyperbolic tree applet to parse multiple small XML documents and tie them to the XTM tree
Xerces	A reliable full-featured parser with reasonable performance	Used by DOM and SAX to extract topics and to transform XML documents from Dublin Core to BiblioML
Xalan	XSLT processor for transforming XML documents into HTML	Used to convert XML documents to HTML
XTM	Specification is an XML syntax for the expression and interchange of topic maps	Used to represent the ACM tree for the hyperbolic visualizer
DOM	Defines a programmatic interface to process smaller XML documents	Used to access and extract data from smaller XML documents
SAX	Defines a programmatic interface to process larger XML documents	Used by the harvester program to process large OAI XML documents
JDOM	Defines an easier version of DOM and SAX for JAVA programmers	Used in the arXiv harvester program
XSL-FO	A language for reformatting XML documents into visually pleasing formats	Used by Xalan to format HTML documents
XSLT	A language for transforming XML documents	Used to convert Dublin Core to BiblioML
Xpath	An expression language used by XSLT to access or refer to parts of an XML document	Used in XSLT code
Xlink	Allows elements to be inserted into XML documents in order to create and describe links between resources	Used in most XML documents to describe HTML links

Brief Commentary on Information Visualization

A goal of the interface should be to support learning. The user, while interacting with the interface over time, can learn more about the underlying structure of the document collection. This idea is supported by the hyperbolic interface (Pirolli, Card, & Wege, 2000, 2001) that we use in our OCS prototype; we use the effective hyperbolic interface in the top frame while providing detailed results using XML and XSL in the bottom frame. Referring to Figure 1, the OCS shows an example of focus- and context-information visualization (Green, Marchionini, Plaisant, & Shneiderman, 1997; Leung & Apperley, 1994; Plaisant, Carr, & Shneiderman, 1995). This type of visual organization has been shown to lead to faster and more intuitive user navigation (Börner & Chen, 2002; Pirolli et al., 2000, 2001). The focus and context frames give the users clues in both windows to help locate information more quickly to "reduce the cost structure of information" (Card, Mackinlay, Shneiderman, & Iacovou, 1999, pp. 14-15).

Putting It All Together

Now that we have discussed the design and implementation of the conversion and visualization features, it is time to present the overall OCS architecture. This can be seen in Figure 5.

Figure 5. The OCS architecture

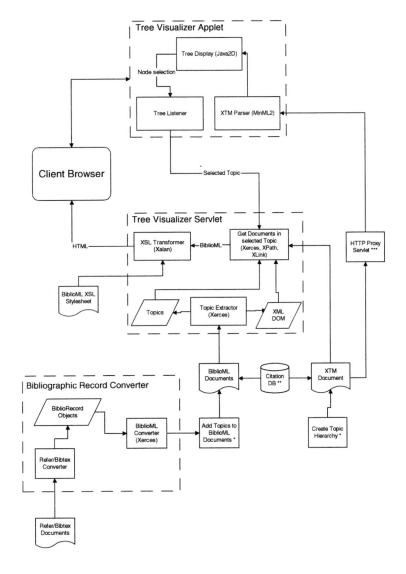

Some commentary on Figure 5 is required. The processes flagged with asterisks are, at present, either manual or not yet implemented. Thus, it was a manual effort to create the ACM's XML topic-map file using their classification scheme found at http://www.acm.org/class/1998/. The encoding of individual BiblioML citation records to tie them into a given topic map was also done manually. Another piece that is not yet implemented is a citation database (DB) that can grow dynamically as new conversion requests come in, as featured in the Stanford Interbib project (Paepcke, 2003). In our current prototype, we visualize 55 citation test records from an MIS PhD thesis that was constructed in LaTeX/BibTeX. The BibTeX records were converted to BiblioML by OCS converters.

FUTURE WORK AND
CONCLUDING REMARKS

Currently, OCS handles all aspects of parsing and creating BiblioML records but does not contain any facility for categorizing them: This is left as a manual process. One difficulty is that there is no standard categorization scheme across all bibliographic formats. Some formats do not even provide a way to encode subjects or categories. For the purposes of the initial prototype of OCS, BiblioML files are manually categorized. Tools for categorizing BiblioML records should be considered for future versions of OCS. BiblioML files are manually categorized using the Note element provided by UniMARC. A Note element of type SubjectAccess can be used to hold any type of reference to information about the record's subjects. OCS uses this element to hold an XLink URL to a topic in a topic-map file. This element must be manually added. One can also add a Subjects element to provide further clarification of what the topic is that the Note element refers to. For the tree visualizer to work properly, all of the BiblioML records that will be displayed should be categorized in this manner. This suggests an extension to the work: to analyze the citation metadata, or the full document text, using classifier algorithms to position it into one or more nodes of a topic map. Much work has been done in the area of document and citation machine categorization, for example, focused crawling in order to poll a large set of documents and classify them according to a given hierarchy. These documents might or might not be available on the Web via HTTP. For example, Mukherjea (2000) developed the WTMS system for focused topic crawling on the Web. Dumais and Chen (2000) demonstrated a system with a similar goal. Researchers have used data and metadata to classify Web documents (Chung & Clarke, 2002) using such diverse algorithms as naïve Bayes classifiers (Mitchell, 1997), Rocchio feedback (Joachims, 1997), and support vector machines (SVMs; Joachims, 2001). In addition, a linear function categorizer has recently shown promise (Wibowo & Williams, 2002). If a suitable approach is found, the OCS can offer a tool kit to allow researchers to create their own custom topic maps and then visualize their bibliographic citations using the new maps.

Another inflexible piece of the current visualizer is the fact that the topic-map file is not user configurable. It is hard coded as an applet parameter. Referring to Figure 5, the presence of the HTTP proxy servlet component means that we can improve the system by allowing the user to designate any networked XML topic-map file to be used by the applet. If the user wishes to change tree views in the middle of the session, there should be supporting facilities to recategorize the BiblioML citations accordingly.

For maximum flexibility, the users should be allowed to select subsets of BiblioML citation collections as well. These collections might reside on one server or they might be distributed across many servers. Ideally, the users can switch topic-map files and also BiblioML collections during a session. An indexed database may be necessary to support scalability as the underlying citation collections become larger.

Finally, the user interface is an interesting piece of the visualizer that can be improved. One natural extension would be to represent the population of each node graphically before the user clicks on the given node to explore its results. Density information would be a valuable additional piece of metadata to help the researcher in the information-foraging (Pirolli et al., 2000, 2001) task. Currently, we are working on this enhancement, experimenting with various approaches such as color coding, 3-D emulation (to provide a thickness to the node), and simple population text labeling. Another extension is to use the OCS framework to group and visualize citations fetched by the OAI-MPH protocol from the arXiv technical-report collection. The arXiv collection classifies each report using the 1998 ACM classification tree (Coulter et al., 1998), and thus the entire content of a digital library can be conveniently fetched and passed to our visualizer, making use of a well-known de facto standard ontology. To generalize to a generic networked environment, our framework is most suited to a situation where entities can be classified using a well-established classification tree and where the entities' metadata are conveniently exposed via open protocols such as OAI-MPH. It is less suited to situations where the underlying classification models are highly fluid or where the metadata are opaque and hard to divine.

To sum up, the Open Citation System has been an interesting exercise in combining XML and Java-component tools. The components fit well together in both conversion and citation visualization. We have a stable proof of concept, but there is much work to be done, particularly in providing the user maximum flexibility in selecting and modeling his or her citations, before it significantly helps address the problem of incompatible citation-management platforms. The task of attaining a worldwide digital library (Fox & Marchionini, 1998) is no easy matter, but it is our hope that the open-standard citation-management architecture supplies an important piece toward its realization.

ACKNOWLEDGMENTS

The author wishes to acknowledge Damien Daspit, who coded much of the initial OCS prototype under the author's supervision as part of his master-of-science degree at the University of Arizona's MIS department. Nhan Nguyen, also an MS-MIS student, contributed code and useful technical documentation. Subsequently, Britton Watson and Jeffrey Keippel contributed code as part of an arXiv digital-library visualization project that was based primarily on the OCS platform. The author also wishes to acknowledge the helpful comments of the anonymous reviewers.

REFERENCES

Anonymous. (1994). *UNIMARC manual: Bibliographic format 1994.* Retrieved June 12, 2003, from http://www.ifla.org/VI/3/p1996-1/sectn1.htm

Auillans, P. (2002). *Toward topic maps processing and visualization.* Paper presented at the Sixth International Conference on Information Visualization, London.

Börner, K., & Chen, C. (2002). *Visual interfaces to digital libraries.* Paper presented at the Second ACM/IEEE-CS Joint Conference on Digital Libraries.

Breeding, M. (2002). The emergence of the open archives initiative. *Information Today, 19*(4), 46.

Brown, G. (2002). *BibML: A markup language for bibliographical references.* Retrieved December 10, 2002, from http://www.cix.co.uk/~griffinbrown/bibMLwp.pdf

Cameron, R. D. (1997, February). A universal citation database as a catalyst for reform in scholarly communication. *First Monday.* Retrieved September 2005, from http://www.firstmonday.dk/issues/issue2_4/cameron

Canós, J. H. (2001). A bibliography manager for Microsoft Word. *ACM crossroads.* Retrieved February 1, 2003, from http://www.acm.org/crossroads/xrds6-4/bibword.html

Card, S. K., Mackinlay, J. D., Shneiderman, B., & Iacovou, N. (1999). *Information visualization: Using vision to think.* San Francisco: Morgan Kaufmann.

Chua, C. E. H., Cao, L., Cousins, K., Mohan, K., Straub, D. W., Jr., & Vaishnavi, V. K. (2002). IS bibliographic repository (ISBIB): A central repository of research information for the IS community. *Communications of the Association for Information Systems, 8,* 0-0.

Chung, C., & Clarke, C. L. A. (2002). *Topic-oriented collaborative crawling.* Paper presented at the Conference on Information and Knowledge Management (CIKM), McLean, VA.

Cole, T. W., Habing, T. G., Mischo, W. H., Prom, C., & Sandore, B. (2002). *Implementation of a scholarly information portal using open archives initiative metadata harvesting protocols: Interim report to the Andrew W. Mellon Foundation.* University of Illinois at Urbana-Champaign.

Coulter, N., French, J., Glinert, E., Horton, T., Mead, N., Rada, R., et al. (1998, January). Computing classification system 1998: Current status and future maintenance report of the CCS update committee. *ACM Computing Reviews, 39*(1), 1-24.

Cover, R. (2001). *BiblioML: XML for UNIMARC bibliographic records.* Retrieved December 10, 2002, from http://xml.coverpages.org/biblioML.html

Dekkers, M., & Weibel, S. L. (2002). Dublin core metadata initiative progress report and workplan for 2002. *D-Lib, 8*(2).

Dumais, S. T., & Chen, H. (2000). *Hierarchical classification of Web content.* Paper presented at the ACM-SIGIR International Conference on Research and Development in Information Retrieval, Athens, Greece.

Farrell, J., & Saloner, G. (1993). Converters, compatibility, and the control of interfaces. *Journal of Industrial Economics, 40*, 9-35.

Fox, E. A., & Marchionini, G. (1998). Toward a worldwide digital library. *Communications of the ACM, 41*, 29-32.

Goossens, M., Mittelbach, F., & Samarin, A. (1993). *The LaTeX companion.* Reading, MA: Addison-Wesley.

Green, S., Marchionini, G., Plaisant, C., & Shneiderman, B. (1997). *Previews and overviews in digital libraries: Designing surrogates to support visual information seeking* (CS-TR-3838). College Park: University of Maryland CS.

Herzog, G., & Huwig, C. (1997). *LIDOS: Literature information and documentation system.* German Research Center for Artificial Intelligence GmbH. Retrieved February 1, 2003, from http://www.dfki.uni-sb.de/imedia/lidos/

Iamnitchi, A. (2001). *BibTeX citations.* University of Chicago, Department of Computer Science. Retrieved February 1, 2003, from http://people.cs.uchicago.edu/~anda/citatations.html

ISO/IEC 13250 topic maps (2000). (ISO/IEC JTC 1/SC34).

Joachims, T. (1997). *A probabilistic analysis of the Rocchio algorithm with TFIDF for text classification.* Paper presented at the 14th International Conference on Machine Learning.

Joachims, T. (2001). *A statistical learning model of text classification for support vector machines.* Paper presented at the 24th Annual International ACM SIGIR Conference on Research and Development in Information Retrieval.

Khare, R., & Rifkin, A. (1997). XML: A door to automated Web applications. *IEEE Internet Computing, 1*, 78-87.

Krechmer, K. (2002). *Cathedrals, libraries and bazaars.* Paper presented at the ACM Symposium on Applied Computing, Madrid, Spain.

Lagoze, C., & Sompel, H. V. d. (2001). *The open archives initiative: Building a low-barrier interoperability framework.* Paper presented at the First ACM/IEEE-CS Joint Conference on Digital Libraries.

Lamport, L. (1986). *LaTeX: A document preparation system.* Addison-Wesley.

Lawrence, S. (2001a). Access to scientific literature. In D. Butler (Ed.), *The nature yearbook of science and technology* (pp. 86-88). London: Macmillan.

Lawrence, S. (2001b). Online or invisible? *Nature, 411*, 521.

Lawrence, S., Bollacker, K., & Giles, C. L. (1999). *Indexing and retrieval of scientific literature.* Paper presented at the Eighth International Conference on Information and Knowledge Management (CIKM 99), Kansas City, MO.

Lawrence, S., & Giles, C. L. (1999a). Digital libraries and autonomous citation indexing. *IEEE Computer, 32*(6), 67.

Lawrence, S., & Giles, C. L. (1999b). Searching the Web: General and scientific information access. *IEEE Communications, 37*, 116-122.

Leung, Y. K., & Apperley, M. D. (1994). A review and taxonomy of distortion-oriented presentation techniques. *ACM Transactions on Computer-Human Interaction, 1*(2), 126-160.

Mitchell, T. (1997). *Machine learning.* New York: McGraw Hill.

Mukherjea, S. (2000). *WTMS: A system for collecting and analyzing topic-specific Web information.* Paper presented at the Ninth International World Wide Web Conference.

Paepcke, A. (2003). *InterBib: General information.* Stanford University. Retrieved January 31, 2003, from http://www-interbib.stanford.edu/~testbed/interbib/interbibInfo.html

Paepcke, A., Chang, C. -C. K., Garcia-Molina, H. e., & Winograd, T. (1998). Interoperability for digital libraries worldwide. *Communications of the ACM, 41*, 33-43.

Pejtersen, A. M. (1998). Semantic information retrieval. *Communications of the ACM, 41*, 90-92.

Petrie, C. J. (1998). The XML files. *IEEE Internet Computing*, 4-5.

Pirolli, P., Card, S. K., & Wege, M. M. V. D. (2000). *The effect of information scent on searching information: Visualizations of large tree structures.* Paper presented at the Working Conference on Advanced Visual Interfaces.

Pirolli, P., Card, S. K., & Wege, M. M. V. D. (2001). Visual information foraging in a focus + context visualization. *ACM Conference on Human Factors in Computing Systems, CHI Letters, 3*, 506-513.

Plaisant, C., Carr, D., & Shneiderman, B. (1995). Image-browser taxonomy and guidelines for designers. *IEEE Software, 12*(2), 21-32.

Raymond, E. S. (1998). The cathedral and the bazaar. *First Monday, 3*(3).

Sévigny, M., & Bottin, M. (2000). *BiblioML project.* Retrieved February 1, 2003, from http://www.culture.fr/BiblioML/en/index.html

Sugimoto, S., Baker, T., Nagamori, T., Sakaguchi, T., & Tabata, K. (2001). *Versioning the Dublin Core across multiple languages and over time.* Paper presented at the Symposium on Applications and the Internet (SAINT) Workshops, San Diego, CA.

Wibowo, W., & Williams, H. E. (2002). *Strategies for minimising errors in hierarchical Web categorisation.* Paper presented at the Conference on Information and Knowledge Management (CIKM), McLean, VA.

Wuestner, E., Hotzel, T., & Buxmann, P. (2002). *Converting business documents: A classification of problems and solutions using XML/XSLT.* Paper presented at the Fourth IEEE International Workshop on Advanced Issues of E-Commerce and Web-Based Information Systems, Newport Beach, CA.

ENDNOTES

[1]   TeTeX is a popular TeX implementation in various Unixes, including many Linux distributions. See http://www.tug.org/teTeX/ for more information.

[2]   See http://www.gnu.org/software/emacs/emacs.html for a description and history of Emacs.

[3]   See http://www.nongnu.org/auctex/ for more information. AUC TeX provides keyboard shortcuts and additional menus for LaTeX and BibTeX document preparation.

APPENDIX A: OCS PLATFORM TECHNICAL DETAILS

The Open Citation System was implemented using a wide range of Java-, Web-, and XML-related technologies. J2SE 1.3 was the primary platform that was used to build the system's components. Web-application development leveraged many of the technologies available from the Java platform, such as servlets and applets. The use of Java allowed for quick prototyping, code portability, and seamless Web integration. The Apache XML Project was the primary source of XML-related software packages to facilitate the use of XML technologies for the Java platform. The Apache Software Foundation (http://www.apache.org) provides many different commercial-quality standards-based XML solutions, including Xerces and Xalan. Xerces 2, a suite of Java XML parsers, was used to parse and build XML documents using DOM Level 2 interfaces. Xalan-Java 2, an XSLT processor, was used for transforming XML to other markup-language formats and for XPath processing. Both of these packages implemented the Java API for XML Processing 1.1 (JAXP) standard, a set of implementation-independent interfaces for XML processing in Java, and therefore allowed for easy integration into the Java-based system. MinML2, a small SAX Level 2 XML parser, was used for parsing on the client side so that the user would not have to have a large, fully functional XML parser implementation installed, such as Xerces. All of the technologies used are freely available and widely used solutions for Web-application development. Code listings for sample ACM-classified BiblioML records, and the ACM XML topic-map file, are available from the project Web server at http://louvain.bpa.arizona.edu/projects/ocs/docs/biblio_ml.html.

Section VI

The Economic Perspective

Chapter XIV

Standardization and Network Externalities

Sangin Park, Seoul National University, Republic of Korea

ABSTRACT

The standardization issue in the ICT industry is mainly compatibility in the presence of network externalities. The compatibility in Economics usually means interoperability between competing products. For instance, the VHS VCRs and the Betamax VCRs are incompatible in the sense that tapes recorded in one format (e.g., VHS) could not be played in the other format (e.g., Betamax). Hence, in the ICT industry, standardization mainly signifies achieving compatibility. Standards can be achieved by mandatory or voluntary measures as well as by de facto standardization. It is an important policy issue whether the government should mandate a standard (or impose compatibility), let the stakeholders (especially, firms) decide a standard, or enforce sponsoring firms to compete in the market, which has substantial impacts on consumer (or end user) well-being as well as business strategies in R&D, technology sponsorship, and competition in the product market. Ultimately, the impacts of standardization policies should be analyzed in terms of costs and benefits of firms (i.e., profit analysis) and the society (i.e., welfare analysis). In this chapter, we suggest an analytical framework to provide a consistent review of theoretical and empirical models of firms' and consumers' (or end users') incentives and behavior under different standardization policies. The chapter is organized as follows. In section 2, we will discuss the Katz and Shapiro model which analyzes how compatibility (or standardization) affects firms' optimization behavior in the product market and whether private incentives for compatibility are consistent with the social incentive. Section 3 will shift our focus onto the consumer's adoption decision of new technology over old technology. We will discuss the pioneering Farrell and Saloner model which studies whether consumers' adoption decision of incompatible new technology is socially optimal. Then we will proceed to introduce several important extensions of the model. The dynamics of standardization process will be explored in section 4. Based on the empirical study of Park (2004a), the de facto standardization of the VHS format in the U.S. home VCR market will be analyzed and further utilized to understand strategic aspects of standardization. Despite recent economists' attentions to the issue of standardization and network externalities, the literature itself still lags behind reality. In section 5, we will examine ongoing and future research issues requiring further cost-benefit analyses based on economic models. Section 6 will conclude.

INTRODUCTION

Basic Concepts

The functionality of standards may be specified in two different contexts. In many cases, standards are a way of certifying product quality in the presence of asymmetric information between buyers and sellers. Examples include processed food and health drugs, professional services, and complex electronic products. As discussed in Akerlof (1970), only "lemons" (bad-quality goods) will be sold in the market with the extreme asymmetric information about the quality of goods. Standards certified by a trustworthy third party will mitigate this problem. In many other cases, standards mainly signify *interoperability*. We can further consider two types of interoperability: interoperability among the components of a product (e.g., the interface between speakers and the CD [compact disc] players of an audio system) and interoperability between competing products. The second type of interoperability is usually called *compatibility* in economics. For instance, the VHS VCRs (videocassette recorders) and the Betamax VCRs are incompatible (or not interoperable) in the sense that tapes recorded in one format (e.g., VHS) cannot be played in the other format (e.g., Betamax). From the viewpoint of social welfare (i.e., the sum of all the stakeholders' net benefits), a trade-off exists between the degree of interoperability and the variety of different products (or systems).[1] On top of this trade-off, the compatibility issue is typically related to *network externalities*. Indeed, the standardization issue in the ICT industries is mainly compatibility in the presence of network externalities.

Network externalities can be understood as positive consumption externalities. In the presence of network externalities, an increased number of users of a product raises the consumer's utility level and hence the demands for that product. As a result, the overall economic value to the society exceeds the economic benefit enjoyed by a consumer from purchasing a hardware product. In the literature on network externalities, the number of users is called *network size*, and the user's benefit from the network size is called *network benefit*.[2] Network externalities are categorized as either direct or indirect. In the presence of direct network externalities, the benefit to each user increases directly with the network size. A good example is a communications network, such as the network of e-mail users, the network of fax machines, or the network of people who exchange MS (Microsoft) Word files. Indirect network externalities are caused by the linkage of some strongly complementary products, whose collection is called a hardware-software system. The bundle of a PC (personal computer; such as a MS-Windows-equipped PC or Macintosh) and various application software programs is an example of this traditional hardware-software paradigm of indirect network externalities. When people choose between a MS-Windows-equipped PC and Macintosh, they consider not only prices and product characteristics, but also the variety of available application software. Software programmers will make more application software products available for a PC if more people use that particular PC. Hence, an increase in the demand for a PC will increase the variety of available software, which in return raises demands for that PC. These positive feedback effects are called indirect network externalities. Hence, in the presence of indirect network externalities, a consumer purchasing a hardware product raises the value of this hardware product for the other consumers indirectly since an increase in demands for the hardware product (e.g., a PC) typically induces a greater variety of available software products (e.g., application software programs).[3] In this sense, the indirect network externalities are understood as positive consumption externalities.[4] Indirect network externalities are significant in many industries, such as the computer industry, the broadcasting industry, and some consumer electronics industries.[5]

Network Externalities and De Facto Standardization

It is widely believed that network externalities give rise to tipping toward a certain technology and thus affect business strategies and government policies on standard setting. Since the study by David (1985) of the adoption of the QWERTY typewriter keyboard, the possible lock-in to inferior technology by historical accidents has attracted economists' attentions. Arthur (1989), based on a simple nonstrategic (or biological) model, demonstrates that random historical (small) events may lead to this type of lock-in with the existence of increasing returns to adoption. Network externalities have been considered a source of these increasing returns to adoption.

In this subsection, we will briefly review the competition between Betamax and VHS in the U.S. home-VCR market, which is one of the best known examples of tipping and de facto standardization in the presence of network externalities. Sony and Matsushita competed to develop a home video system. Finally, Sony introduced the Betamax format to the U.S. home-VCR market in February 1976. One and a half years later, Matsushita launched the VHS format into the U.S. market, which was incompatible with the Betamax format in the sense that tapes recorded in one format could not be played in the other format. The evolution of the U.S. home-VCR market up to the de facto standardization in 1988 can be divided into two phases.

The First Phase (1976-1980)

Although Sony first introduced the Betamax VCRs, Matsushita's VHS immediately took the lead in the market right after entry. In 1977, 140,000 Betamax VCRs were sold in the entire year, while 80,000 VHS VCRs were sold in only 5 months. In every year from 1978, VHS outsold Betamax. This was mainly because VHS could switch speeds to provide 4 hours of recording time, while Betamax could provide only 2 hours of recording time.[6] At that time, consumers used VCRs mainly to record TV programs.[7] Yet many programs, such as football games, tennis matches, and many movies, required more than 2 hours of recording time.

Sony tried to catch up with Matsushita by developing a 5-hour-recording-time Betamax in 1979. Matsushita immediately introduced a 6-hour-recording-time VHS. The intense competition between Sony and Matsushita in the VCR market led to the steep reduction of prices and the rapid upgrade of performance and features. In 1977, RCA introduced its VHS model at $995, $300 below the price of the Betamax. Zenith and Sony immediately dropped the prices of their Betamax models to $995 and $1,095, respectively. As cited in Klopfensten (1985), the price of the average VCR dropped from $2,165 to $600 between 1976 and 1983. As a result, consumers enjoyed unexpectedly rapid innovations, and VCR producers obtained a very high household penetration rate.[8]

The Second Phase (1981-1988)

In 1978, RCA, a producer of VHS, allying with Magnetic Video Corporation of America (MV), gave each VCR purchaser two free MV program cassettes and a free MV club membership. In 1979, Sony, the sponsor of Betamax, joined forces with Video Corporation of America. The British firm, Granada, began opening its rental shops in 1980. Video-rental shops began to expand in the early

1980s. Sales and rentals of movie titles began to grow exponentially, doubling each year from 1982 to 1986.[9]

In the second phase, there was almost no significant difference between the two formats in performance, features, or prices. Both Betamax and VHS VCRs provided recording time long enough for TV programs. According to the tests of *Consumer Reports* published in May 1982, neither Betamax nor VHS was superior, and the low-end models generally did as well as the high-end models, regardless of brand.[10] The price data from *Orion Blue Book: Video & Television 1993* and *Consumer Reports* show that Betamax (equipped with features similar to those of VHS) set a little bit lower prices on average.

However, a dramatic increase occurred in the relative sales of VHS to Betamax in this second phase, except in 1983. Many new VHS-producing firms entered while many Betamax-producing companies exited. The Korean electronics companies began to export VHS VCRs to the U.S. market in 1985. Zenith and Emerson decided to stop producing Betamax in 1983, and Sanyo and Toshiba followed suit in 1986. The total sales of Betamax decreased from 1985. VHS became dominant around 1985 and 1986, occupying 87.6% of the market share in 1985 and 90.5% in 1986 (see Figure 1). Sony, the only important producer of Betamax since 1986, finally announced in January of 1988 that it would produce VHS, thus ending the competition between VHS and Betamax. De facto standardization was achieved.

Road Map

In the ICT industry, standards can be achieved by mandatory or voluntary measures as well as by de facto standardization. As discussed, the VHS format became the standard in the home-VCR market through de facto standardization. However, the standard for the DVD- (digital video disc) player format was set

Figure 1. Sales of VHS and Betamax

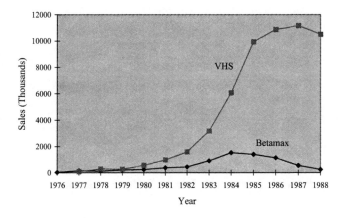

by the voluntary cross-licensing agreement between two consortia that had competed to develop incompatible formats, while the CDMA was mandated by the government as the standard of mobile communications in Korea. It is an important policy issue whether the government should mandate a standard (or impose compatibility), let the stakeholders (especially firms) decide a standard, or enforce sponsoring firms to compete in the market, which has substantial impacts on consumer (or end user) well-being as well as business strategies in R & D (research and development), technology sponsorship, and competition in the product market. Ultimately, the impacts of standardization policies should be analyzed in terms of the costs and benefits of firms (i.e., profit analysis) and the society (i.e., welfare analysis).

In this chapter, we suggest an analytical framework to provide a consistent review of theoretical and empirical models of firms' and consumers' (or end users') incentives and behavior under different standardization policies. Specifically, we begin by defining the consumer's utility function for a product in the presence of network externalities as the sum of a *stand-alone benefit* and a *network benefit*.[11] The stand-alone benefit of a product is a function of the product's attributes, while the network benefit is specified as an increasing nonnegative real function of the network size of the compatible products. In many examples of network externalities, products such as VCRs are durable goods. In the case of durable goods, the consumer may take into account not only the current utility, but also the expected future utilities derived from the use of a product. This dynamic concern of the consumer will be represented by the consumer's *valuation function*, which is composed of the average (of the present value of) network benefit and the average (of the present value of) stand-alone benefit of the product in current and future periods. If a model considers this dynamic concern explicitly, it will be called a *dynamic model* in this chapter. Otherwise, the model is named a *static model*.

The chapter is organized as follows. In the following section, we will discuss the Katz and Shapiro model, which analyzes how compatibility (or standardization) affects firms' optimization behavior in the product market and whether private incentives for compatibility are consistent with the social incentive. Then we will shift our focus onto the consumer's adoption decision of new technology over old technology. We will discuss the pioneering Farrell and Saloner model, which studies whether consumers' adoption decision of incompatible new technology is socially optimal. Then we will proceed to introduce several important extensions of the model. The dynamics of the standardization process will be explored next. Based on the empirical study of Park (2004a), the de facto standardization of the VHS format in the U.S. home-VCR market will be analyzed and further utilized to understand strategic aspects of standardization. Despite recent economists' attentions to the issue of standardization and network externalities, the literature itself still lags behind reality. Next, we will examine ongoing and future research issues requiring further cost-benefit analyses based on economic models. Finally, we will conclude.

STANDARDIZATION AND COMPETITION

Katz and Shapiro Model

One of the pioneering works in the literature of network externalities is Katz and Shapiro (1985), which examined two key questions: (a) whether compatibility is socially desirable and (b) whether the private incentives for compatibility are consistent with the social incentive.

The Model

To answer these questions, Katz and Shapiro (1985) developed a two-stage static model in which all the products are homogeneous and a consumer of type r's utility level for product j, say, U_{rj}, is the sum of the stand-alone benefit $r - p_j$ and the network benefit $k(Q_j^e)$, where p_j is the price of product j, Q_j^e is the expected network size of firm j, and $k(\cdot)$ is a network-benefit function.[12]

$$U_{rj} = r - p_j + \kappa(Q_j^e). \tag{1}$$

Let q_j^e denote the expected sales (or the number of consumers) of firm j. Then if the first m firms' products (including product j itself) are compatible with firm j's product, we have $Q_j^e = \sum_{i=1}^{m} q_i^e$. Furthermore, the network benefit function is assumed to satisfy the following typical properties: $\kappa(0) = 0$, $\kappa' > 0$, $\kappa'' < 0$, and $\lim \kappa'(Q) = 0$ as $Q \to \infty$. It is also assumed that the consumer type r is uniformly located between minus infinity and A (> 0). The timing of the model is as follows. In the first stage, consumers form expectations about the network sizes. Then in the second stage, the symmetric firms play a Cournot output game, taking consumers' expectations as given, and consumer r purchases one of the products if $r - p_j + \kappa(Q_j^e) \geq 0$ for at least one j.

In this two-stage static model, two firms, i and j, will both have positive sales only if $p_i - \kappa(Q_i^e) = p_j + \kappa(Q_j^e) \equiv \phi$ where ϕ is called the hedonic price. Then, for a given ϕ, there are A − ϕ demands for the products. Let z denote total outputs, that is, $z = \sum q_j$. Then, A - ϕ = z, or A + $\kappa(Q_j^e) - p_j = z$ for all j. Hence, firm j's inverse demand function is

$$p_j = A + \kappa(Q_j^e) - z. \tag{2}$$

We consider two types of costs: costs of production and costs of achieving compatibility. For simplicity, let production costs be 0. Firm j will earn profits equal to $\pi_j = q_j \times (A - z + \kappa(Q_j^e))$. From the first-order conditions of profit maximization of the Cournot game, we obtain a vector of the equilibrium sales, $(q_1^*, \ldots, q_j^*, \ldots, q_n^*)$, such that for $j = 1, 2, \ldots, n$,

$$q_j^* = A + \kappa(Q_j^e) - \sum_{i=1}^{n} q_i^* = \{(A + n\kappa(Q_j^e)) - \sum_{i \neq j} \kappa(Q_i^e)\} / (n+1). \tag{3}$$

As indicated in Equations 2 and 3, a firm's equilibrium output level is equal to the price in this model, and thus we obtain firm j's profits in equilibrium as $\pi_j = (q_j^*)^2$. Since a type r consumer will purchase one of the products if $r + z - A \geq 0$, consumer surplus, say, CS, and social welfare (gross of the fixed costs of compatibility), say, SW, are derived as follows.

$$CS = \int_{A-z}^{A} (\rho + z - A) d\rho = \frac{z^2}{2}, \text{ and } SW = \sum_{i=1}^{n} \pi_i + CS = \sum_{i=1}^{n} q_i^2 + \frac{z^2}{2} \tag{4}$$

Results

When all the products are compatible, we have $Q_j^e = z^e$. Then Equation 3 becomes $q_j^* = (A + \kappa(z^e))/(n+1)$. In the rational expectations equilibrium without uncertainty, we have $q_j^e = q_j^*$. Hence, in the rational expectations equilibrium, we obtain $z^c = (n / (n + 1)) (A + \kappa(z^c))$. On the other hand, under incomplete compatibility (i.e., when at least two products are not compatible to each other), we obtain $z^I = (nA + \sum_j \kappa(Q_j)) / (n+1)$. Since $z > Q_j$ for at least one firm, we obtain $z^C > z^I$ as shown in Figure 2.[13]

Hence, we have the following proposition.

Proposition 1. The level of total output is greater under industry-wide compatibility than in any other equilibrium with less-than-complete compatibility.

Figure 2. Complete vs. incomplete compatibility

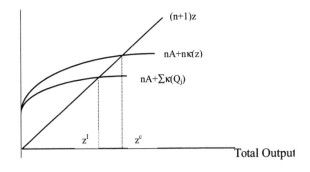

Equation 3 indicates that the equilibrium price will be higher under complete compatibility than those of symmetric equilibrium under incomplete compatibility. However, consumer surplus will be higher under complete compatibility since it increases with the total output level as indicated in Equation 4. In general, the compatibility among the competing technologies dampens competition in the product market but is likely to increase consumers' network benefits. In the Katz and Shapiro (1985) model, the positive effects of the increased network benefits dominate the negative effects of the increased price, and thus compatibility raises consumer surplus.

When side payments are feasible among all firms,[14] private incentives for compatibility are given by the changes in industry-wide profits, say, $\Delta\pi$. The social incentive is determined by the changes in social welfare, say, ΔSW, which is the sum of the changes in industry-wide profits and the change in consumer surplus, say, ΔCS. That is, $\Delta SW = \Delta\pi + \Delta CS$. Therefore, when compatibility costs are purely fixed costs, any move to complete compatibility that raises industry profits is socially beneficial (since $\Delta CS > 0$, $\Delta W = \Delta p + \Delta S > \Delta\pi$). However, if the industry-wide costs of compatibility, say, F, satisfy $\Delta\pi < F < \Delta W$, then the private firms will fail to adopt a socially desirable system of compatibility. Hence, (collective) private incentives for compatibility are not excessive. The private-standardization rule will be more stringent if side payments are infeasible.[15]

Extensions

We now discuss several directions of the extensions from the Katz and Shapiro (1985) model. In general, the compatibility among the competing technologies dampens competition in the product market but is likely to increase consumers' network benefits. Hence, the private incentives for standardization may be reversed under different modeling (for example, price competition is more intense than quantity competition). Katz and Shapiro (1986a) posited a three-stage model and allowed exogenous technological progress. In contrast to Katz and Shapiro (1985), Katz and Shapiro (1986a) found that firms often have excessive (collective) private incentives for compatibility. In the early stages of industry evolution, producers of incompatible technologies may engage in extremely intense price competition in order to get ahead of its rivals by building up an installed base. Hence, firms may use compatibility as a means of reducing competition among themselves.

Incentives for compatibility may differ across firms. Katz and Shapiro (1986a) showed that the producer of an inferior technology in the present (for example, an old technology) may prefer compatibility with superior technologies in the future (for example, new technologies), whereas the producer introducing a superior technology in the future is biased against compatibility. Since the producer of a superior technology in the future is biased against compatibility, voluntary compatibility between these two producers will reduce consumer surplus.

Compatibility can be partially achieved through converters (or emulators, adapters) as well. Farrell and Saloner (1992) found that market outcomes without converters are often inefficient, but the availability of converters can worsen the matter. Furthermore, when one of the technologies is supplied only by a single firm, that firm may have an incentive to make conversion costly, which implies anticompetitive disruption of interface standards.

TECHNOLOGY-ADOPTION TIMING

In the previous section, we discussed compatibility and the firm's optimization behavior in the product market in which the consumer's adoption decision is simply concerned about which product to purchase. In this section, we will focus on the timing of the consumer's adoption decision. The products we consider in the literature of network externalities are typically durable products, and thus consumers may delay their adoption decision, especially faced with the introduction of new technology. A pivotal question is whether the timing of the adoption of incompatible technologies is socially efficient or not. We will begin with the adoption of new technology over old technology as discussed in Farrell and Saloner (1986a). Then we will extend the model to take account of the firm's strategic decision on the introduction timing of new products and the value of consumers' waiting.

Farrell and Saloner Model

The adoption of new technology over old technology will be affected by the installed base of the old technology in the presence of network externalities. Indeed, the existence of an installed base causes a disparity between the social incentives for the adoption of new technology and the private incentive. For example, consider the case in which the new technology is adopted by only new users.[16] Then, there are two externalities induced by the existence of the installed base of the old technology. New consumers may be unwilling to adopt a new technology since the network benefit of the new technology may not be large enough for early adopters. In other words, most of the benefits of switching will accrue to late adopters, and early adopters would bear disproportionate switching costs. Hence, the new technology may be adopted too slowly (excess inertia). This is more likely to happen if the installed base is large, or if the new technology is not sufficiently superior. On the other hand, new consumers do not take account of the loss of the users of the old technology when new consumers decide to adopt the new technology. Hence, the users of the old technology may be stranded (or orphaned). In this case, the new technology may be adopted too early (excess momentum).[17]

Farrell and Saloner (1986a) provided a model that formulates these two externalities and the possibility of the inefficient adoption of new technology. In the following, we will briefly review this model in Farrell and Saloner (1986a). In the model, a new technology, say V, becomes available unexpectedly: Before time T*, only technology U is available, and at T*, the new technology becomes available. The technologies are assumed to be competitively supplied (i.e., prices are exogenously given), and consumers are infinitesimal and arrive continuously to the market over time with the arrival rate of $n(t) \geq 0$. Let $N(t) = \int_0^t n(t)\,dt$, and let U_{ut} and U_{vt} denote a (representative) consumer's flow of utilities from technologies U and V, respectively. More specifically, the model considers the linear case

$$n(t) = 1,\ N(t) = t,\ U_{ut} = u(N(t)) = a + bt, \text{ and } U_{vt} = v(N(t)) = c + dt, \quad (5)$$

where a and c represent the stand-alone benefit of U and V, and bt and ct indicate the network benefits of U and V.[18]

There are two extreme possibilities: Everyone from T* onward adopts the new technology V (adoption outcome), or nobody adopts V (nonadoption outcome). In the case of the nonadoption of V, the valuation function (i.e., the present-value payoff) of a user who adopts U at time T is

$$\bar{u}(T) = \int_T^\infty u(N(t))e^{-\varphi(t-T)}\,dt = \frac{a+bT}{\varphi} + \frac{b}{\varphi^2}, \quad (6)$$

where φ is the consumer's discount rate. The first term is the present value of benefits if the network sizes remain unchanged, and the second term is the present value of benefits from the future growth of the network. Similarly, the valuation function of a user who is the last to adopt the old technology U at T is

$$\tilde{u}(T) = u(N(T))\int_T^\infty e^{-\varphi(t-T)}\,dt = \frac{a+bT}{\varphi}. \quad (7)$$

Suppose that all new adopters use V after T*. Then

$$\bar{v}(T) = \int_T^\infty v(N(t) - N(T^*))e^{-\varphi(t-T)}\,dt = \frac{c+d(T-T^*)}{\varphi} + \frac{d}{\varphi^2} \text{ for } T \geq T^*. \quad (8)$$

Also, $\tilde{v}(T)$ can be defined and given by the first term of $\bar{v}(T)$ as in the case of $\tilde{u}(T)$.

The adoption of the new technology is a subgame perfect equilibrium (SPE) if and only if $\bar{v}(T^*) \geq \tilde{u}(T^*)$, whereas nonadoption is an SPE if and only if $\bar{u}(T^*) \geq \tilde{v}(T^*)$. At least one of these conditions must hold so that equilibrium exists. They may hold simultaneously, in which case there will be multiple equilibrium.

If V is adopted, each user gains (or loses) $\bar{v}(T) - \bar{u}(T)$ for $T > T^*$. However, the old users (installed base) lose $\bar{u}(T^*) - \tilde{u}(T^*)$. Hence, the present value of the net gain in social welfare from the adoption of V is

$$G \equiv \int_{T^*}^{\infty} [\bar{v}(t) - \bar{u}(t)]e^{-\varphi(t-T^*)}dt - \{\bar{u}(T^*) - \tilde{u}(T^*)\} = \frac{2(d-b) - 2rbT^* + r(c-a)}{\varphi^3}. \quad (9)$$

Therefore, there are four possible cases: (a) unique nonadoption equilibrium and excess inertia ($d > 2b$), (b) multiple equilibria and possible excess inertia ($b/2 < d < 2b$), (c) if adoption is efficient, then there is unique equilibrium ($d < b/2$), and (d) both ($c-a$) and bT^* are large and there is possible excess momentum.

Farrell and Saloner (1986a) further added three extensions to this simple setting. First, if potential users learn in advance via a product preannouncement that a new technology V will become available, the technology V will be adopted when it would not have been adopted otherwise. However, the preannouncement may reduce social welfare. Second, it may be both feasible and profitable for the incumbent to prevent the adoption of new technology by engaging in predatory pricing if the existing technology is provided by a monopolist. If the excess momentum is important enough, this strategic pricing may enhance social welfare. Hence, it would be extremely difficult to frame a legal rule regarding predatory pricing in the presence of network externalities. Lastly, in the case that old users must switch to adopt a new technology (i.e., there are no new users), there still exists the possibility of excess momentum and excess inertia. On one hand, when a user switches, the other users lose some network benefits, but the switching user ignores this in his or her calculation. On the other hand, even if users unanimously favor a switch, each user may prefer the other to switch first (penguin effect).

Extensions

Katz and Shapiro (1992) extended the model in Farrell and Saloner (1986a) to allow the endogenous introduction of a new technology as well as strategic pricing, and confirmed the possibilities of excess inertia and excess momentum. However, Katz and Shapiro assumed perfect foresight equilibria in which all consumers have perfectly accurate predictions of the future evolution of prices, network sizes, and the date of any subsequent product introductions. Hence, the consumers' option of waiting was not explicitly analyzed in Katz and Shapiro since consumers never exercise their option to wait in perfect foresight equilibria.

Choi (1994) considered a two-period model in which the option of waiting is valuable with stochastically evolving technologies. Typically, second-generation consumers live only for the second period while first-generation consumers live for two periods in a two-period model (see, for example, Choi; Katz & Shapiro, 1986a, 1986b), in which it is usually assumed that consumers can coordinate their choices within the period to end up with a Pareto-optimal choice. Since adoption decisions are made in different periods by two different generations of consumers, there may be two externalities in the adoption of technologies. Since technologies evolve stochastically, the early adoption of a technology deprives second-generation consumers of an opportunity to coordinate efficiently based on better information (forward externality). On the other hand, second-generation consumers do not take account of the loss of early adopters (backward externality). In other words, early adopters can be stranded inefficiently. Choi showed that forward externality dominates backward externality and thus there exist excessive private incentives for early adoption. This is because the occurrence of forward externality is more pervasive than that of backward externality.

Katz and Shapiro (1986b) focused on the effects of strategic pricing on the adoption of technologies. They showed that the technology that will be superior tomorrow has a strategic advantage since the firm with today's superior technology cannot credibly promise to price below the marginal cost in the second period. However, the firm with tomorrow's superior technology can price below the marginal cost in the first period.

STANDARDIZATION AND STRATEGIC MANEUVERING

Despite the rich theoretical literature, there have been few empirical studies that discuss statistical evidence on the significance of network externalities. These few empirical studies, such as Brynjolfsson and Kemerer (1996) and Gandal (1994), were often based on the estimation of the hedonic price equations. However, as discussed in Park (2004a), the existence and the significance of network externalities may not be appropriately tested by using format dummy variables in the hedonic price equation, especially when the products we consider are differentiated with heterogeneous producers. In this section, we will discuss a dynamic structural model in Park that is applied to quantitatively analyze the extent to which network externalities contributed to the de facto standardization of the VHS format in the U.S. home-VCR market during the years 1981 to 1988. The structural model in Park can be applied to simulation studies of various policies related to standardization. An example of these simulations will be further discussed.

Quantitative-Analysis Model

The Model

    In the presence of network externalities, we have specified the consumer's utility function for a product as the sum of a stand-alone benefit of the product and a network benefit. In this model, the stand-alone benefit of a product is further specified by nested logistic assumptions as in Berry (1994), Cardell (1997), and McFadden (1973) in order to reflect the consumer's idiosyncratic tastes for differentiated individual products (e.g., VCRs) and formats (e.g., VHS or Betamax). Let product jt denote the durable product introduced by producer j in period t. Let $(X_{jt}, c_{1t}, \xi_{jt})$ denote a vector of the characteristics of product jt, where X_{jt} is observable (to researchers), and (c_{1t}, ξ_{jt}) is unobservable.[19] Let p_{jt} be the price of product jt in period t. Let ζ_{igt} denote consumer i's idiosyncratic taste for format g, and ε_{ijt} denote consumer i's idiosyncratic taste for product jt. Then the utility level of consumer i, if he or she purchases product jt of format g in period t, is given by

$$U_{ijt} = \kappa(Q_{gt}) - \lambda p_{jt} + X_{jt}\alpha + c_{1t} + \xi_{jt} + \zeta_{igt} + (1-\sigma)\varepsilon_{ijt}, \tag{10}$$

where $\kappa(\cdot)$ is a network benefit function, Q_{gt} is the network size of format g in period t, and $0 \leq \sigma < 1$. In the nested logit model, the parameter σ is called the within-format correlation coefficient. As σ approaches 1, the within-format correlation of utility levels goes to 1, and as σ approaches 0, the within-format correlation goes to 0.

    In the durable-goods case, the network size of format g in period t is the sum of the installed base, say, B_{gt}, and the expected sales in this period, say, q_{gt}, and consumers take into account not only the current utility but also the expected future utilities from the use of a product. We assume that the consumer's expected idiosyncratic taste in the future is the currently perceived one. We also assume that the expected stand-alone benefits of a product in the future are the same as the current ones.[20] Note that the network benefit of a product changes over periods since the network size of a format changes over periods. We define the consumer's valuation function of a product as the average present value of utilities derived from the use of that product forever.[21] Let φ be the consumer's discount rate. Since the consumer who purchases a product in period t does not have to pay the price again after period t, the consumer's valuation function, say, V_{ijt}, will be:

$$V_{ijt} = N_{gt} - \varphi\lambda p_{jt} + X_{jt}\alpha + c_{1t} + \xi_{jt} + \zeta_{igt} + (1-\sigma)\varepsilon_{ijt}, \tag{11}$$

where $N_{gt} = \varphi\sum_{s\geq t}(1-\varphi)^{s-t}E_t[\kappa(B_{gs}+q_{gs})]$.[22] We call N_{gt} the network effect, which is the average of (the present value of) the consumer's network benefits in current and future periods. We will consider uncertainty in the future by assuming that the vector of exogenous-state variables, say, Ω_t, including product and cost characteristics, evolves stochastically as a Markov process. Then, due to the rational expectations assumption, we have $N_{gt} = N_g(\Omega_t, B_t)$, where B_t denotes the vector of B_g in period t. In other words, in rational expectations, the network effect can be projected by current information such as exogenous-state variables and installed bases.

Then, with further simplifying assumptions on the product offerings, the consumer's option of waiting, installed bases, and the market size, Park (2004a) obtained the market-share function of product j, say, S_{jt}, and the market-share function of format g, say, S_{gt}, in period t as follows.

$$S_{jt} = \frac{e^{(-\varphi\lambda p_{jt}+\delta_{jt}+N_{gt})/(1-\sigma)}D_{gt}^{-\sigma}}{1+\sum_g D_{gt}^{(1-\sigma)}} \text{ and } S_{gt} = \frac{D_{gt}^{(1-\sigma)}}{1+\sum_g D_{gt}^{(1-\sigma)}}, \tag{12}$$

where $\delta_{jt} = c_t - \varphi\lambda p_{jt} + X_{jt}\alpha + \xi_{jt}$, and $D_{gt} = \sum_{j\in J_g}\exp[(-\varphi\lambda p_{jt}+\delta_{jt}+N_{gt})/(1-\sigma)]$ with J_g denoting the set of all the products of format g. The demand function for product j in period t, say, q_{jt}, is $q_{jt} = M_t \cdot S_{jt}$, where M_t is the market size (the number of potential buyers).

For the above demand function, we assume that in each period, producers set prices in the oligopolistically competitive market, which is characterized by dynamic oligopoly due to durability. Let subscript j indicate a vector of firm j's variables, and subscript $-j$ indicate a vector of variables of firm j's rivals. Hence, for example, $\Omega_t' = (\Omega_{jt}', \Omega_{-jt}')$. Then, under some regularity conditions,[23] firm j's optimal decision rule solves for the Bellman equation:

$$v(\Omega_{jt}, \Omega_{-jt}, B_{jt}, B_{-jt}) = \sup_{p_{jt}}\{ \Pi_j(\Omega_t, B_t, p_t; \theta_0) + \beta \int v(\Omega_{jt+1}, \Omega_{-jt+1}, B_{jt+1}, B_{-jt+1}) P_B(dB_{t+1}|B_t, p_t) P_W(d\Omega_{t+1}|\Omega_t) \}, \tag{13}$$

where $\Pi_j(\cdot)$ is a single-period profit function, β is the discount factor, $P_\Omega(\cdot|\cdot)$ is the conditional distribution of the next-period exogenous-state vector Ω_{t+1} given the current-period exogenous-state vector Ω_t, and $P_B(\cdot|\cdot)$ is the conditional distribution of the next-period endogenous-state vector B_{t+1} given the current-period endogenous-state vector B_t and the current decision vector p_t.

Assuming that the marginal cost is the sum of the hedonic cost function, say, $\Gamma(X_j)$, and an unobserved cost characteristic, say, ω_j: $mc_j = \Gamma(X_j) + \omega_j$, Park (2003) derived the pricing equation as follows:

$$p_j = (1-\sigma)/([1-\sigma S_j/S_g - (1-\sigma)S_j]\varphi\lambda) + \phi_j(B,\Omega) + \omega_j, \qquad (14)$$

where $\phi_j(B,\Omega) = \Gamma(X_j) - H_j(B,\Omega)$, and $H_j(B,\Omega)$ is the dynamic part of the margin that reflects that the current price affects future network sizes and thus the stream of the firm's future profits.[24] Note that the first term of the right-hand side of Equation 14 is the standard margin of the nested logit model. The dynamic part of the margin may have a positive or a negative sign depending on the intensity of competition and the stage of the product cycle. Hence, strong network externalities do not necessarily imply that the firm's price increases with its market share as interpreted in Brynjolfsson and Kemerer (1996).

Results

In order to quantitatively analyze network externalities, Park (2004a) derived a system of estimating equations from the market-share functions in Equation 12 and the pricing equation in Equation 14, and then obtained consistent and asymptotically normal estimates of the parameters such as $\varphi\lambda$, σ, α, and $\{N_{gt} + c_t\}_{g,t}$. For this estimation, Park used unbalanced firm-level panel data of sales, price, and product characteristics for the U.S. home-VCR market from the year 1981 to the year 1988 except the year 1985.[25]

Based on these estimates, Park (2004a) proceeded to analyze the de facto standardization toward VHS. Let v denote VHS and b denote Betamax. From the market-share functions of formats in Equation 12, the logarithm of the relative sales (log relative sales, hereafter) of the two formats can be decomposed into the sum of two differences as follows.

$$(1-\sigma)[\ln(\sum_{j\in J_v} e^{\delta_j/(1-\sigma)}) - \ln(\sum_{j\in J_b} e^{\delta_j/(1-\sigma)}] + [N_v - N_b] \qquad (15)$$

The first term of the right-hand side in Equation 15, if it is positive, represents the price-quality advantage (or stand-alone-benefit advantage) of VHS over Betamax, while the second term stands for the network advantage of VHS over Betamax. The estimated ratios of for the years 1981 to 1988 (except the year 1985) indicate that the network advantage of VHS explains 70.3% to 86.8% of the log relative sales of VHS to Betamax in each year (see Table 1). Hence, during these years, the network advantage of VHS was the key reason that VHS outsold Betamax in the U.S. home-VCR market. Furthermore, we can confirm that the yearly change of the log relative sales was mainly due to the yearly change in network advantage. Hence, the increase in the network advantage of

Table 1.

Year		$[N_v - N_b]/\ln(q_v/q_b)$	
		Estimate	Standard Error*
1981		0.703	0.095
1982		0.751	0.080
1983		0.818	0.058
1984		0.747	0.058
1986		0.807	0.019
1987		0.868	0.065
1988		0.811	0.009

Standard errors are calculated by a bootstrapping procedure with a resampling size of 3,000.

Table 2.

Year	$\kappa(B_v - B_b)/[N_v - N_b]$
1981	0.058
1982	0.088
1983	0.163
1984	0.314
1986	0.688
1987	0.722
1988	0.818

VHS was an engine of the tipping toward and the de facto standardization of VHS in the U.S. home-VCR market.

With a linear network-benefit function, $NB = \kappa B_{gt} + \kappa E_t[q_{gt}]$, where the parameter k is a positive constant,[26] Park (2004a) further decomposed the network effect as the sum of the installed-base effect (i.e., the average network benefit if the network size remains unchanged) and the expected-growth effect (i.e., the average network benefit from the expected growth of the network size). As shown in Table 2, 5.8% to 81.8% of VHS's network advantage can be explained by its installed-base advantage. Moreover, in the early stage of competition, the expected-growth advantage was more important in the adoption of the VHS format, but as the household penetration rate of VCR jumped in the mid-1980s, the installed-base advantage became dominant. As discussed, VHS and Betamax had no significant difference in qualities and features in the 1980s, but VHS had a larger installed base and a bigger lineup of producers in the beginning of the 1980s due to its longer recording time in the late 1970s. We may infer that a larger installed base of VHS and a bigger lineup might help consumers to form expectations for greater expected-growth advantages of VHS.[27]

Based on the above analyses, we may explain the de facto standardization process as an amplification-reinforcement process in the following sense. The small difference in the initial installed bases, which were accumulated prior to the network externalities in action (via the consumer's use of prerecorded videocassettes such as movie titles), amplified the difference in sales between VHS and Betamax via network externalities, leading to a bigger advantage with more installed bases. Moreover, this amplification-reinforcement process in the presence of network externalities was reinforced by the expected future dominance of VHS, and this expectation itself became the main reason for the network advantage of VHS in the early stage of competition.

Strategic Maneuvering

Apparently, the interpretation of the previous section is consistent with the popular idea of the *critical mass*: The adoption of a technology with network externalities takes off to be self-sustaining if it reaches the critical mass (see, for example, Allen, 1987; Economides & Himmelberg, 1995; Lim, Choi, & Park, 2003; Mahler & Rogers, 1999; Markus, 1988). However, extending the results in Park (2004a), this section will highlight the importance of strategic maneuvering in the technology-adoption process.

In contrast to the model in Arthur (1989), technologies are typically sponsored by firms that engage in strategic maneuvering in the adoption process.[28] In fact, the tipping and de facto standardization process in the U.S. home-VCR market was affected by the strategic maneuvering of sponsoring firms. For instance, there was a surge (an increase of the market share) of Betamax in 1983 in the process of the de facto standardization toward VHS. Park (2004c), based on the estimates of Park (2004a), further analyzed the strategic maneuvering of Betamax in the de facto standardization process. We begin by further decomposing the log relative sales of the VHS format to the Betamax format as follows. Henceforth, it is understood that all the variables are indexed by period t.

$$\ln(q_v/q_b) = [N_v - N_b] + [\delta_v - \delta_b] - \varphi\lambda[\bar{p}_v - \bar{p}_b] + (1-\sigma)\ln[\#(J_v)/\#(J_b)], \quad (16)$$

where δ_v (δ_b) is the average quality of VHS (Betamax) VCRs, $\bar{p}_v$ $(\bar{p}_b)$ is the quality-adjusted average price of the VHS (Betamax) format, and $\#(J_v)$ $(\#(J_b))$ is the number (or the lineup) of VHS (Betamax) producers. Refer to Park (2004c) for the derivation of Equation 16. On the right-hand side of Equation 16, the first term is called the network advantage of VHS, the second term the average-quality advantage of VHS, the third term the average-price advantage (APA) of VHS, and the last term the lineup advantage of VHS. The sum of the average-quality advantage and APA (i.e., $[\delta_v - \delta_b] - \varphi\lambda[\bar{p}_v - \bar{p}_b]$) is called the average-price and -quality advantage (APQA), which measures the strategic advantage of VHS. The sum of APQA and the lineup advantage of VHS is called price-quality advantage (PQA) in this chapter.

Figure 3 illustrates VHS's PQA, APA, and APQA in each year for 1981 to 1988 except 1985. Since the distance between APQA and APA indicates the average-quality advantage, we can infer that there was almost no difference in average quality between Betamax and VHS during these years except for 1987, although Betamax maintained slight advantages until 1986.[29] However, VHS had a visible price advantage on average except in 1983, 1986, 1987, and 1988.[30] Overall, the APQA of VHS had negative values in these four years, which implies that Betamax had strategic advantages in those years. Figure 3 implies that the surge of Betamax in 1983 was mainly due to the APA of Betamax. However, a more aggressive strategic maneuvering of Betamax in 1987 (primarily due to its average-quality advantage) did not make any apparent interruption in the de facto standardization toward VHS (see Figure 1).[31]

Although the strategic advantages of Betamax in 1983 and 1987 were relatively substantial (compared with those in 1986 and 1988) and resulted in the decreases in PQA of VHS from the previous levels, VHS's advantage in the lineup of producers dominated Betamax's APQA even in 1983 and 1987. Note that the difference between the PQA and APQA of VHS represents the lineup advantage of VHS in Figure 4. The lineup advantage of VHS was generated by the licensing agreements between the sponsors and the licensees that allow entry into the VHS lineup and exit from the Betamax lineup. However, the reason why more producers (or licensees) have positive effects on the format sales is the consumers' preferences for the variety of the products that are reflected in the

Figure 3. Decomposition of price-quality advantage of VHS

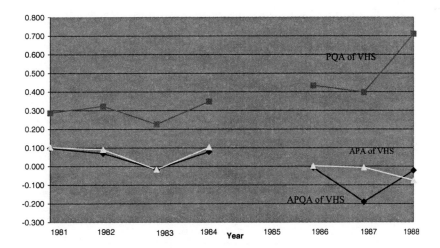

nested logit specification of the utility function. Different producers of the same-format VCRs may still be differentiated by consumers' idiosyncratic tastes for the reliability and the brand images of the producers. Indeed, the within-format correlation coefficient s is estimated to be 0.805 in the U.S. home-VCR case, which indicates that the VCRs of the same format are much closer substitutes to each other but still are perceived as differentiated products. Hence, the sponsor will have an incentive to license for product differentiation. On the other hand, more licensees might join the lineup of VHS in the tipping and de facto standardization process, which was mainly caused by its huge network advantages. Therefore, the lineup advantage of VHS reflects indirect contribution of the network advantage of VHS via increased product varieties.

Based on the calibrations of the network advantage and the lineup advantage, Park (2004c) proceeded to conduct several simulations to study the impacts of strategic maneuvering in price-quality advantages on de facto standardization in the presence of network externalities. Since we are mainly interested in the effects of hypothetical strategic advantages in price and product quality on the de facto standardization process, the strategic advantage (APQA) is considered a control variable in the simulations.[32] Since we do not have information on the sell-off value of the sponsors, we will assume that the de facto standardization of VHS is achieved if $\ln(q_v/q_b)$ gets bigger than 3.749, which was the value of $\ln(q_v/q_b)$ when the de facto standardization of VHS was fulfilled in 1988. Consistently, we will assume that the de facto standardization of Betamax is done if $\ln(q_v/q_b)$ gets smaller than -3.749.

Figure 4. 1987's APQA of Betamax in 1983

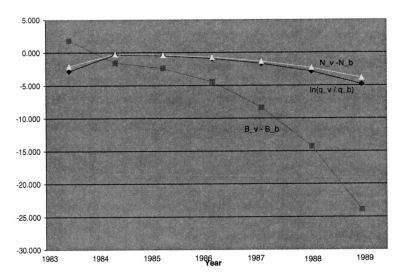

As indicated in Figure 1, VHS seems to have achieved the critical mass around 1982 and 1983. The one-shot strategic maneuvering of Betamax in 1983 created only a temporary interruption in the tipping process toward VHS. In 1984, the yearly increase in the relative sales of VHS to Betamax had recovered, and the tipping process was back on track. The simulation in Figure 4, however, shows that the sponsor of Betamax could reverse the tipping process toward VHS if it had the same amount of strategic advantages in 1983 as it did in 1987. In this case, the sales of Betamax would surpass those of VHS immediately from 1983. The log relative sales of VHS to Betamax, however, would rise in 1984 and then decrease from 1985 along with the rapid reversed tipping process toward Betamax. The strategic advantage was sufficient to reverse the installed-base advantage from 1984, and the installed-base advantage of Betamax became the main force of the de facto standardization toward Betamax from 1984 via the amplification-reinforcement process. The de facto standardization of Betamax would occur in 1989 under this scenario. Indeed, our simulations indicate that 42% of 1987's strategic advantage of Betamax is the minimal advantage (i.e., critical advantage) to reverse the de facto standardization process.[33] Hence, the seemingly critical mass of VHS in 1983 did not necessarily lead to the further rate of adoption to be self-sustaining.

FURTHER RESEARCH ISSUES

The influences of standardization policies certainly go further beyond the product market. Standardization policies will affect not only standards-setting procedures and standards implementations, but also innovative activities such as R & D investments and organizational forms of R & D competition and cooperation.[34] In the context of innovations and standardization, the role of IPRs (intellectual property rights) and public policies such as antitrust measures should be carefully reviewed and reevaluated. The recognition of standards tends to be a technical barrier to trade adopted strategically by individual nations, which raises an issue of international trade and standardization. All these issues are interconnected (see Figure 5). In this section, we will briefly review recent developments in these areas of research and discuss some further issues.

R&D and Intellectual Property Rights

Compatibility regimes affect R&D competition. Kristiansen (1998) analyzed the R&D competition for the timing of the introduction of new technologies in the presence of network externalities. Kristiansen extended the two-period model of Choi (1994) to a three-period model in which two rival firms decide on the timing of the introduction of new products in the first period. Kristiansen showed that the increased profitability from having an installed base in the third period exceeds the loss from more intense competition in the second period.

Figure 5. Standardization issues

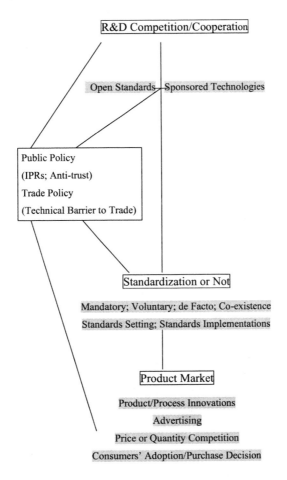

Hence, network externalities may induce one or both firms to introduce new technologies early. Since R&D costs increase with early introduction, the early introduction of new technologies is socially harmful in the analysis of Kristiansen. As compatibility dampens the competition in the product market as shown in Katz and Shapiro (1986a), so does compatibility dampen the R&D competition of new-technology introduction in Kristiansen998). In contrast to Katz and Shapiro, compatibility increases social welfare by delaying the introduction of new technologies. However, Kristiansen's model lacks in taking account of incentives for R&D investments across different regimes of compatibility.

If a certain standard is not mandated by the government, firms will choose to coordinate through standard-setting committees or to compete in the market. Farrell and Saloner (1988) studied the case in which firms prefer any standard to incompatibility but disagree on which standard is better. Farrell and Saloner found that the committee outperforms de facto standardization, although it is slower.[35] Farrell and Saloner, however, did not consider that firms' expectations for the choices between committees and market competition affect innovations and R & D competition. Furthermore, incentives for joining the standard-setting process may be affected by R & D intensities and the firm's perspective of the superiority of its technology. Blind and Thumm (2004) showed that the higher the patent intensities of firms, the lower their tendency to join the standardization process. An economic-analysis framework should be further developed to evaluate the relationship between standard-setting processes and R & D incentives.

Often in the ICT industry, several firms control IPRs. Hence, blocking patents are not unusual, especially in a standard-setting process. Indeed, most standard setting takes place through formal standard-setting bodies. There are hundreds of official standards-setting bodies in the world, such as the International Telecommunications Union (ITU), the Institute of Electronics and Electronic Engineers (IEEE), the National Institute of Standards and Technology (NIST), the American National Standards Institute (ANSI), the International Organization for Standardization (ISO), and the International Electrotechnical Commission (IEC). On top of these, there is any number of unofficial groups such as various special-interest groups (SIGs). To clear blocking patents, participants to the standard-setting process are usually required to license all of its patents essential to implementing the standard on fair, reasonable, and nondiscriminatory terms. For this purpose, cross-licenses and patent pools are usually employed.[36] However, as pointed out in Shapiro (2000), it is still an unanswered question whether this cooperation leads to efficient standardization and increased social welfare. Naturally, collusive motivations and antitrust concerns can be pointed out. On the other hand, it is also argued that patents to interfaces to achieve interoperability foster disagreement and slow down innovation, and cross-licensing tends to favor the older and larger firms rather than the newer, smaller, possibly more innovative firms (see, for example, Krechmer, 2004). Hence, it all raises a new challenge in public policies, especially in the IPRs and antitrust policies.[37]

A mandatory policy of standardization will affect a firm's R & D decision and the evolution of the industry. An example of the impact of governmental standardization policy is found in the European and the U.S. cellular-communications-equipment industries. The U.S. government adopted a laissez-faire policy, resulting in the three competing cellular standards while the European Union pursued a single unified standard, GSM. Consequently, the two European equipment manufacturers, Nokia and Ericsson, pulled far ahead of their rival, Motorola, the U.S. firm. However, during the same period, a new U.S. equipment firm, Qualcomm, grew dramatically by promoting a different standard, CDMA, for cellular communications. This example indicates that standardization policy has a great impact on industry evolution and social welfare, but there may not exist a simple rule to evaluate the standardization policy.

We need to develop a more comprehensive analytical framework that takes account of the interconnected impacts among the standardization policy, the choice between open standards and the sponsorship of technologies, the R & D competition and cooperation, the compatibility and standard-setting process, and the competition in the product market.

Strategic-Trade Policy

As implied in the above example of the cellular communications industry, standardization policies often have impacts on international trade. However, the economic analysis of the strategic-trade policy on standardization is only in the beginning stage. In this subsection, we review some studies in this context.

Gandal and Shy (2001) considered standard recognition as a strategic-trade policy and analyzed governments' incentives to recognize foreign standards when there exist both network effects and conversion costs. In a three-country, three-firm, horizontal differentiation model, two countries can increase their welfare by forming a standardization union that does not recognize the standard of the third country if conversion costs are relatively large, but all the countries mutually recognize all standards if network effects are significant. The model in Gandal and Shy, however, is short of taking account of the dynamic aspects of standard recognition such as preemptive recognition and the strategic delay of recognition for domestic industry protection.

Kende (1991) and Shy (1991) considered the effect of compatible international standards on licensing and the incentives for R & D investments, respectively. Those papers, however, did not proceed to take account of strategic-standardization trade policy.

CONCLUSION

In this chapter, we have reviewed the economic literature of network externalities and standardization, and discussed further research issues. We focused on the two major contributions by Katz and Shapiro (1985) and Farrell and Saloner (1986a). The Katz and Shapiro model analyzes whether private incentives for compatibility are consistent with the social incentive, while the Farrell and Saloner model studies whether consumers' adoption decision of incompatible new technology is socially optimal. Compatibility has a trade-off in social welfare since it dampens competition but increases network benefits. Adoption timing involves two counteracting externalities since early adopters would have disproportionate switching benefits that mostly accrue to late adopters, while late adopters (or new consumers), when they decide to switch, do not take account of the loss of the users of the old technology. Then we proceeded to discuss several important extensions of these models. We shifted our attentions to the empirical evidence of network externalities. Based on the study of Park (2004a), the de facto standardization of the VHS format in the U.S. home-VCR market was quantitatively analyzed in terms of network advantages and stand-alone-benefit advantages. The dynamics of the standardization process can be understood as an amplification-reinforcement process. The dynamic structural model was further utilized to understand the strategic aspects of standardization. Next, we reviewed ongoing research in standardization domestic and trade policies, standard-setting processes, IPRs, and antitrust policy issues, and discussed future research issues requiring cost-benefit analyses based on economic models.

Despite recent economists' attentions to the issue of standardization and network externalities, we are only in the beginning of this growing field of study. Cooperation among various but relevant disciplines may be necessary to enhance our understanding and to develop evaluation frameworks. In the process, we may also experience standardization or interoperability between jargons of different disciplines.

REFERENCES

Akerlof, G. A. (1970). The market for "lemons:" Quality uncertainty and the market mechanism. *Quarterly Journal of Economics, 84*, 488-500.

Allen, D. (1988). New telecommunications services: Network externalities and critical mass. *Telecommunications Policy, 12*, 257-271.

Allen, R. H., & Sriram, R. D. (2000). The role of standards in innovation. *Technological Forecasting and Social Change, 64*, 171-181.

Arthur, B. (1989). Competing technologies, increasing returns, and lock-in by historical events. *Economic Journal, 99*, 116-131.

Auriol, E., & Benaim, M. (2000). Standardization in decentralized economies. *American Economic Review, 90*, 550-570.

Bensen, S. M., & Farrell, J. (1994). Choosing how to compete: Strategies and tactics in standardization. *Journal of Economic Perspectives, 8*, 117-131.

Berg, S. V. (1988). Duopoly compatibility standards with partial cooperation and standards leadership. *Information Economics and Policy, 3*, 35-53.

Berry, S. (1994). Estimating discrete choice models of product differentiation. *RAND Journal of Economics, 25*, 242-262.

Berry, S., Levinsohn, J., & Pakes, A. (1995). Automobile prices in market equilibrium. *Econometrica, 63*, 841-890.

Blind, K., & Thumm, N. (2004). Intellectual property protection and standardization. *International Journal of IT Standards and Standardization Research, 2*, 61-75.

Brynjolfsson, E., & Kemerer, C. F. (1996). Network externalities in microcomputer software: An econometric analysis of the spreadsheet market. *Management Science, 42*, 1627-1647.

Cardell, N. S. (1997). Variance components structures for the extreme value and logistic distributions with applications to models of heterogeneity. *Econometric Theory, 13*, 185-213.

Choi, J. P. (1994). Irreversible choice of uncertain technologies with network externalities. *RAND Journal of Economics, 25*, 382-401.

Church, J., & Gandal, N. (1992). Integration, complementary products, and variety. *Journal of Economics and Management Strategy, 1*, 651-675.

Church, J., & Gandal, N. (1996). Strategic entry deterrence: Complementary products as installed base. *European Journal of Political Economy, 12*, 331-354.

Cusumano, M., Mylonadis, Y., & Rosenbloom, R. (1992). Strategic maneuvering and mass market dynamics: The triumph of VHS over BETA. *Business History Review, 66*, 51-94.

David, P. A. (1985). Clio and the economics of QWERTY. *American Economic Review Proceedings, 75*, 332-337.

David, P. A., & Greenstein, S. (1990). The economics of compatibility standards: An introduction to recent research. *Economics of Innovation and New Technology, 1*, 3-41.

De Palma, A., & Leruth, L. (1995). Variable willingness to pay for network externalities with strategic standardization decisions. *European Journal of Political Economy, 12*, 235-251.

Economides, N. (2000). *The Microsoft antitrust case* (Stern School of Business Working Paper 2000-09). New York: New York University.

Economides, N., & Himmelberg, C. (1995). Critical mass and network evolution in telecommunications. In G. Brock (Ed.), *Toward a competitive telecommunication industry: Selected papers from the 1994 Telecommunications Policy Research Conference.* Mahwah, NJ: Lawrence Erlbaum Associates.

Egyedi, T. M., Krechmer, K., & Jakobs, K. (Eds.). (2003). *Proceedings of the Third IEEE Conference on Standardization and Innovation in Information Technology.* Delft University of Technology, The Netherlands.

Ericson, R., & Pakes, A. (1995). Markov perfect industry dynamics: A framework for empirical work. *Review of Economics Studies, 62*, 53-82.

Farrell, J., & Saloner, G. (1985). Standardization, compatibility, and innovation. *RAND Journal of Economics, 16*, 70-83.

Farrell, J., & Saloner, G. (1986a). Installed base and compatibility: Innovation, product preannouncement, and predation. *American Economic Review, 76*, 940-955.

Farrell, J., & Saloner, G. (1986b). Standardization and variety. *Economics Letters, 20*, 71-74.

Farrell, J., & Saloner, G. (1988). Coordination through committees and markets. *RAND Journal of Economics, 19*, 235-252.

Farrell, J., & Saloner, G. (1992). Converters, compatibility, and the control of interfaces. *Journal of Industrial Economics, 40*, 9-35.

Gandal, N. (1994). Hedonic price indexes for spreadsheets and an empirical test of the network externalities hypothesis. *RAND Journal of Economics, 25*, 160-170.

Gandal, N., Kende, M., & Rob, R. (2000). The dynamics of technological adoption in hardware/software systems: The case of compact disc players. *RAND Journal of Economics, 31*, 43-61.

Gandal, N., & Shy, O. (2001). Standardization policy and international trade. *Journal of International Economics, 53*, 363-383.

Gilbert, R. (1992). Symposium on compatibility: Incentives and market structure. *Journal of Industrial Economics, 40*, 1-8.

Katz, M. L., & Shapiro, C. (1985). Network externalities, competition, and compatibility. *American Economic Review, 75*, 424-440.

Katz, M. L., & Shapiro, C. (1986a). Product compatibility choice in a market with technological progress. *Oxford Economic Papers, 38*, 146-165.

Katz, M. L., & Shapiro, C. (1986b). Technology adoption in the presence of network externalities. *Journal of Political Economy, 94*, 822-841.

Katz, M. L., & Shapiro, C. (1992). Product introduction with network externalities. *Journal of Industrial Economics, 40*, 55-84.

Katz, M. L., & Shapiro, C. (1994). Systems competition and network effects. *Journal of Economic Perspectives, 8*, 93-115.

Kende, M. (1991). *Strategic standardization in trade with network externalities* (INSEAD Working Paper 92/50/EP).

Kende, M. (1998). Profitability under an open versus a closed system. *Journal of Economics and Management Strategy, 7*, 307-326.

Klopfensten, B. C. (1985). *Forecasting the market for home video players: A retrospective analysis.* Unpublished doctoral dissertation, Ohio State University.

Krechmer, K. (2004). Standardization and innovation policies in the information age. *International Journal of IT Standards and Standardization Research, 2*, 49-60.

Kristiansen, E. G. (1998). R&D in the presence of network externalities: Timing and compatibility. *RAND Journal of Economics, 29*, 531-547.

Liebowitz, S. J., & Margolis, S. E. (1990). The fable of the keys. *Journal of Law and Economics, 33*, 1-25.

Liebowitz, S. J., & Margolis, S. E. (1995). Are network externalities a new source of market failure? *Research in Law and Economics, 17*, 1-22.

Lim, B., Choi, M., & Park, M. (2003). The late take-off phenomenon in the diffusion of telecommunication services: Network effect and the critical mass. *Information Economics and Policy, 15*, 537-557.

Mahler, A., & Rogers, E. (1999). The diffusion of interactive communication innovations and the critical mass: The adoption of telecommunications services by German banks. *Telecommunications Policy, 23*, 719-740.

Markus, M. L. (1987). Toward a "critical mass" theory of interactive media: Universal access, interdependence and diffusion. *Communications Research, 14*, 491-511.

McFadden, D. (1973). Conditional logic analysis of quantitative choice behavior. In P. Zarembka (Ed.), *Frontiers in econometrics*. New York: Academic.

Pakes, A., & McGuire, P. (1994). Computation of Markov Perfect Nash Equilibria I: Numerical implications of a dynamic product model. *RAND Journal of Economics, 25*, 555-589.

Pakes, A., & McGuire, P. (2001). Stochastic Algorithm, Symmetric Markov Perfect Equilibrium, and the "curse" of dimensionality. *Econometrica, 69*, 1261-1282.

Park, S. (2003). Semiparametric instrumental variables estimation. *Journal of Econometrics, 112*, 381-399.

Park, S. (2004a). Quantitative analysis of network externalities in competing technologies: The VCR case. *Review of Economics and Statistics, 86*, 937-945.

Park, S. (2004b). Some retrospective thoughts of an economist on the Third IEEE Conference on Standardization and Innovation in Information Technology. *International Journal of IT Standards & Standardization Research, 2*, 76-79.

Park, S. (2004c). *Strategic maneuvering and standardization: Critical advantage or critical mass?* Unpublished manuscript, Seoul National University, Republic of Korea.

Park, S. (2004d). *Website usage and sellers' listing in Internet auctions.* Unpublished manuscript, Seoul National University, Republic of Korea.

Park, S. (2005a). Integration between hardware and software producers in the presence of indirect network externality. *EURAS Yearbook of Standardization, 5*, 47-70.

Park, S. (2005b). *Network benefit function and indirect network externalities.* Unpublished manuscript, Seoul National University, Republic of Korea.

Rysman, M. (2000). *Competition between networks: A study of the market for yellow pages.* Unpublished manuscript, Boston University.

Saloner, G., & Shepard, A. (1995). Adoption of technologies with network effects: An empirical examination of the adoption of automated teller machines. *RAND Journal of Economics, 26*, 479-501.

Shapiro, C. (2000). *Setting compatibility standards: Cooperation or collusion?* Unpublished manuscript, University of California at Berkeley.

Shapiro, C. (2003). Antitrust limits to patent settlements. *RAND Journal of Economics, 34*, 391-411.

Shapiro, C., & Varian, H. R. (n.d.). *Information rules: A strategic guide to the network economy.* Boston: Harvard Business School Press.

Shy, O. (n.d.). The economics of compatibility standards. In *Industrial organization: Theory and applications* (chap. 10, pp. 253-277). Cambridge, MA: The MIT Press.

Shy, O. (1991). *International standardization protection.* Unpublished manuscript, Tel Aviv University, Israel.

ENDNOTES

1   See Farrell and Saloner (1986b) and Gilbert (1992) for details.

2   For surveys of network externalities, see David and Greenstein (1990) and Katz and Shapiro (1994).

3   The integration between hardware and software producers is an interesting issue, which the historical antitrust case of the Microsoft centered around. See Economides (2000). For the strategic aspects of this integration issue, refer to Park (2005a), Church and Gandal (1992, 1996), and Kende (1998).

4   Some researchers consider indirect network externalities as positive production externalities (Liebowitz and Margolis, 1995). In this case, an increase in demands for a hardware product causes a decline in the prices of software products and thus input costs of that hardware product.

5   Examples include a home video game system and video game cartridges, DVD players and DVD movie titles, HDTV sets and HDTV programs, and 3G mobile phones and applications programs.

6   Cusumano, Mylonadis and Rosenbloom (1992) argued that Sony's reluctance to build VCRs for its licensees was a key reason for Matsushita's lead. However, *Fortune* (pp 110 - 116, July 16, 1979) indicated that the key reasons for Matsushita's lead were the longer recording time and lower price, citing the Vice President of marketing for Zenith, "The longer playing time turned out to be very important, . . .". *Consumer Reports* (September, 1978) provided the VCR format lineups in 1978: 8 companies supported the Betamax format, while 13 companies produced the VHS format. VHS's longer recording time affected the lineups as well, making RCA and Hitachi join the ranks of VHS instead of Betamax.

7   Surveys in *Broadcasting* (April 2, 1979) and *The New York Times* (March 19, 1979) show that the majority of consumers used VCRs to record TV programs.

8   In 4 years after its introduction, the VCR had a 1.3 percent household penetration ratio, while it took 5 years before color TV accounted for 1 percent of the sets in use (*Fortune*, July 16, 1979).

9   Klopfensten (1985) cited a survey in *Newsweek* Research Report, 1983, finding that in mid-1983, 60 percent of those surveyed had used prerecorded cassettes.

[10] In July 1983, *Consumer Reports* provided similar test results.

[11] The stand-alone benefit and the network benefit are sometimes called 'similarity standards' and 'compatibility standards', respectively. See Krechmer (2004).

[12] Consumers may have different tastes for the network benefit. De Palma and Leruth (1995) proposed a model in which consumers have heterogeneous willingness to pay for network externalities. The welfare consequences of compatibility in De Palma and Leruth (1995), however, remained the same as in Katz and Shapiro (1985).

[13] When an increased compatibility does not result in complete compatibility, total output may not rise. See Katz and Shapiro (1985) for details.

[14] Side payments may take the form of licensing fees or compensation for the expenses of making the products compatible.

[15] Auriol and Benaim (2000) showed that compatibility outcomes depend on adopters' attitudes to problems caused by in compatibility. If individuals display aversion to incompatibility, official action can be useful to quickly achieve sensible standardization. Otherwise, regulation may be a poor alternative.

[16] This is a reasonable assumption if old users switch slowly, compared to the arrival rate of new users.

[17] Farrell and Saloner (1985) found that symmetric excess inertia (a Pareto-superior new technology not being adopted) could not occur with complete information when all consumers have the opportunity to adopt the new technology at the same time.

[18] For instance, for the user of technology U, the network benefit function is $k_U(N(t)) = k_U(t) = bt$.

[19] The observable product characteristics, X_j, include high picture quality (*HQ*), number of programmable events (*events*), on-screen display (*osd*), multichannel TV sound decoder (*mts*), stereo, and hi-fi.

[20] This assumption can be mitigated in such a way that consumers expect that the utility from the product characteristics and idiosyncratic tastes is depreciated *proportionally* as time passes.

[21] Our model can be extended to the case in which a consumer plans to use the product "for finite periods" instead of "forever".

[22] Note that this network benefit function is a reduced form in the case of indirect network externalities since the network benefit reflects the consumer's benefit from available software (e.g., movie titles) induced by the use of the (hardware) product (e.g., VCRs). Park (2005b) shows that this reduced-form network benefit function can be derived when indirect network externalities are modeled as positive feedback effects between the number of (hardware product) users and a variety of available software. These positive feedback effects are empirically studied in the CD players' case by Gandal, Kende and Rob (2000), in market for the

Yellow Pages by Rysman (2000) and in the Internet auctions by Park (2004d).

23   For the regularity conditions, refer to Ericson and Pakes (1995) and the literature cited there.

24   Hence, network externalities are (partially) internalized by producers' pricing behavior.

25   Relevant data are not available in 1985. For the sources of these data, refer to Park (2004a).

26   As shown in Park (2005b), the network benefit function can be a linear function of the network size under certain specific form of the consumer's utility function for hardware/software systems.

27   Often in the previous empirical analyses, expected growth effects are abstracted. See, for example, Saloner and Shepard (1995) and Economides and Himmelberg (1994).

28   For instance, the two competing technologies in the home VCR market (the Betamax format and the VHS format) were sponsored by Sony and Matsushita, respectively. Katz and Shapiro (1986a) shows that the pattern of adoption depends on whether technologies are sponsored or not.

29   This conclusion is consistent with the perception discussed in Park (2004a) and published articles such as in *Consumers Reports* and others during 1980s.

30   Note that during the years in question, many new VHS producers entered the U.S. market and they usually had focused on low-end models with relatively inexpensive prices.

31   Considering that the installed base advantage of VHS became much bigger in 1987, these observations suggest the significance of the *applications barrier to entry* in the presence of (indirect) network externalities.

32   Indeed, the strategic advantage is an equilibrium outcome in pricing and R&D games. In principle, we can adapt the computational algorithms in Pakes and McGuire (1994, 2001) to our simulations to solve the equilibrium. However, it will induce tremendous computational burden.

33   However, the big strategic advantage of Betamax in 1987 might be possible because the VHS producers did not aggressively respond to Betamax's deep price cut in the final stage of the de facto standardization. On the other hand, the Betamax producers might have had incentives to sell as many Betamax VCRs as possible prior to the de facto standardization.

34   Standardization policies will also affect internal (within a firm) standards setting procedures and standards implementations. Françoise Bousquet in Egyedi, Krechmer and Jakobs (2003) emphasized the importance of standardizers in the internal procedures.

35   Refer to Bensen and Farrell (1994) for overviews of this area of research.

36   Sometimes, the precise requirements imposed by a standard-setting group may be unclear. For instance, there could exist hidden IPRs.

37   Shapiro (2003) proposed a specific antitrust rule limiting patent settlements.

Chapter XV

Network Effects and Diffusion Theory:
Extending Economic Network Analysis

Tim Weitzel, Johann Wolfgang Goethe University, Germany

Oliver Wendt, Johann Wolfgang Goethe University, Germany

Daniel Beimborn, Johann Wolfgang Goethe University, Germany

Wolfgang König, Johann Wolfgang Goethe University, Germany

ABSTRACT

In this chapter, some of the main results of an interdisciplinary research project on standards are presented and integrated into a single framework of technology diffusion. Based on network-effect theory and diffusion theory, we present an agent-based simulation model that extends the traditional economical network perspective by incorporating structural determinants of networks (centrality, topology and density) from sociology and geography and individual standard decision making under realistic informational assumptions. The main finding is a substantial impact of network topology on standard diffusion. The model has so far served as a tool both for developing and evaluating network strategies in practical applications like EDI networks or corporate directory service planning, as well as for providing theoretical insights into standardization problems and possible solutions.

INTRODUCTION

The use of common standards generally makes possible or simplifies transactions carried out between actors or eases the exchange of information between them (David & Greenstein, 1990). Examples are DNA, natural languages, and currency or metric standards, as well as communication protocols (IP [Internet protocol], TCP (transmission-control protocol], HTTP [hypertext transfer protocol]) and syntactic and semantic standards such as XML (extensible markup language) and EDI. In information systems such as intranets, standards provide for compatibility and are a prerequisite for collaboration benefits (Besen & Farrell, 1994). More generally speaking, standards can constitute networks. Inherent in standards, the commonality property deriving from the need for compatibility implies coordination problems. From a theoretical perspective, the existence of network effects as a form of externality that is often associated with communication standards renders efficient neoclassical solutions improbable (Weitzel, Wendt, & Westarp, 2000).

Based upon the final public report of the interdisciplinary research project *Economics of Standards in Information Networks* (Weitzel, Wendt, & Westarp, 2002) and its successor project *IT Standards and Network Effects*, an extended economic framework for network analysis is presented in this chapter. Responding to the challenge of incorporating findings from other noneconomic disciplines into a single framework of network analysis, two network-simulation models are introduced and merged as a first step toward extending a pure economic perspective. The *standardization model* focuses on individuals deciding about the use of standards and providing a first best standardization solution for any given network, while the *diffusion model* describes the diffusion of technological innovations from a vendor's viewpoint as a basis for deriving pricing strategies, for example, in software markets. As the main outcome of this chapter, these two models are merged, combining the consideration of individual anticipatory decision behavior with a consideration of the topological properties of networks. The authors are indebted to the German National Science Foundation (DFG) for their funding of the project.

ECONOMIC APPROACHES TOWARD NETWORK ANALYSIS

Network-Effect Theory

Standardization problems, or more generally economic network analysis, are often based upon the theory of positive network effects (Besen & Farrell, 1994). Network effects describe a positive correlation between the number of users of a standard and its utility (demand-side economies of scale; Farrell & Saloner, 1985; Katz & Shapiro, 1985). They imply multiple equilibria (Arthur, 1989) and hence possibly unfavorable outcomes. The pattern of argument in network-effect theory is always the same: The discrepancy between private and collective gains in networks under increasing returns leads to possibly Pareto-inferior results (market failure, unexploited network gains; David & Greenstein, 1990). With incomplete information about other actors' preferences, *excess inertia* (a start-up problem) can occur as no actor is willing to bear the disproportionate risk of being the first adopter of a standard or technology and then becoming stranded in a small network if all others eventually decide in favor of another technology (Farrell & Saloner, 1986). This renowned start-up problem prevents any standardization at all even if it is preferred by everyone (see Figures 1 and 2 for the start-up problem in the standardization model). From an economic perspective, this is not surprising: In traditional neoclassical economics, there is no difference between local and global efficiency (private and collective gains) if the validity of the fundamental theorems of welfare economics (Hildenbrand & Kirmann, 1976) can be proven. This is the case when certain premises are fulfilled, especially the absence of externalities. Unfortunately, network effects as constituting particularities in networks are a form of externality, thus disturbing the automatic transmission from local to global efficiency (Weitzel, Wendt, et al., 2000). Accordingly, standardization problems that are characterized by the existence of strong network effects transcend large parts of traditional economics. Additionally, since the network metaphor and therefore most practical standardization problems are strongly influenced by factors outside the premises mentioned above, it has proven difficult to find empirical evidence for, for example, start-up problems that is not too ambiguous (Liebowitz & Margolis, 1995). The reason is that many of the traditional findings owe their particularities in large part to implicit premises like infinitely increasing network effects or homogeneous network agents. Thus, while the traditional models contributed greatly to the understanding of a wide variety of general problems associated with the diffusion of standards, much research is still needed to make the results applicable to real-world problems (Liebowitz & Margolis, 1994). Additionally, the specific interaction of standards adopters within their personal socioeconomical environment and the potential decentralized coordination of network efficiency (as has long been demanded by sociologists and institution theorists) are neglected. As a result, important phenomena of modern network-effect markets such as the coexistence of different products despite strong network effects or the fact that strong players in communication networks force other participants to use a certain solution cannot be sufficiently explained by the existing approaches (Liebowitz & Margolis, 1994; Weitzel, Wendt, et al.). For an extensive overview, see Economides (2000) and Weitzel (2004).

Diffusion of Innovation Theories

While the theory (or theories) of positive network effects analyzes the specific characteristics of markets for network-effects goods (such as software products), models of the diffusion of innovations focus on explaining and forecasting the process of the adoption of innovations over time. The term diffusion is generally defined as "the process by which an innovation is communicated through certain channels over time among the members of a social system" (Rogers, 1983, p. 5). In particular, the question of which factors influence the speed and specific course of diffusion processes arises (Weiber, 1993). Traditional diffusion models are based on similar assumptions: Generally, the number of new adopters in a certain period of time is modeled as the proportion of the group of market participants that have not yet adopted the innovation. Most of the traditional approaches aim at revealing the relationship between the rate of diffusion and the number of potential adopters according to the nature of the innovation, communication channels, and social system attributes (Mahajan & Peterson, 1985). The three most common types of diffusion models (Lilien & Kotler, 1983; Mahajan & Peterson, 1985; Weiber, 1993) are exponential (assumes that the number of new adopters is determined by influences from outside the system, e.g., mass communication), logistic (assumes that the decision to become a new adopter is determined solely by the positive influence of existing adopters, e.g., word of mouth), and semilogistic (considers both influences) diffusion models. A famous example of the latter is the Bass model, which has been used for forecasting innovation diffusion in various areas such as retail service, industrial technology, and agricultural, educational, and pharmaceutical markets (Bass, 1969; Mahajan, Muller, & Bass, 1990).

For modeling the diffusion of network-effect products, three areas of deficit are eminent: Critical-mass phenomena are not sufficiently analyzed, real-life diffusion processes cannot be explained, and the interaction of potential adopters within their socioeconomic environment is not sufficiently elaborated (Schoder, 1995). Therefore, logistic and semilogistic approaches are primarily found in areas where innovations have only small consumer interdependencies, where the acceleration of the adoption is characteristically slow, and where the diffusion function is similar to normal distribution.

A long, mostly empirical, research tradition exists in the area of network models of the diffusion of innovations. Network analysis in this context is an instrument for analyzing the pattern of interpersonal communication in social networks (for geographical network analysis, see Haggett, Cliff, & Frey, 1977; for sociological network analysis, see Jansen, 1999). Two early studies can be seen as the starting point for network-diffusion analysis (Valente, 1995). Collecting network and diffusion data in 1955 and 1956, Coleman, Menzel, and Katz (1957) studied the diffusion of a drug innovation (tetracycline) in four towns. With the aim of determining the role of social networks, doctors were asked to name other doctors from whom they most frequently sought discussion, friendship, and advice. For example, the diffusion of the new drug was faster among doctors with an integrated position in the network than among isolated doctors. In general, network-diffusion models can be divided into *relational models* and *structural models*. Relational models analyze how direct contacts between participants in networks influence the decision to adopt or not adopt an innovation. In contrast, structural models focus on the pattern of all relationships and show how the structural characteristics of a social system determine the diffusion process.

THE STANDARDIZATION PROBLEM: A BASIC MODEL

The basic concept of a decentralized standardization model is summarized below. See Buxmann (1996) for a centralized model. The standardization model was developed by incorporating some of the findings summarized above. In the following, costs and benefits associated with standardization are generically modeled and can be adapted to many particular application domains (e.g., EDI networks, electronic markets, corporate intranets). For practical applications and for empirical data based on the model, see Weitzel, Son, & König (2001; an implementation of directory services in a network of six banks and insurance companies), Weitzel (2004; empirical data for large and small EDI users), and Beck, Beimborn, & Weitzel (2003; an application to the German mobile commerce market, i.e., WAP vs. iMode). For extensions of the standardization model (e.g., network topology and density, agent size), see Weitzel (2004).

Typically, standardization costs include the costs of hardware, software, switching, introduction or training, and operations. Furthermore, the interdependence between individual standardization decisions can yield coordination costs of agreeing with market partners on a single standard (Kleinemeyer, 1998). Standardization benefits are direct savings resulting from decreased information costs due to cheaper and faster communication (Kleinemeyer), and also more strategic benefits like avoiding media discontinuities, eliminating errors and costs, creating less converting and friction costs (Braunstein & White, 1985; Thum, 1995), and associating the automation and coordination of different business processes, for example, enabling just-in-time production (Picot, Neuburger, & Niggl, 1993).

Using a computer-based simulation model rather than an analytical approach implies associated disadvantages like smaller analytical transparency and results that cannot be presented in the form of an equation. But these shortcomings are compensated for by the ability to analyze more complex and dynamic structures and discrete choice scenarios as typically found in networks. See Tesfatsion (2002b) for the paradigm of computational economics.

A Basic Standardization Model

Let K_i be the standardization costs of agent i, and c_{ij} be i's network-effect benefit induced by j (i.e., standardization benefits modeled as information costs between agents i and j that can be saved by mutual standardization). Then, the standardization condition of agent i is described by Equation 1.

$$\sum_{\substack{j=1 \\ j \neq i}}^{n} c_{ij} - K_i > 0 \tag{1}$$

However, given autonomous agents and the availability of a realistic information set, it is not clear that all partners actually standardize. If it is assumed that all agents i know K_j and the communication costs directly associated with them (i.e., c_{ij} and c_{ji}) but no further data, like the information costs between other agents, from an individual perspective, the standardization problem is mainly a problem of anticipating the partners' ($j \in \{1,...,n\}$; $i \neq j$) strategic standardization decisions. As in most models, we assume that the agents are risk-neutral decision makers. Note that additional information about agents' individual preferences, standard endowments, or cost parameters can and should be incorporated if available. Other risk preferences, for example, can easily be incorporated by reducing p_{ij} the higher an agent's risk aversion is. The agent's anticipation calculus can now be approximated according to Equation 2, where p_{ij} describes the probability with which agent i believes that j will standardize. If $E[U(i)] > 0$, then agent i will standardize.

$$E[U(i)] = \sum_{\substack{j=1 \\ j \neq i}}^{n} \frac{c_{ji}(n-1) - K_j}{c_{ji}(n-1)} c_{ij} - K_i = \sum_{\substack{j=1 \\ j \neq i}}^{n} p_{ij} c_{ij} - K_i \tag{2}$$

Given the informational assumptions, p_{ij} can be heuristically computed as in Equation 2 (expected utility ex ante) using all data presumed available to the deciding agent. The numerator describes the maximum net savings possible through the standardization for node j. The denominator normalizes the fraction for nonnegative K_j as a value from 0 to 1. Should the fraction have a negative value ($[c_{ji} (n-1)] < K_j$), then $p_{ij} = 0$ holds. We call this the decentralized standardization model since it focuses on describing the individual perspective of agents deciding on adopting standards by determining local costs and benefits associated with that decision, like, for example, autonomous business units in corporate intranets.

As long as the individual agents are unable to influence the standardization decisions of their communications partners, they can do no more ex ante than estimate the probability that their partners will standardize. The decentralized model allows the prediction of standardization behavior in a network, thereby creating a basis for predicting the effects of various concepts of coordination, including influencing the development of expectations regarding the future spread of standards (installed base) and providing a proving ground for forms of cooperation to jointly reap standardization profits through the (partial) internalization of network effects.

To determine the efficacy of solution strategies, as a diametrical concept, the centralized standardization model by Buxmann (1996) is used as a benchmark of the maximum possible coordination quality, assuming that agency or coordination problems are resolved (at zero cost) and that there is a central manager who can determine and implement the optimum result network-wide.[1] Therefore, the first best allocation of standards in a network given there is an all-knowing planner can be determined using the mixed-integer problem formulation in Equation 3 (the basic standardization model in centralized networks; Buxmann).

The binary indicative variable x_i takes on a value of 1 if agent i has standardized, and 0 if not. The binary variable y_{ij} equals 0 if both i and j are standardized ($x_i=1 \wedge x_j=1$). For a multiperiod, multistandard extension of the centralized standardization model, see Buxmann.

$$OF = \sum_{i=1}^{n} K_i \, x_i + \sum_{i=1}^{n} \sum_{\substack{j=1 \\ j \neq i}}^{n} c_{ij} \, y_{ij} \quad \rightarrow \quad Min!$$

s.t.:

$$x_i + x_j \geq 2 - M \, y_{ij} \qquad\qquad \forall \, i, j; \, i \neq j$$

$$x_i, x_j, y_{ij} \in \{0,1\} \qquad\qquad \forall \, i, j; \, i \neq j \qquad (3)$$

Simulation Design

Individual benefits E_i to agent i ($i \in \{1, \dots, n\}$) from implementing a standard are:

$$E_i = \sum_{\substack{j=1 \\ j \neq i}}^{n} c_{ij} \cdot x_j - K_i .$$

(4)

The binary variable x_i is again 1 if i standardizes. GE denotes aggregate network-wide savings resulting from standardization, that is, the horizontal aggregation of all individuals' benefits (network-wide ex post savings).

$$GE = \sum_{i=1}^{n} E_i = \sum_{i=1}^{n}\sum_{\substack{j=1 \\ j \neq i}}^{n} c_{ij} \cdot (1 - y_{ij}) - \sum_{i=1}^{n} K_i \cdot x_i = \underbrace{\sum_{i=1}^{n}\sum_{\substack{j=1 \\ j \neq i}}^{n} c_{ij}}_{\text{ex ante costs}} - \left(\underbrace{\sum_{i=1}^{n} K_i \cdot x_i + \sum_{i=1}^{n}\sum_{\substack{j=1 \\ j \neq i}}^{n} c_{ij} \cdot y_{ij}}_{\text{ex post costs}} \right)$$

(5)

During the simulations, first a network is initialized assigning approximately[2] normally distributed random values for K_i ($K \sim ND(K_i | \mu, 1{,}000^2)$) to all agents i, and c_{ij} ($c \sim ND(c_{ij} | 1{,}000\mu, 200^2)$) to all communications relations <ij>. Networks with other distributions and sizes have yielded analogous results. Having generated a network, the centralized solution is determined according to the centralized model. For this purpose, a linear program with all network data is formulated and solved using JAVA packages drasys.or by DRA Systems (http://www.opsresearch.com) and lp.solve 2.0 by M. Berkelaar (http://siesta.cs.wustl.edu/~javagrp/help/LinearProgramming.html). Afterward, the decentralized decisions are computed until an equilibrium is reached. The whole simulation process is repeated 50 times before reducing $\mu(K)$ by 125 and starting anew. The following figures consist of 4,500 simulation runs each. All simulation results were generated using JAVA 1.2 applications. For data analysis, SPSS 9.0 was used.

Results of Simulating the Standardization Process: The Standardization Gap

To compare the quality of decision making in centrally and decentrally coordinated networks, Figure 1 shows the results from randomly generated networks according to the parameter values described above. The graphs show the decision quality under centralized and decentralized (background) coordination at different standardization costs in periods t=1, ..., 7. GE is graphed against decreasing standardization costs on the abscissa, GE(dz) denotes the total decentralized savings, and GE(z) denotes the total centralized savings. To get a better understanding of the dynamics behind the diffusion of standards, we will first look at the results from the first period and then extend the analysis to t periods (here, the system always found a stationary state within seven periods). For the multiperiod simulations, it is assumed that an agent can only decide to implement a standard once, and that after this decision, he or she is tied to it. This is a common setting in network-effect models (e.g., Arthur, 1989). See Weitzel (2004) for other settings. Incomplete but perfect information is assumed: If an agent can see that a partner has standardized, the uncertainty about this partner's action vanishes (p_{ij}=1). Thus, the state of the network (as binary 1*n-vector consisting of all agents' x_i) and therefore the available information set can change up to n-1 times; no later than in period n, all agents will eventually have made a decision in favor of standardization or they will no longer do so.

In a centrally coordinated network, all agents standardize if $\mu(K) \leq 34{,}000$. In a decentrally coordinated network, agents standardize much later,[3] that is, only at significantly lower standardization costs do they consider standardizing to be an advantageous strategy. Uncertainty about their partners' standardization behavior and thereby about their ability to reap network effects implies a start-up problem (excess inertia) since no agent is willing to bear the disproportionate risk of being the first adopter of a standard. In the first period (front graph), at lower K ($\mu(K)$=[19,000; 16,700]), some agents decide in favor of standardizing, but not all their expected partners do, which results in negative savings subsequently (i.e., wrong decisions ex post). In contrast, wrong decisions are impossible under centralized coordination as the ex post results cannot deviate

Figure 1. Standardization gap (background: centralized solution)

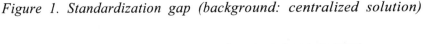
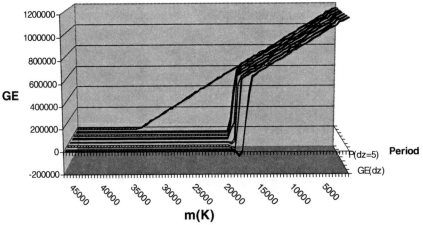

Figure 2. A standardization gap (magnified with no averaging)

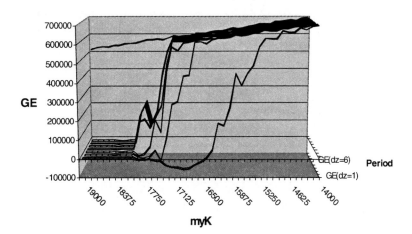

from those planned ex ante. Measured against centralized coordination, too few actors standardize.

In t=1, for $\mu(K)<14,000$, decentralized decisions equal those in centrally coordinated networks. Thus, there is an intuitive discrepancy between centralized and decentralized networks that takes on the form depicted in Figure 1 and is magnified in Figure 2 for the basic standardization problem. We call this perpendicular distance between GE(z) and GE(dz) the *standardization gap* quantifying the magnitude of the standardization problem. The risk of incorrect decisions under decentralized coordination leads to less willingness to standardize within the moderate range of values. Measured in cumulative network-wide costs, the accuracy of the solution achieved in the centralized model cannot be attained by the decentralized network. On the other hand, there are no coordination costs in the decentralized model. Network-wide savings attainable through centralized coordination determine the critical value for the costs of coordination above which a centralized solution is no longer advantageous. This value corresponds to the (vertical) size of the standardization gap.

To accentuate the dynamics of diffusion, Figure 2 shows a magnification of the standardization gap. One can see that the initial negative values of GE(dz, t=1) are neutralized over time. Uncertainty for later adopters is reduced by the costly standardizations of a few early adopters in Period 1. Then, they also standardize. The first movers' additional costs are neutralized and later yield benefits. This phenomenon is often called the *penguin effect* in network-effect literature: "Penguins who must enter the water to find food often delay doing so because they fear the presence of predators. Each would prefer some other penguin to test the waters first" (Farrell & Saloner, 1986, p. 943). Figure 2 gives a detailed impression of the dynamics behind the penguin effect and shows the speed of the standardization process (the graph in the background again is the centralized solution). The first results of an extension of the model to Q standards show a sudden quite-surprising second standardization gap at very low standardization costs resulting from a locally persistent inefficient heterogeneity of standards.

NETWORK-DIFFUSION MODEL OF THE SOFTWARE MARKET

In contrast to the standardization model presented above that focuses on standard users, the diffusion model draws from diffusion theory and is aimed at deriving strategies for network-effect-goods vendors like software producers. Accordingly, while the standardization model will be used to develop and evaluate coordination strategies for standards users' problems, the diffusion model will focus on vendor-side challenges such as optimal pricing and communication strategies. Again, the adoption decision is modeled discretely, meaning that it is not rational to buy or use more than one unit of the same product or even of different products in the same category. This is a common assumption in network models. As confirmed by empirical research, the network effects in the utility function are modeled only as being dependent on the decision behavior of the direct communication network of the potential buyer and not on the installed base (Westarp, 2003; Westarp, Buxmann, Weitzel, & König, 1999). Including topology into the network analysis, triggered by interdisciplinary research cooperations with most notably sociologists and geographers, turned out to provide one of our key improvements in explaining standardization problems: Topological determinants of network behavior have long been highlighted in social network analysis, but have not yet been incorporated into economic models. First simulations with an extended standardization model also confirm the hypothesis of a surprisingly small influence of the aggregate installed base (especially at low standardization costs). One main hypothesis is that the (macro) dynamics of networks as multiagent systems not only depend on the individual (micro) decisions of the participants, but also on personal neighborhood structures reflecting the institutional patterns of networks.

A Basic Diffusion Model

The terminology of our diffusion model is similar to that of Katz & Shapiro (1985) but uses differently interpreted terms. Let r denote the stand-alone utility of a network-effect product (i.e., x=0) and $f(x)$ denote the additional network-effect benefits (i.e., the value of the externality when x is the number of other adopters). For reasons of simplification, we assume that all network participants have the same function $f(x)$. This unrealistic premise can be overcome by integration with the standardization model, thereby also describing asymmetries as, for example, found in EDI networks with large enterprises reaping most of the network benefits whereas the smaller players seem to be left with costs only (Weitzel, 2004). We also assume that $f(x)$ increases linearly. The willingness to pay for a software product can then be described by $r+f(x)$. Let p be the price or the cost of a certain software product. A consumer buys the solution if $r+f(x)-p>0$. In case of v competing products in a market, the consumer buys the product with the maximum positive surplus:

$$\max_{q\in\{1,...,v\}}\left\{r_q + f(x_q) - p_q\right\}. \tag{6}$$

Unlike most of the existing network models, we interpret a network not as homogeneous but as relational; that is, the buying decision is not influenced by the overall installed base within the whole network, but rather by the adoption decisions within the personal communication network. The significance of this proposition is demonstrated by

Figure 3 showing the communication network of Consumer A deciding on adopting Standard 1 or 2. We assume that both standards are free and have identical functionality so that the buying decision solely depends on network effects.

In traditional models that base the standardization decision on the size of the installed base, Consumer A would choose Standard 2 since the installed base with four existing adopters is larger. If we use the relational network approach and therefore only focus on the relevant communication partners of A, the consumer will decide to use Standard 1 since the majority of his or her direct communication partners uses this solution. In the following, we will systematically prove the importance of the personal network structure.

Simulation Design

Our simulations are based on the assumption that network structure, consumer preferences, and the prices of the software are constant during the diffusion process. Price changes, for example, resulting from penetration pricing strategies by vendors or the influence of marketing efforts changing standardizers'

Figure 3. Communication network environment of Consumer A

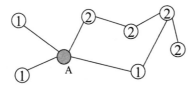

preferences, can be incorporated using the extensions developed in Westarp (2003). All the results presented below are based on a network size of 1,000 consumers. We also tested our simulations for larger and smaller network sizes, again without significant differences in the results. All entities of our model were implemented in JAVA 1.1, and their behavior was simulated on a discrete-event basis.

Network Structure

First, the n consumers are distributed randomly on the unit square; that is, their x- and y-coordinates are sampled from a uniform distribution over [0; 1]. Then, the network's structure is generated by either choosing the c closest neighbors measured by Euclidean distance (*close topology*) or selecting c neighbors randomly from all n-1 possible neighbors (*random topology*). This distinction is made to support our central hypothesis, namely, Ceteris paribus (e.g., for the same network size n and connectivity c), the specific neighborhood structure of the network strongly influences the diffusion processes.

The graphs in Figure 4 give examples of randomly sampled cases of the close topology (exemplary for 100 consumers and a connectivity c of 2, 5, and 10). As we see, a low number of neighbors may lead to a network structure that is not fully connected; that is, agents can only experience network externalities within their local cluster. The standardization processes in individual clusters cannot diffuse to any agents from a different cluster. These subpopulations

Figure 4. Typical networks with two, five, or ten closest neighbors (close topology)

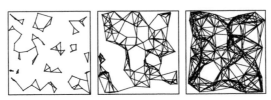

Figure 5. Typical networks with two, five, or ten random neighbors (random topology)

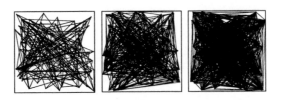

evolve in total separation, and it is therefore rather unlikely that all the isolated regions will evolve to the same global standard. With increasing connectivity (c=5 or 10), the chances that a network is not connected become rather small; that is, any subgroup of consumers agreeing on a specific product may convince their direct neighbor clusters to join them. The domino effect might finally reach every consumer even in the most remote area of the network. However, the number of dominoes that have to fall before a standard that emerged far away in a certain area of the network reaches the local environment of an actor and therefore influences the decision to adopt is typically much higher than in the corresponding graph with random topology. Speaking more formally, the average length of the shortest path connecting two arbitrarily chosen vertices of the graph is smaller for the same connectivity if the graph has a random topology.

In Figure 5, the graphs with the same connectivity (2, 5, and 10) but random topology are shown. The visual impression of a higher connectivity (which is an illusion) results from the fact that within the close topology, a symmetric neighborhood is more common, and here the two links are plotted on top of each other. Evidently, most real-world networks represent an intermediate version of these extreme types, but since the costs of bridging geographical distance have become less and less important the more information technology evolves, the tendency is clear. Electronic markets will tend to resemble the random type of structure (since we select our partners by other criteria than geographical distance), while in markets for physical goods (or face-to-face communication), physical proximity is still a very important factor for selecting business partners and therefore the close topology will be a good substitute for the real-world network structure. Certainly, there are other types of distances as we have learned, for example, from an analysis of proximity measures in geography and sociology. While those disciplines mostly do not consider network effects explicitly yet, they are an important addition to our framework. See Westarp (2003) for the diffusion model incorporating distances based on opinion leadership, group pressure, and centrality.

Preferences, Prices, and Network Effects

Regardless of topology, in our simulation, every consumer can choose from all existing standards and knows all their prices. Initially, all consumers are (randomly) equipped with one standard, which may be considered their legacy software.

The direct utility that each consumer draws from the functionality of the v different products is then sampled from a uniform random distribution over the interval $[0; util]$. For each consumer and every standard we use the same interval. Thus, a value of $util=0$ leads to homogeneous direct preferences (of 0), while the higher the exogenously given value of $util$, the more heterogeneous the preferences of the consumers become. The weight of the positive network externalities deriving from each neighbor using the same software has been set to an arbitrary (but constant) value of 10,000 (for every consumer and every run). In order to isolate the network externalities and heterogeneity of consumer preferences from other effects, we decided to fix all prices for the software products to a constant value and all marketing expenditures to 0 for the simulations presented here; that is, consumers decide solely upon potential differences of direct utility and the adoption choices of their neighbors (see term above).

Dynamics of the Decision Process

In each iteration of the diffusion, every consumer decides whether to keep his or her old software, or whether to replace it with a new one based on the decision rationale described above. The adoption decisions are made in a sequential order. Although we have not yet established a formal proof, for our simulations, this decision process always converged toward an equilibrium in which no actor wanted to revise his or her decision anymore.

Results of Simulating the Diffusion Process

First, a total number of 3,000 independent simulations were run with 1,000 consumers and 10 different software products. The distribution reached in equilibrium was then condensed into the Herfindahl[4] index used in industrial economics to measure market concentration (Tirole, 1993). In the following diagrams, every small circle represents one observation.

Figure 6 illustrates the strong correlation (0.756) of connectivity and equilibrium concentration for close topology. Despite this strong correlation, it can clearly be seen that even in networks with 200 neighbors per consumer (i.e., a connectivity of 200), the chances that one standard will completely dominate the market are still very low. For random topologies, an even stronger correlation (0.781) is obtained. Note that all the correlations illustrated in this chapter are

Figure 6. Strong correlation of connectivity and concentration

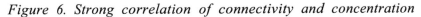

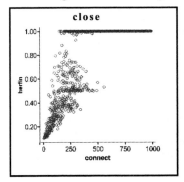
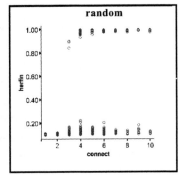

Figure 7. Equilibria in different topologies

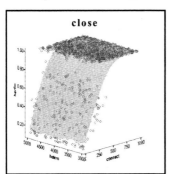 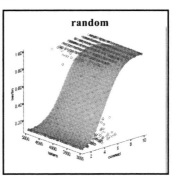

significant on the 0.01 level. Figure 6 clearly shows that the likelihood of the total diffusion of only one standard is very high in a random topology network even for very low connectivity (here, there is a high frequency of maximum concentration, that is, one standard; note the different abscissa scale).

In Figure 7, we additionally consider the heterogeneity of preferences as a third dimension. We did not find any significant dependency of the sampled equilibria on this factor for close topologies (Figure 7, left). However, sampling random topologies, we found a slight but significant negative correlation of heterogeneity and concentration (-0.141). Note that the axis for connectivity is again scaled differently. It can clearly be seen that for 10 neighbors per consumer (1% of the total population), it is already almost certain that only one standard will eventually take over the whole market (Figure 7, right).

This comparison strongly supports our hypothesis that for a given connectivity, the indirect domino effects are much stronger for random-topology networks, and thus the diffusion process shows much higher tendencies toward standardization. To test this statistically, we ran a Kolmogorov-Smirnov test (Hartung, 1989) rejecting the hypothesis that the concentration indices obtained for close and random topologies follow the same distribution on a significance level better than 0.0005 (KS-Z of 2.261). This result substantiates our findings statistically.

A second interesting phenomenon can be seen in the fact that, although the mean concentration for a random-topology network of connectivity 5 is about 0.5, there are hardly any equilibria with concentration indices between 0.2 and 0.8; that is, either the diffusion process leads to one strong product or many will survive. This is different for close-topology models where intermediate solutions with two or three strong products can be stable equilibria, obviously being the result of subgroups of consumers (with strong intragroup communication and fewer links to other groups) collectively resisting external pressure to switch their selected standard.

Figure 8. Typical diffusion processes for 1,000 consumers and connectivity of 5 depending on topology and the heterogeneity of preferences

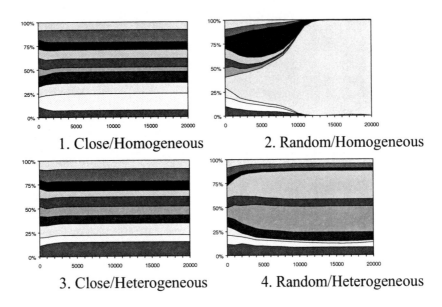

1. Close/Homogeneous   2. Random/Homogeneous

3. Close/Heterogeneous   4. Random/Heterogeneous

Summarizing our findings so far, we display four typical patterns for diffusion processes toward an equilibrium depending on network topology and the heterogeneity of preferences (Figure 8). The *x*-axis shows the number of iterations with every consumer deciding once per iteration. The *y*-axis illustrates the market shares of the 10 software products.

The influence of topology on the diffusion of innovations in networks is obvious. While close topology is generally the basis for a greater diversity of products since clusters or groups of consumers can be relatively independent from diffusion processes in the rest of the market, random topology tends to market the dominance of one or few products.

FUSION OF THE TWO MODELS

Although we used a different notation, the strong similarities of the standardization model and the diffusion model are obvious: The stand-alone utility r_q of the consumer's utility function $\max_{q \in \{1,\dots,v\}} \{r_q + f(x_q) - p_q\}$ may directly be deducted from the price, which then corresponds to the individual standardization costs:

$$\underbrace{-(p_q - r_q) + f(x_q)}_{\text{Diffusion}} = \underbrace{x_q \left[\sum_{q=1}^{v} K_q - \sum_{j=1}^{n} c_{ij} prob_{ijq} \right]}_{\text{Standardization}} \rightarrow \max_q \qquad (7)$$

The network benefits of the diffusion model, however, are modeled as a function of the number of partners using the same product, which is furthermore restricted to being linear and homogeneous for all consumers. In contrast to that, the standardization model allows for differentiating these benefits c_{ij} to be a function of the individual link from agent i to a specific partner j. To adapt the decision calculus of the standardization model to the diffusion model, we relax the assumption that the ex post observability of other agents' standardization decisions is restricted to the agent's direct neighbors and instead assume that each agent i also knows the identity and the standards used by the direct neighbors of his or her own neighbors (called *second-degree neighbors*). Thus, we may use their current endowment for a more precise forecast of actor i's direct neighbors' decisions. One possible way to do this is to replace actor i's decision criterion, "installed number of product q used in my direct neighborhood," with "expected number of product q used in my direct neighborhood," and to try to approximate this number indirectly by the observable "installed number of product q used in my second-degree neighborhood," of course normalized to the number of direct neighbors, since we will not receive 10 times the positive network gains, even when each of our 10 neighbors has 10 distinct neighbors himself or herself. Formally, we may express this expected installed base of an actor i by Equation 8, with x_{kq} being the Boolean variable indicating whether actor k currently uses product q. The decision of each actor is then made as before according to the decision function used in the diffusion model.

Figure 9. The effect of anticipative consumer behavior on market concentration

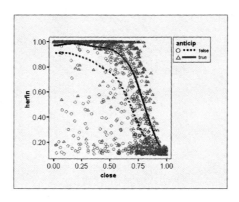

$$E_i(base_q) = \sum_{j \in neighbors(i)} prob_{ijq} \quad \text{with} \quad prob_{ijq} = \frac{\sum_{j \in neighbors(i)} x_{kq}}{|neighbors(j)|} \tag{8}$$

Figure 9 shows that this anticipative behavior (anticip=true) from the adapted decentralized standardization model leads to a significantly higher expected market concentration than in the original network-diffusion model. This effect is of course stronger in close-topology networks, since random-topology networks already exhibited a strong tendency toward market concentration with the nonanticipatable-decision criterion. This observation is not very surprising if we consider the question of choosing an operating system for your new PC (personal computer): Although you may run into some trouble by installing a Windows 2000 workstation if your company still runs NT 4.0 servers, it may be wise to do so if you observe your company's suppliers and customers switching to Windows 2000 and thus expect your system administrator to be urged to follow them soon.

Figure 10 also illustrates that this anticipative behavior really pays off compared to the ex post observation of direct neighbors: The figure shows the difference in the cumulative benefit to all consumers measured by the aggregated utility (direct and external effects) minus the total cost of buying the products, and maybe buying a new product in the next period since we have to revise the decision made in the last period.

Figure 10. The effect of anticipative consumer behavior on consumer welfare

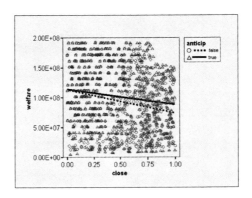

IMPLICATIONS AND FURTHER RESEARCH

Two approaches aimed at transcending the traditional scope of economic network analysis have been presented and integrated, incorporating among others structural elements of networks into the analysis. Next research steps will include the development of higher order anticipatory decision functions, the incorporation of additional information about the standards at choice, and altering the anticipation of partner behavior according to their decision history. These aspects will then allow modeling the influence of vendor strategies (e.g., intertemporal pricing, marketing, bundling) and in turn the interplay with topology determinants such as centrality, group pressure, or opinion leadership. Requirements concerning the modeling power of an eventual interdisciplinary theory of networks have been proposed in Weitzel, Wendt, et al. (2000) and include the integration and explanation of social and economic interaction of human actors and automated agents. Among others, considering the bounded rationality of deciding agents and their institutional embeddedness, evolutionary system dynamics and the emergence of system components and links should allow one to model the emergence of completely new network participants and their decease and relations. However, since assuming bounded rationality—a primary proposition of social network analysis—usually implies the impossibility of determining analytical (ex ante) results (existence and/or efficiency of equilibria), a recourse toward empirical and simulative approaches seems unavoidable. Therefore, we must rather rely on the simulation of system dynamics and analysis of the observed behavior of the simulation model.

What does that mean for future standardization research? We consider the research paradigm of agent-based computational economics (ACE) a promising way to cope with the inevitable system complexities associated with standard diffusion (Weitzel & König, 2003). According to Tesfatsion (2002a, pp. 55-56), "Agent-based computational economics is the computational study of economies modeled as evolving systems of autonomous interacting agents." It evolved as a specialization to the economics of the basic complex adaptive-systems paradigm (Vriend, 1996, 1999).

One principal concern of ACE researchers is to understand why certain global regularities have been observed to evolve and persist in decentralized market economies despite the absence of top-down planning and control...The challenge is to demonstrate constructively how these global regularities might arise from the bottom up, through the repeated local interactions of autonomous agents. (Tesfalsion, 2002b)

And for normative network analysis, ACE uses "computational laboratories within which alternative socioeconomic structures can be studied and tested with regard to their effects on individual behavior and social welfare" (Tesfatsion,

2002b). Three recent special issues of international journals (Tesfatsion, 2001a, 2001b, 2001c) show the scope of possible ACE applications:

(i) Learning and the embodied mind; (ii) evolution of behavioral norms; (iii) bottom-up modeling of market processes; (iv) formation of economic networks; (v) modeling of organizations; (vi) design of computational agents for automated markets; (vii) parallel experiments with real and computational agents; and (viii) building ACE computational laboratories. (Tesfatsion 2001a)

In the context of an information-systems view on the economics of standards, the strand on the formation of economic networks is particularly interesting as it focuses on the selection of transaction partners forming cooperation networks between these agents. Thereby, the IS community can contribute to the perennial debate in economics (Haye) on the complex mutual feedback between micro- and macrostructures by providing building blocks to the decentralized coordination of networked agents (Tesfatsion, 2002a).

Applications of the proposed standardization model to derive pricing strategies for software markets, and empirical evidence for the model's validity for predicting diffusion results in software markets, EDI networks, and other systems subject to network effects (e.g., cellular phone market, SME value-chain integration, X.500 services) are presented in Weitzel (2004) and Westarp (2003). For an application of the model on the standards battle between iMode and WAP mobile commerce standards, see Beck et al. (2003). An elaboration of the coevolution of agents' expectations could be based on the interesting work on self-referential predictions about the future choice of other network participants as presented in Arthur's (1995) model of an ecology of coevolving expectations. The extended models will then be used to develop and evaluate solution strategies to standardization problems, and to try to contribute to the ongoing discussion about if and how we need standardization policies that are different from traditional industry policies. After all, the implications and underlying principles within social and economic networks and their intended and unintended influence on existing systems are not yet well understood, and "an economic theory of norms and standards…is still lacking" (Knieps, Müller, & Weizsäcker, 1982, p. 213).

REFERENCES

Arthur, W. B. (1989). Competing technologies, increasing returns, and lock-in by historical events. *The Economic Journal, 99*, 116-131.

Arthur, W. B. (1995). Complexity in economic and financial markets. *Complexity, 1*(1).

Bass, F. M. (1969). A new product growth for model consumer durables. *Management Science, 15*(5), 215-227.

Beck, R., Beimborn, D., & Weitzel, T. (2003). The German mobile standards battle. *Proceedings of 36th Hawaii International Conference on System Sciences (HICSS-36)*. Retrieved from http://www.is-frankfurt.de/publikationen/publikation423.pdf

Besen, S. M., & Farrell, J. (1994). Choosing how to compete: Strategies and tactics in standardization. *Journal of Economic Perspectives, 8*(2), 117-131.

Braunstein, Y. M., & White, L. J. (1985). Setting technical compatibility standards: An economic analysis. *Antitrust Bulletin, 30*, 337-355.

Buxmann, P. (1996). *Standardisierung betrieblicher informationssysteme.* Wiesbaden, Germany: DUV.

Coleman, J. S., Menzel, H., & Katz, E. (1957). The diffusion of an innovation among physicians. *Sociometry, 20*, 253-270.

David, P. A., & Greenstein, S. (1990). The economics of compatibility standards: An introduction to recent research. *Economics of Innovation and New Technology, 1*, 3-41.

Economides, N. (2000). *An interactive bibliography on the economics of networks and related subjects.* Retrieved from http://www.stern.nyu.edu/networks/biblio.html

Farrell, J., & Saloner, G. (1985). Standardization, compatibility, and innovation. *RAND Journal of Economics, 16*, 70-83.

Farrell, J., & Saloner, G. (1986). Installed base and compatibility: Innovation, product preannouncements, and predation. *The American Economic Review, 76*(5), 940-955.

Haggett, P., Cliff, A. D., & Frey, A. (1977). *Location models* (Vols. 1 & 2, 2nd ed.). London: Oldenbourg.

Hartung, J. (1989). *Statistik: Lehr und handbuch der angewandten statistik.* München, Germany.

Hildenbrand, W., & Kirman, A. P. (1976). *Introduction to equilibrium analysis.* Amsterdam: North-Holland Publishing.

Jansen, D. (1999). *Einführung* in die Netzwerkanalyse. Opladen: Leske und Budrich.

Katz, M. L., & Shapiro, C. (1985). Network externalities, competition, and compatibility. *The American Economic Review, 75*(3), 424-440.

Kleinemeyer, J. (1998). *Standardisierung zwischen kooperation und wettbewerb.* Frankfurt, Germany: Lang.

Knieps, G., Müller, J., & Weizsäcker, C. C. v. (1982). Telecommunications policy in West Germany and challenges from technical and market development. *Zeitschrift für Nationalökonomie, 2*(Suppl.), 205-222.

Liebowitz, S. J., & Margolis, S. E. (1994). Network externality: An uncommon tragedy. *The Journal of Economic Perspectives, 8*(2), 133-150.

Liebowitz, S. J., & Margolis, S. E. (1995). Path dependence, lock-in, and history. *Journal of Law, Economics and Organization, 11*, 205-226.

Lilien, G. L., & Kotler, P. (1983). *Marketing decision making: A model building approach*. New York: Harper & Row.

Mahajan, V., Muller, E., & Bass, F. M. (1990). New product diffusion models in marketing: A review and directions for research. *Journal of Marketing, 54*, 1-26.

Mahajan, V., & Peterson, A. P. (1985). *Models for innovation diffusion*. Newbury Park, CA: Sage Publications.

Picot, A., Neuburger, R., & Niggl, J. (1993). Electronic data interchange (EDI) und lean management. *Zeitschrift für Führung und Organisation, 1*, 20-25.

Rogers, E. M. (1983). *Diffusion of innovations* (3rd ed.). New York: The Free Press.

Schoder, D. (1995). *Erfolg und mißerfolg telematischer innovationen*. Wiesbaden, Germany: Gabler edition Wissenschaft.

Tesfatsion, L. (Guest Ed.). (2001a). Agent-based computational economics. Special issue. *Computational Economics, 18*(1).

Tesfatsion, L. (Guest Ed.). (2001b). Agent-based computational economics. Special issue. *Journal of Economic Dynamics and Control, 25*(3-4).

Tesfatsion, L. (Guest Ed.). (2001c). Agent-based modelling of evolutionary economic systems. Special issue. *IEEE Transactions on Evolutionary Computation, 5*(5).

Tesfatsion, L. (2002a). *Agent-based computational economics*. Retrieved from http://www.econ.iastate.edu/tesfatsi/ace.htm

Tesfatsion, L. (2002b). Agent-based computational economics: Growing economics from the bottom up. *Artificial Life, 8*, 55-82.

Thum, M. (1995). *Netzwerkeffekte, standardisierung und staatlicher regulierungsbedarf*. Tübingen, Germany: Mohr.

Tirole, J. (1993). *The theory of industrial organization* (6th ed.). Cambridge, MA: MIT Press.

Valente, T. W. (1995). *Network models of the diffusion of innovations*. Cresskill, NJ: Hampton Press.

Vriend, N. (1996). Rational behavior and economic theory. *Journal of Economic Behavior and Organization, 29*, 263-285.

Vriend, N. (1999). *Was Hayek an ACE?* Working Paper 403. London: University of London, Queen Mary and Westfield College.

Weiber, R. (1993). Chaos: Das ende der klassischen diffusionsforschung? *Marketing ZFP, 1*, 35-46.

Weitzel, T. (2004). *Economics of standards in information networks*. New York: Springer/Physica.

Weitzel, T., Beimborn, D., & König, W. (2003). Coordination in networks: An economic equilibrium analysis. *Information Systems and e-Business Management* (ISeB), *1*(2), 189-211.

Weitzel, T., & König, W. (2003). Computational economics als Wirtschaftsinformatischer Beitrag zu Einer Interdisziplinären Netzwerktheorie. *Wirtschaftsinformatik, 45*(5), 497-502.

Weitzel, T., Son, S., & König, W. (2001). Infrastrukturentscheidungen in Vernetzten Unternehmen: Eine Wirtschaftlichkeitsanalyse am Beispiel von X.500 Directory Services. *Wirtschaftsinformatik, 4*, 371-381.

Weitzel, T., Wendt, O., & Westarp, F. v. (2000). Reconsidering network effect theory. *Proceedings of the 8th European Conference on Information Systems (ECIS 2000)*. Retrieved from http://www.wiwi.uni-frankfurt.de/~tweitzel/paper/reconsidering.pdf

Weitzel, T., Wendt, O., & Westarp, F. v. (2002). Modeling diffusion processes in networks. In K. Geis, W. Koenig, & F.v. Westarp (Eds.), *Networks: Standardization, infrastructure, and applications* (pp. 3-31). Berlin, Germany/New York: Springer.

Westarp, F. v. (2003). *Modeling software markets: Empirical analysis, network simulations, and marketing implications.* Unpublished doctoral dissertation, Institute of Information Systems, Frankfurt, Germany.

Westarp, F. v., Buxmann, P., Weitzel, T., & König, W. (1999). *The management of software standards in enterprises: Results of an empirical study in Germany and the U.S.* (SFB 403 AB-99-7). Retrieved from http://www.wiwi.uni-frankfurt.de/~tweitzel/report.pdf

ENDNOTES

[1]   Thus the individual implications of standardization decisions are irrelevant for the centralized model. In a game theoretical equilibrium analysis it becomes clear that the centralized solution is a Kaldor-Hicks optimum determining the biggest savings yet to be (re)distributed [Weitzel/Beimborn/König 2003].

[2]   No negative cost values are used.

[3]   "Later" in this context does not describe a later period in time regarding one particular diffusion process, of course, but rather "at parameter constellations farther to the right of the abscissa" (as mostly lower K).

[4]   The Herfindahl index is the sum of the squared market share per vendor. If all market shares are evenly distributed among our ten alternative products, we get the minimal concentration index of $10*(0.1)^2 = 0.1$ while we get a maximal concentration index of $1*1^2+9*0^2 = 1$ if the diffusion process converges to all consumers using one standard.

Chapter XVI

How High-Technology Start-Up Firms May Overcome Direct and Indirect Network Externalities

Mark Pruett, George Mason University, USA

Hun Lee, George Mason University, USA

Ji-Ren Lee, National Taiwan University, Taiwan

Donald O'Neal, University of Illinois – Springfield, USA

ABSTRACT

This chapter presents a conceptual model of strategic choice for high-technology start-up firms in the face of network externalities—the strength of the market's preference for standardized technology. Our model suggests that the commercialization strategies followed by such a firm will depend on the type of network externalitites—direct vs. indirect—as well as the degree of appropriability—the firm's ability to retain the value of innovation. We offer a number of propositions generated by the model and discuss their implications.

INTRODUCTION

A particularly vexing barrier for some start-up firms is how to overcome network externalities that may exist in their markets. Similarly, another hurdle for many start-up firms is how to appropriate value from an innovation. The model in this chapter suggests that the distinction between direct and indirect network externalities, and the degree of appropriability, will determine whether the firm's commercialization strategy focuses on internal resources and decision variables or on interactions with its competitive environment.

High-technology start-up firms may be particularly sensitive to network externalities and appropriability since many such firms are introducing products based on technologies for which there are yet no market standards for compatibility (Hill, 1997) and facing particularly uncertain appropriability conditions that affect their ability to grow and survive (Shane, 2001). Not all new technologies, and not all start-up firms, face network externalities or appropriability issues. For those that do, however, overcoming these barriers to commercialization is crucial as these barriers may influence the firm's strategy for commercialization, growth, and survival.

This chapter models strategic choice for start-up high-technology firms in the face of network externalities—the strength of the market's preference for standardized or compatible technology (Farrell & Saloner, 1985; Katz & Shapiro, 1985). It suggests that commercialization strategies will depend on appropriability—the firm's ability to retain the value of an innovation (Arrow, 1962; Teece, 1986)—and the type of network externality—direct vs. indirect. Following prior researchers (Katz & Shapiro, 1985; Kotabe, Sahay & Aulakh, 1996), direct network externalities refers to a direct relationship between the number of users of a product and the product's quality or utility, while indirect network externalities refers to the indirect effects from the price and availability of goods and services that complement a product. The chapter is rooted in streams of research from technology and innovation literature on how technologies become commercialized (Lee, O'Neal, Pruett & Thomas, 1995; Tushman & Rosenkopf, 1992), organization research on technological discontinuities (e.g., Anderson & Tushman, 1990; Tushman & Anderson, 1986; Tushman & Rosenkopf, 1992), and literature from strategy and economics focused on the impact of technological standards (e.g., Farrell & Saloner, 1987; Garud & Kumaraswamy, 1993; Hill, 1992, 1997; Katz & Shapiro, 1986; Majumdar & Venkataraman, 1998; McGrath & McGrath, 2000).

These streams have posed long-standing questions for researchers and for firms. How can standards be established? How can a new entrant compete? What roles do switching costs, first or second mover advantage, regulation, and intra-industry cooperation play? In competitive strategy, how can a firm profitably commercialize its own technology if the technology poses a network externality for customers and there is the potential for competition from other firms, either through imitation or through alternative technologies? In particular, our model suggests that direct network externalities will lead firms to pursue strategic choices centered on internally controllable decision variables. Indirect network externalities, on the other hand, will lead firms toward efforts to manage their competitive environment by cooperating with outside actors. In addition to theory-building, these questions also pose the need for additional work to empirically quantify the relationship between strategic choice, appropriability, and the two distinct forms of network externalities. The model developed in this chapter offers a basis for subsequent empirical study.

Significance to High-Technology Industries

This topic is particularly significant in a global economy that is evolving rapidly in the area of so-called "high" technology, a term encompassing a variety of industries focused on newer technologies. The most prominent may be telecommunications and information technology (IT) hardware, software, and services. There are other significant areas as well that often are placed under the high-technology umbrella, including biotechnology, advanced manufacturing technologies, and advanced product technologies. These industries have moved to the forefront of business activity and change in the last decade, in part because of their impact on traditional economic sectors, but also in part because of their own growing economic significance.

The growth and evolution of these industries has not been entirely pain-free, however. We have witnessed in the past several years a dramatic weeding-out process in the IT sector. The "dot-com" collapse in the United States may be the central example in this process. Hundreds of IT-related firms have, in a remarkably brief span of time, declared bankruptcy, been sold off, experienced sales declines, laid off large numbers of employees, redefined their missions and strategies, or simply closed their doors. It is worth noting, however, that many continue to survive, even prosper.

Although some firms in IT are well-established (such as the traditional telecom and mainframe computer companies), many IT-related firms are remarkable for their youth and their independent emergence. Many of these firms, whether failures or survivors, are relatively young companies, having come into being in the 1990s, particularly in the second half of the decade. Financed by a healthy venture capital sector and strong responses to public stock offerings, many of these companies were independent start-ups rather than the progeny of large, established firms.

These firms also have faced a highly competitive environment characterized by rapid technological advancement and entry by new firms. In this environment, strong network externalities and widely varying appropriability pose a particularly fascinating mix of competitive issues.

First, network externalities are common in IT hardware, software, and services. That is, the value of many IT products to a user depends on the number of other users. New firms have strong incentives to build a critical mass of customers for their product quickly. At the same time, many high-technology products may be unfamiliar to existing markets, so new firms also must educate customers about new product technologies and their uses.

Second, firms have faced varying degrees of appropriability. Appropriability is the ability of a firm to capture the economic rents generated by its activity. It may depend on financing or other resources, on technological barriers to imitation, and/or on legal barriers to imitation. The degree of appropriability is a major factor in a firm's ability to sustain a competitive advantage.

The perception that these new areas could offer substantial economic rents has created strong incentives for competitors to arise. Certainly, venture capital for new IT firms is now harder to obtain than in the last decade, but we have seen a variety of new would-be competitors arriving in fairly short order in many segments of information technology. In some segments these new entrants compete with incumbents, yet other segments are new areas of business populated by *de novo* firms or by incumbents entering from other arenas.

Entry has been facilitated by the intangible nature of intellectual property. The high job mobility of workers with technical knowledge and other intellectual capital has facilitated start-ups. Further, since innovation and competition have moved rapidly, it has been risky for firms to rely heavily on legal protection for their intellectual property. Legal protection may be definitive, but it is not necessarily swift. In a highly competitive environment, relying primarily on the law for protection may lead a firm to "win the battle but lose the war."

STRATEGIC CHOICES BASED ON APPROPRIABILITY AND THE TYPE OF NETWORK EXTERNALITIES

A start-up firm with an innovative product facing network externalities has a difficult situation. It needs a substantial customer base in the short term to gain a first mover advantage.

From a consumer's point of view, network externalities echo the supplier's minimum efficient scale. For the supplier, it may be uneconomical to produce below some given level. For the consumer, demand may be absent below some level, regardless of price. Figure 1 presents a stylized example of hypothetical demand curves for two products, one with a network externality and one without. This is a simplified model, not a depiction of a typical situation.

D_0 is a demand curve for a product without a network externality. It represents the traditional normal good—demand is an inverse function of price,

Figure 1. Hypothetical product demand curves with and without network externality

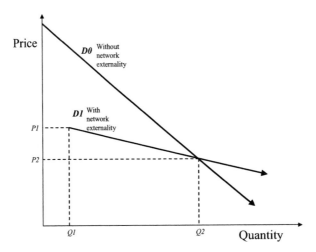

and an individual consumer's product utility does not depend on aggregate demand. D_1 is a stylized demand curve for a product with a network externality. We have presented it in this form to highlight three network externality issues that are particularly interesting from a competitive standpoint for start-up firms.

First, demand does not begin at zero—there is no demand below quantity *Q1*. This reflects the basic nature of a network externality. In a world with only one telephone, there should be no demand for that phone, regardless of price. From a competitive perspective, this illustrates the start-up firm's initial sales issue—how to achieve sales of quantity *Q1*. Customer expectations about the product's likely popularity with other users will influence whether the firm can achieve any sales at all.

Second, by depicting demand beginning at price *P1*, it suggests that low levels of demand for a product with a network externality may be satisfied only at a price less than that for a similarly priced product without a network externality. Stated another way, if two similarly priced competing products are introduced, one with a network externality and one without, initial demand may be higher for the product without. Strategically, this also suggests that, *ceteris paribus,* resolving the network externality is the central competitive issue for the start-up.

Third, the flatter slope of the demand curve for the product with a network externality suggests that the market may be more price-sensitive for this product. It highlights that at prices below some level, denoted here as *P2*, the product's rising marginal utility may actually induce the market to prefer the externality-driven product. Stated another way, it suggests that a firm facing a network externality for its product can gain market dominance and earn economic rents if it can stimulate demand to and beyond the point *Q2*.

Figure 2 introduces the subsequent discussion, in which we develop the logic of strategic choices that a firm is likely to pursue in the face of network externalities. Although the model focuses on start-up firms (entrepreneurial or corporate), we recognize that an incumbent can influence the choice and success of these strategies. For example, an incumbent can block entry through their unique and inimitable resources and capabilities, and through market signals for a reputation of retaliation or by making irreversible investments (Afuah, 1999). Alternatively, an incumbent can encourage entry to allow rivals to fill out product lines that it cannot provide themselves (Porter, 1980) to persuade buyers to adopt products and services provided by monopolists (Garud & Kumaraswamy, 1993), and to help win a standard/dominant design (Afuah, 1999). With this understanding, our model, however, focuses on start-up firms (entrepreneurial or corporate) and emphasizes the distinction between two types of network externalities—direct and indirect—and differing conditions of appropriability. The following discussion develops propositions for each of the four cells in Figure 2.

Figure 2. Strategic choices depending on appropriability and type of network externalities

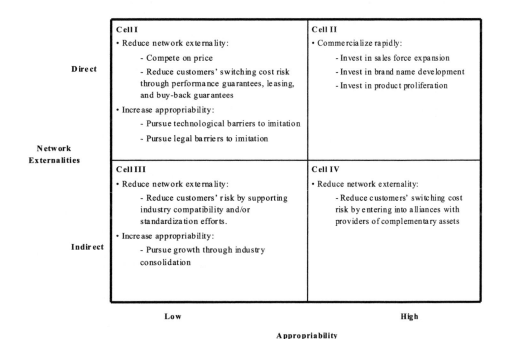

Cells 1 and 2: Direct Network Externalities

As noted earlier, *direct network externalities* are driven by a direct relationship between the number of users of a product and the product's quality or utility (Katz & Shapiro, 1985; Kotabe, Sahay & Aulakh, 1996). Direct network externalities often are found in products that facilitate human interaction. An obvious example, the telephone, has greater consumption utility as more users join the network (Katz & Shapiro, 1985). (As one reviewer noted, direct network externalities may be found in other areas as well, such as the issue of machine-machine interoperability in production or operations environments.)

Cell 1 illustrates the situation of direct network externalities in conditions of low appropriability. The firm faces customers whose marginal utility rises with the number of other users, yet the firm lacks the resources and barriers with which it could profit. Accordingly, the firm's strategy should be concerned with reducing the significance of the network externality to customers, and with increasing barriers to entry/imitation.

One way to reduce the significance of the network externality is to reduce a prospective customer's out-of-pocket investment. Although network externalities are known to increase a customer's willingness to pay, lowering prices can also increase demand (Kaufmann & Wang, 2001). If the innovation provides less utility for users at lower levels of demand, then it must be priced accordingly. Thus:

Proposition 1a. Direct network externalities and low appropriability will lead the firm to compete on price.

A significant factor underlying initial demand may be the expected switching costs for users should they stop using the innovation or change to an alternative design or supplier. As a result, network externalities typically create greater switching costs. Accordingly, the firm may seek to reduce these potential costs. One obvious way is to increase standardization and/or compatibility with competitors' products. This may lower customers' incentives to switch, and it may help those who do switch to sell more easily their now-unneeded technology.

However, increasing standardization may not be a viable option in at least two instances. First, standardization may require technological changes that compromise the distinctive strengths of the firm's product. Second, the firm may be a first-mover without competitors. There are ways to reduce switching costs that do not affect the product development process or require coordination with competitors. Instead, the firm may simply shift switching costs from users to the firm by providing performance guarantees, offering leasing, and/or buy-back guarantees. Thus:

Proposition 1b. Direct network externalities and low appropriability will lead the firm to reduce customers' switching cost.

In addition to the above steps to make the innovation more attractive to the market by addressing the network externality, the firm may wish to raise barriers to entry to forestall competitors if it is to profit from its efforts. To a large degree in high-technology industries, appropriability depends on both technological barriers and legal barriers to imitation (Teece, 1998). Barriers to entry in this instance center on preventing imitation of the innovation, for without this protection the firm has little prospect of profiting. Technological barriers to imitation include secrecy practices (not sharing the existence or details of a new technology or a firm's start-up plans) and technological complexity (offering sophisticated, multi-featured designs). Examples of legal barriers to imitation include patent enforcement, securing a multiplicity of patents, and the use of non-compete and nondisclosure agreements with employees. Thus:

Proposition 1c. Direct network externalities and low appropriability will lead the firm to pursue technological barriers to imitation.

Proposition 1d. Direct network externalities and low appropriability will lead the firm to pursue legal barriers to imitation.

Cell 2 illustrates the situation of direct network externalities in conditions of high appropriability. The firm still faces customers whose marginal utility rises with the number of other users, but the firm possesses the resources necessary to secure a first-mover advantage. Accordingly, the firm's strategy should be concerned with putting those assets to their best use through rapid commercialization to lock out new entrants. By pursuing marketing barriers, firms can prolong first-mover advantages that can lead to differentiation and sustainable competitive advantage (Porter, 1980). Consider, for example, the long-standing success of Intel's personal computer (PC) processors. The firm built on its early success with several interesting moves. First, it made its processors available to any firm wanting to build PCs. Second, it took the remarkable step of creating brandname recognition for its component among the buyers of finished computers through its "Intel Inside" advertising. Third, it reinvested its high returns in the rapid development of new product offerings.

These traditional marketing barriers to entry—intensive selling efforts, developing brandname recognition, and product proliferation—are feasible solutions for a firm with high network externalities in conditions of high appropriability. For instance, several researchers have argued that marketing assets, such as promotional development, can enhance the appropriability of an innovation (Rao & Klein, 1994; Vinod & Rao, 2000) and can preempt competitors by dominating a market (Hill, 1998). Similarly, Utterback and Suarez (1995) argued that product variety or proliferation can facilitate a dominant design and can provide a competitive advantage in a technologically uncertain environment. Thus:

Proposition 2. Direct network externalities and high appropriability will lead the firm to pursue marketing barriers to entry through intense early efforts to expand sales, develop its brandname, and proliferate products.

Cells 3 and 4: Indirect Network Externalities

Indirect network externalities are dependent on indirect effects of using a product. More specifically, they are found in products that require distinct scale-driven complementary assets (Katz & Shapiro, 1985; Kotabe, Sahay & Aulakh, 1996). For example, the utility of software support or of automobile repair service depends on the number of other users in the same user group or "network," since in these cases the unit costs and availability of these complementary products are highly related to the number of other users (Arthur, 1989; Chou & Shy, 1990; Teece, 1986). A car owner may have a piston-driven engine or a rotary one, and a videocassette recorder owner may have a VHS-format VCR or a Betamax one. The utility of the car does not directly depend on whether anyone else has the same design, but it does depend on the availability of repair services and parts for that particular engine technology. Similarly, the utility of the VCR depends on the availability of videotapes to fit that particular VCR design. In fact, an innovation can be rejected in instances where there exists a lack of complementary assets (Schilling, 1998). The faster rate of growth in the availability of VHS-format videotapes vs. that of Betamax was one factor in the eventual demise of Sony's consumer-market Betamax VCRs.

In the case of direct network externalities, the firm's strategy highlights elements that are essentially internal to the firm, or firm-specific. In contrast, a firm facing indirect network externalities will turn its attention to the competitive environment. More specifically, the decisions that will drive the firm's success will center on interactions and relationships with other firms.

As Moner-Colonques and Sempere-Monerris (2000) observe, cooperation between firms is often considered in terms of its anti-competitive impact. However, research already argues that appropriability is one driver of such dynamics (Gemser, Leenders, & Wijnberg, 1996; Gemser & Wijnberg, 1995) and of vertical integration (Krickx, 1995). Similarly, the importance of complementary assets possessed by incumbent firms is one explanation for alliances between incumbents and new entrants to an industry (Rothaermel, 2001). If the network externality is indirect, the situation becomes more complex—the higher the indirect externality, the greater the need for some form of compatibility or standardization among competing designs, if customers are to be willing to invest.

Ceteris paribus, we may see the emergence of product standards, whether formal or *de facto*.

Cell 3 illustrates the situation of indirect network externalities in conditions of low appropriability. The firm lacks resource, technological, and/or legal barriers to imitation. Further, the firm's product requires complementary assets that are distinctive from those required by its competitors, but economies of scale mean the producers, distributors, and/or consumers of those assets prefer to use only one version of the complementary asset. Complementary asset producers will invest and exploit economies of scale, thus lowering their customers' fully loaded cost of use.

Certainly, firms may find it convenient to conduct product development/ R&D processes within a variety of existing standards, many of which may simply be taken for granted as constraints in the development process. Software products may depend on standardized programming languages and may be developed for existing hardware, electronic products may conform to existing technological standards and may use some degree of standardized components, and production processes may be designed to use already-available machinery.

In this case, the firm may be less likely to invest heavily in unique technology and may be more likely to pursue standardization in its industry, whether before initiating research and development, during the R&D process, or during the initial phases of competition. Consider for example the myriad computer manufacturers that arose in the late 1980s and early 1990s. Few created highly novel designs—the majority followed the IBM PC format. They did so partly because of scale economies achievable in standardized, off-the-shelf components, and partly so that the crucial complementary asset—software—would work seamlessly across computer brands. Thus:

Proposition 3a. Indirect network externalities and low appropriability will lead the firm to pursue industry standardization and compatibility.

Certainly, an alternative solution in this case would be for the firm is to pursue industry consolidation by merging with competitors and then to standardize within the merged firm's market area and customer base. Thus:

Proposition 3b. Indirect network externalities and low appropriability will lead the firm to pursue industry consolidation.

Cell 4 illustrates the situation of indirect network externalities in conditions of high appropriability. The firm possesses the financial or other resources, technological expertise, and/or legal barriers needed to capitalize on its product, but the complementary asset question remains unresolved. Customers remain sensitive to the switching costs associated with incompatible forms of complementary assets. Such conditions may lead to stronger relations between firms (Boschetti & Marzocchi, 1998). This case offers another form of capitalizing on a first-mover advantage. Although cooperative efforts with other firms may create appropriability hazards (Oxley, 1997), and although the asset specificity of those complementary assets is likely to mold the form of alliances between firms (Brousseau & Quelin, 1996), it has been argued that the lack of needed complementary assets is clearly tied to the extent of a firm's alliances with other industry actors (Arora & Gambardella, 1994). The firm is likely to encourage development, availability, and links to providers of the needed complementary assets, since those assets comprise the only significant remaining barrier to a successful start-up. To build on an earlier example, a dramatic scenario is the successful commercialization of the VHS-format VCR by Matsushita, a relatively small firm. Matsushita's major competitor in VCRs was the far-larger Sony, which possessed greater experience, the resources needed for commercialization, and a superior VCR technology (BetaMax). Matsushita's decision to license its design to multiple firms stimulated market growth, fostered entry by firms, encouraged product innovation, and spurred the widespread availability of pre-recorded VHS videotapes. The combination of these factors drove Sony and its technology out of the consumer VCR market. Thus:

Proposition 4. Indirect network externalities and high appropriability will lead the firm to pursue alliances with providers of complementary assets.

CONCLUSION

This chapter provides an overview of a model and propositions regarding a firm's choices when commercializing a new technology. The model is developed in the context of start-up firms, as the success of these firms is particularly sensitive to the issues of network externalities and appropriability. By distinguishing between direct and indirect network externalities, and by examining their relationship to appropriability, the model advances theory-building in the technology and innovation literature and offers specific insights into how start-up firms may overcome network externalities. In particular, the model suggests that direct network externalities will lead firms to pursue strategic choices centered on internally controllable decision variables. Indirect network externalities, on the other hand, will lead firms toward efforts to manage their competitive environment by cooperating with outside actors.

REFERENCES

Afuah, A. (1999). Strategies to turn adversity into profits. *Sloan Management Review, 40*(2), 99-109.

Anderson, P., & Tushman, M. (1990). Technological discontinuities and dominant designs: a cyclical model of technological change. *Administrative Science Quarterly, 35,* 604-633.

Arora, A., & Gambardella, A. (1994). Evaluating technological information and utilizing it: Scientific knowledge, technological capability, and external linkages in biotechnology. *Journal of Economic Behavior and Organization, 24,* 91-114.

Arrow, K. (1962). Economic welfare and the allocation of resources for invention. In R. Nelson (Ed.), *The rate and direction of inventive activity: Economic and social factors* (pp. 609-619). Princeton, NJ: Princeton University Press.

Arthur, W. B. (1989). Competing technologies, increasing returns, and lock-in by historical events. *The Economic Journal, 99,* 116-131.

Boschetti, C., & Marzocchi, G. (1998). Complementary resources, appropriability and vertical interfirm relations in the Italian movie industry. *Journal of Management and Governance, 2,* 37-70.

Brousseau, E., & Quelin, B. (1996). Asset specificity and organizational arrangements: The case of the new telecommunications services market. *Industrial and Corporate Change, 5,* 1205-30.

Chou, C., & Shy, O. (1990). Network effects without network externalities. *International Journal of Industrial Organization, 8,* 259-270.

Farrell, J., & Saloner, G. (1985). Standardization, compatibility, and innovation. *Rand Journal of Economics, 16*(1), 70-83.

Farrell, J., & Saloner, G. (1987). Competition, compatibility and standards: The economics of horses, penguins and lemmings. In H.L. Gabel (Ed.), *Product standardization and competitive strategy* (pp. 1-21). Amsterdam: North-Holland.

Garud, R., & Kumaraswamy, A. (1993). Changing competitive dynamics in network industries: An exploration of Sun Microsystems' open systems strategy. *Strategic Management Journal, 14*(5), 351-369.

Gemser, G., Leenders, & Wijnberg, N. (1996). The dynamics of inter-firm networks in the course of the industry life cycle: The role of appropriability. *Technology Analysis and Strategic Management, 8,* 439-53.

Gemser, G., & Wijnberg, N. (1995). Horizontal networks, appropriability conditions and industry life cycles. *Journal of Industry Studies, 2,* 129-40.

Hill, C. W. (1992). Strategies for exploiting technological innovations. *Organization Science, 3,* 428-441.

Hill, C. W. (1997). Establishing a standard: Competitive strategy and technological standards in winner-take-all industries. *Academy of Management Executive, 11*(2), 7-25.

Katz, M., & Shapiro, C. (1985). Network externalities, competition, and compatibility. *American Economic Review, 75*(3), 424-440.

Katz, M., & Shapiro, C. (1986). Technology adoption in the presence of network externalities. *Journal of Political Economy, 94*(4), 822-41.

Kaufman, R. J., & Wang, B. (2001). New buyers' arrival under dynamic pricing market microstructure: The case of group-buying discounts on the Internet. *Journal of Management Information Systems, 18*(2), 157-188.

Kotabe, M., Sahay, A., & Aulakh, P.S. (1996). Emerging role of technology licensing in the development of global product strategy: Conceptual framework and research propositions. *Journal of Marketing, 60*(1), 73-88.

Krickx, G. (1995). Vertical integration in the computer mainframe industry: A transaction cost interpretation. *Journal of Economic Behavior and Organization, 26,* 75-91.

Lee, J. R., O'Neal, D., Pruett, M., & Thomas, H. (1995). Planning for dominance: A strategic perspective on the emergence of a dominant design. *R&D Management, 25,* 1-13.

Majumdar, S. K., & Venkataraman, S. (1998). Network effects and the adoption of new technology: Evidence from the U.S. telecommunications industry. *Strategic Management Journal, 11,* 1045-1062.

McGrath, B., & McGrath, R. (2000). Competitive advantage from knowledge spillovers: Implications of the network economy. In R. Bresser, M. Hitt, R. Nixon, & D. Huskel (Eds.), *Winning strategies in a deconstructing world* (pp. 267-288). Chichester, UK: John Wiley & Sons.

Moner-Colonques, R., & Sempere-Monerris, J. (2000). Cooperation in R&D with spillovers and delegation of sales. *Economics of Innovation and New Technology, 9,* 401-420.

Oxley, J. (1997). Appropriability hazards and governance in strategic alliances: A transaction cost approach. *Journal of Law, Economics, and Organization, 13,* 387-409.

Porter, M. E. (1980). *Competitive strategy.* New York: The Free Press.

Rao, P. M., & Klein, J. A. (1994). Growing importance of marketing strategies for the software industry. *Industrial Marketing Management, 23*(1), 28-37.

Rosenberg, N. (1982). *Inside the black box: Technology and economics.* Cambridge: Cambridge University Press.

Rothaermel, F. (2001). Complementary assets, strategic alliances, and the incumbent's advantage: An empirical study of industry and firm effects in the biopharmaceutical industry. *Research Policy, 30,* 1235-51.

Schilling, M. A. (1998). Technological lockout: An integrative model of the economic and strategic factors driving technology success and failure. *Academy of Management Review, 23*(2), 267-284.

Shane, S. (2001). Technology regimes and new firm formation. *Management Science, 47*(9), 1173-1190.

Teece, D. J. (1986). Profiting from technological innovation: Implications for integration, collaboration, licensing and public policy. *Research Policy*, 15, 285-305.

Teece, D. J. (1998). Capturing value from knowledge assets: The new economy, markets for know-how, and intangible assets. *California Management Review, 40*(3), 55-79.

Tushman, M., & Anderson, P. (1986). Technological discontinuities and organizational environments. *Administrative Science Quarterly, 31,* 439-465.

Tushman, M. L., & Rosenkopf, L. (1992). Organizational determinants of technological change: Toward a sociology of technological evolution. *Research in Organizational Behavior, 14,* 311-347.

Utterback, J. M., & Suarez, F. F. (1995). Dominant designs and the survival of firms. *Strategic Management Journal, 16,* 415-430.

Vinod, H. D., & Rao, P. M. (2000). R&D and promotion in pharmaceuticals: A conceptual framework and empirical exploration. *Journal of Marketing Theory and Practice, 8*(4), 10-20.

This chapter was previously published in the Journal of IT Standards & Standardization Research, 1(1), 33-45, January-March, Copyright © 2003.

Section VII

After Standardization

Chapter XVII

In Pursuit of Interoperability

Scott Moseley, Farbum Scotus, France

Steve Randall, PQM Consultants, UK

Anthony Wiles, ETSI PTCC, France

ABSTRACT

Traditionally, conformance testing has been the domain of the telecommunications industry, while interoperability testing has mainly been limited to the Internet world. Many see these as either/or solutions, ignoring the fact that recent experience shows that both approaches have their strengths when used wisely. This paper discusses the merits and shortcomings of each approach and shows how they can usefully be combined to maximise the effectiveness of the testing process. This is especially relevant where testing is being treated as a potential branding issue by various fora. This paper is based on many years of practical experience of writing test specifications at the European Telecommunications Standards Institute (ETSI). It presents ETSI standardisation activities on testing, including the development of a generic interoperability testing methodology and the work being done by the Technical Committee Methods for Testing and Specification (MTS), the ETSI Protocol and Testing Competence Centre (PTCC), and the ETSI Plugtests^{TM} service.

INTRODUCTION

The telecommunications industry requires different kinds of base specifications and standards to ensure that its products function, interoperate with each other, are safe, and comply with regulatory requirements. Any effective standardisation activity requires test specifications to support these base requirements. Without such test specifications, a product fatally risks being dysfunctional, not working with other products, being unsafe, and incurring legal liabilities. The only way to ensure that standards are met is to test products in an effective way using the test specification.

Responsible engineering mandates appropriate and thorough testing. There are not many people willing to cross a bridge at the risk of their lives if they are not certain that it has not been proven to be safe. For this purpose, a bridge is tested before, during, and after its construction in various manners. Similarly, proving that a telecommunications system works correctly requires testing before, during, and after its development.

In the development and maintenance of its products and services, the telecommunications industry uses many kinds of testing such as integration, performance, stress, load, electro-magnetic emissions, electrical safety, mechanical resistance, conformance, and interoperability. In general, protocol conformance testing is appreciated in the telecommunications world, whereas it appears that the IP world, at best, avoids it and, at worst, tolerates it only if imposed by "force majeure."

Here at ETSI, protocol conformance testing specifications have dominated our testing activities and will continue to do so in the future. However, we are seeing new and significant interest in interoperability testing for a variety of reasons, some of which are valid and others, arguably, less so. The increasing success of ETSI's Plugtests™ Service is proof that the concept of interoperability testing is appreciated by our members and the industry. Current thinking includes the view that interoperability testing effectively replaces conformance testing with significantly less cost and time.

In order to discuss this position and develop others, definitions and understandings of conformance and interoperability testing are necessary. There is indeed much confusion on what conformance testing is, what it does, and how much it costs. There is just as much confusion concerning the definition of interoperability testing. An informal survey recently asked seven manufacturers, network operators, and application providers their definition of interoperability. Each of the seven answers bore no relation to the other!

This paper discusses the industry's view of interoperability testing, provides ETSI's definitions and methodologies for both types of testing, and explores their advantages and disadvantages. It shows that interoperability testing produces results that are indeed different from those obtained in conformance testing. This, then, precludes the possibility of interoperability testing replacing conformance testing in toto. In this paper we hope to show clearly that the results of either kind taken separately do not guarantee interoperability, and that good engineering practice requires both kinds of testing to ensure the interoperability intended by the base standards. We go further by daring to assert that conformance testing is necessary in accomplishing effective and rigorous interoperability testing.

THE CURRENT APPROACH OF INTEROPERABILITY TESTING IN THE INDUSTRY

There is actually no commonly agreed definition of interoperability, let alone a common view of interoperability testing. However, interoperability testing has, until recently, been viewed generally as the rather informal interconnection of prototype equipment for the combined purposes of product debugging and technology development. Interop Events, Plug Fests, and bake-offs all fall into this category.

The approach here has been for manufacturers to bring their products together at a central location where appropriate network facilities have been provided. As shown in Figure 1, the equipment is connected to the network and, by mutual arrangement, two or more manufacturers will attempt to make their products communicate to execute particular joint functions. Information on the success or failure of these tests is used to improve the product designs or as feedback to the underlying standardisation activities to advance the technology itself.

Figure 1 shows a typical (though simplified) arrangement for bake-off testing where a number of products from different suppliers have all been connected to a central network. The figure shows a session where the manufacturers of Products 2, 3, and 6 have agreed to check the interoperability of some common functions. Similar testing between Products 1, 2, and 4 might take place in a later session. However, testing between Product 1 and 5 could be carried out in parallel.

Testing of this type is a very powerful method for improving product stability and the technology upon which the products are based. However, it does not provide proof that the products conform to any standard, nor does it show that the products interoperate fully as the standards intend.

Market demand for guarantees of interoperability has led to a growing interest by telecommunications manufacturers and operators in certification and logo schemes based on a more formal approach to testing. Such testing depends not only on the use of predefined suites of interoperability tests covering the full range of possible functions, but also on the conformance testing of products to specific protocol standards as a pre-requisite to interoperability testing.

Figure 1. Typical bake-off connection arrangement

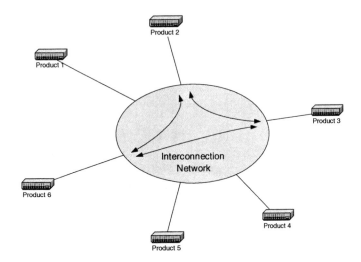

ETSI'S APPROACH TO CONFORMANCE TESTING

The purpose of conformance testing is to determine to what extent an implementation of a particular standard conforms to the individual requirements of that standard. For nearly 15 years ETSI Technical Bodies have been producing conformance test specifications for key technologies such as GSM, UMTS, DECT, INAP, TETRA, ISDN, B-ISDN/ATM, HiperLAN/2, VB-5, FSK, and VoIP (H.323/SIP). ETSI test specifications are developed according to the well-proven ISO/IEC 9646 conformance testing methodology illustrated in Figures 2 and 4.

In the conformance testing architecture of Figure 2, there are two main components: the System Under Test (SUT), which contains the Implementation Under Test (IUT), and the Means of Testing (MOT). The IUT is usually a single protocol, although an SUT contains several protocol layers that may be tested a layer at a time, with one conformance test suite for each layer.

The MOT includes at least one tester depending on the architecture of the IUT and the interfaces that are accessed during testing. It also handles such activities as coordination, logging, and the reporting of test results. Connection of the SUT to the testers is achieved by some Means of Communication (MOC). For example, when testing, say, a layer N protocol with a single tester, the MOC would be a protocol stack of layer N-1.

The testers, which in a real testing environment may be distributed, execute test programs or scripts, which in ISO 9646 terminology are called Test Cases. The entire set of Test Cases is known as a Test Suite.

Figure 2. A generic model for conformance testing (based on ISO/IEC 9646)

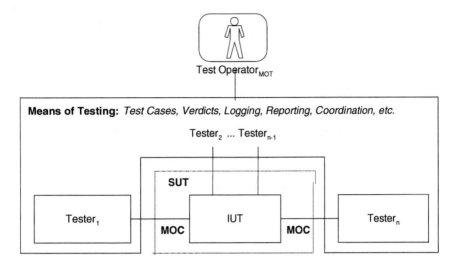

ETSI develops Abstract Test Suites (ATSs) written in the standardised testing language, TTCN, which can be compiled and run on a variety of real test systems. The following illustrations show testing of a network element (A) and terminal equipment (B).

In each case there may be different conformance test suites for the different protocols (components) that make up the products. At no time does an individual test suite test the product as a complete system. The dotted lines in these illustrations indicate that the interfaces at which testing occurs are usually standardised protocol interfaces internal to the product and the test system.

Figure 3. Illustrations of conformance testing

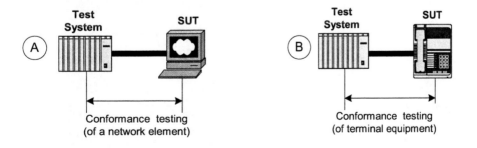

Conformance Testing Methodology

The main element of the ISO/IEC 9646 methodology as applied by ETSI is summarised in Figure 4. While originally targeted at protocol testing, this methodology can be applied to the black-box testing of other reactive systems such as services and APIs. For simplicity, we shall mainly consider conformance testing of protocols.

ISO/IEC 9646 recommends the production of test specifications comprising the following documentation:

- The Test Suite Structure and Test Purposes (TSS&TP) are derived from the relevant base standard. They provide an informal, easy-to-read description of each test, concentrating on the meaning of the test, rather than detailing how it may be achieved. Each Test Purpose focuses on a specific requirement or combinations of requirements identified in the base standard. They are usually defined on a detailed protocol level, possibly referring to protocol messages, states, or capabilities. Finally, the Test Purposes are grouped into a logical Test Suite Structure according to suitable criteria (e.g., basic interconnection, error handling, protocol functionality, etc.).
- The Abstract Test Suite (ATS) is the entire collection of Test Cases. Each Test Case specifies the detailed coding of the Test Purposes, written usually in a test specification language such as the standardized TTCN.
- The Implementation Conformance Statement (ICS) is a checklist of the capabilities supported by the Implementation Under Test embedded in the SUT. It provides an overview of the features, capabilities, functionality, and options that are implemented by that product. The ICS can be used to select and parameterise test cases, and as an indicator for basic interoperability between different products.

Figure 4. Illustration of the ISO/IEC 9646 conformance testing methodology

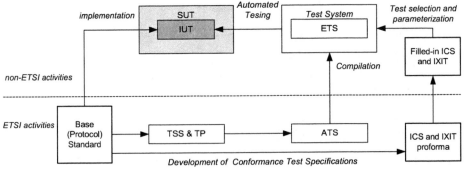

- The Implementation eXtra Information for Testing (IXIT) contains additional information (e.g., specific addresses, timer values, etc.) necessary for testing.
- The Executable Test Suite (ETS) can be quickly and easily implemented from the ATS using the TTCN compilers available on most modern test tool platforms (C++, Java, etc.). Runtime support, such as message encoders/decoders, test control, and adaptation layers, is needed to execute the tests in a real test system.

ISO/IEC 9646 does not just define test specifications as described above. This seven-part standard also covers test realisation (executable tests), requirements on test laboratories, and the development of Protocol Profile test specifications.

Characteristics of Conformance Testing

Because the conformance tester maintains a high degree of control over the sequence and contents of the protocol messages sent to the IUT, it is able to be comprehensive in that it can explore a wide range of both expected and unexpected (invalid) behaviour. Protocol design must take into account events that could occur only once in more than a million instances. Such a low probability event cannot (intentionally) be generated with interoperability testing.

A complete system (product) is often based on a number of different standards, each covering only a part of the entire product. Conformance testing tests those parts individually to a greater or lesser degree depending on the quality and coverage of the test suites. By testing these specific requirements in a controlled and to some extent artificial environment, conformance testing tests in a narrow but deep manner. It does not test the system as whole, and it does not test how well the system interoperates with other systems in the real world.

Figure 5. Basic components of a TTCN-3 test system

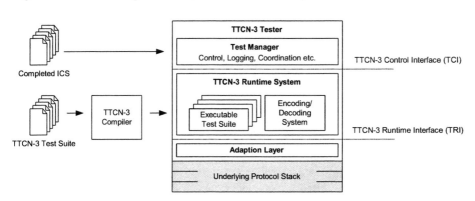

One criticism of conformance testing is that it is expensive. Indeed, that may have been true when it was associated with bureaucratic and inefficient third-party testing schemes. However, that is no longer the case. ETSI has streamlined the process of writing test specifications, and the test suites themselves focus on testing the essentials—the motto being "conformance testing for interoperability." Because the test specifications are produced by teams of experts recruited from the ETSI Technical Bodies, the cost is distributed among the ETSI membership. ETSI test specifications are in increasing demand, often being used by manufacturers as a basis for internal development testing and as an essential complement to interoperability testing. In the words of one 3GPP TG chairman, "The Conformance Test Suites for 3GPP terminals are extremely good value for money. They test crucial requirements ensuring interoperability of terminals and form a core set of tests on which suppliers can extend."

It is true that test systems for radio-based systems are expensive. In a large market such as GSM and UMTS, it is worthwhile to build such test systems. However, for less extensive applications, where testing is considered important but where costs must be kept to a minimum, there is a growing interest in emulating the lower (radio) layers in a non-radio protocol such as IP for testing purposes. This is a very effective but much lower cost alternative to the real thing!

Turning back to the bake-off example in Figure 1, there are three interconnected products (Products 2, 3, and 6), each of which are almost certainly composed of several sub-systems or components. These components may, themselves, have sub-components or sub-systems. Among these components there is likely to be at least one protocol stack having several protocol layers. For the purposes of conformance testing, an IUT could be one of the protocols in the stack over the other underlying protocols or an emulator. Thus, conformance test suites are designed to test one implementation of a specification in a product. Theoretically, conformance testing can be conducted on each implementation of a specification that is present in a product. In other words, conformance testing is specific to a sub-system or component of a product for which there exists a base specification or standard. If there is no base specification, there can be no conformance testing!

Considering only the conformance testing of protocols, it is true to say that a telecommunications system uses a number of protocols, some of which may be "tried and true" and others that are newly specified. If the newer protocols are lower in the stack of protocols and the "tried and true" protocols are higher up on the stack (which often occurs), the reliability in the higher level protocols is not determinable until confidence in the lower levels is established. To put a reliable protocol over an unproven entity is asking for trouble, with the amount of trouble being directly proportional to the complexity of the unproven protocol.

A conformance test for a single protocol can have hundreds of test cases even if it is a relatively simple protocol. The number of messages in each sequence, the number of re-attempts that are permitted, and the number of exception conditions that are catered for can all increase the complexity exponentially.

Although this specific example discusses only the protocol component in our system, the same reasoning is valid for any component in a product. True, protocols demand the most extensive form of conformance testing, but conformance testing could be required in any of the other sub-systems (applications, services, electrical and electronic systems, connectors, and interfaces), particularly if they have multiple ways of accomplishing the same function. Such robustness is a quality factor often designed into transmission and electrical/electronic systems.

Advantages and Disadvantages of Conformance Testing

If comprehensively specified, conformance testing determines whether the behaviour of an implementation conforms to the requirements laid out in its base specification, including the full range of error and exception conditions that can only be induced or replicated by dedicated test equipment.

On the other hand, conformance testing does not prove end-to-end interoperability of functions between the two similar communicating systems, and it does not prove the operation of proprietary features, functions, interfaces, and systems that are not in the public domain. However, these proprietary facilities may be exercised indirectly as part of the configuration or execution of the conformance tests.

ETSI'S APPROACH TO INTEROPERABILITY TESTING

For ETSI, the purpose of interoperability testing is to prove that end-to-end functionality between (at least) two communicating systems is as required by the standard(s) on which those systems are based. Interoperability activities can be loosely divided into three classes: bake-offs, interoperability demonstrations, and interoperability testing.

Bake-offs are used to develop and validate technologies and standards. They are best begun at an early stage of the development cycle when prototypes and early implementations are available. Their value is to move the standards forward and to iterate to the best technical solutions. Feedback to standards bodies and implementers is essential.

As the name implies, interoperability demonstrations are used mainly in a marketing context and do not usually add value to standards development.

It is the third kind of event, interoperability testing, which is the focus of this paper. This activity requires a more rigorous approach than bake-offs, especially if the end purpose is some form of certification or branding. There seems not to exist a universal methodology with well-defined concepts and terminology. In order to fill this gap, ETSI EP TIPHON™ (Telecommunications and Internet Protocol Harmonisation Over Networks) has defined a generic methodology (TS 102 237-1) for the interoperability testing of communications systems in Next Generation Networks (NGNs), as illustrated in Figures 6 and 8. Note that the methodology, which is described in TS 102 237-1, is still in its early stages, and some of the ideas presented in this paper may be subject to change as it matures.

Even if this is not a global view, TIPHON™ sees a benefit in developing this generic approach. It is hoped that with the cooperation of ETSI TC MTS (Methods for Testing and Specification), the methodology will at least develop into being one acceptable to the entire ETSI community.

Like the conformance model of Figure 2, the two main architectural components of the interoperability architecture are the Means of Testing (MOT) and the System Under Test (SUT), illustrated in Figure 6.

However, this is where the similarity ends. The SUT comprises the Equipment Under Test (EUT) and one or more pieces of Qualified Equipment (QE). The EUT is the main focus of testing. The QE, while being part of the entire SUT, does not have the main focus and will have already undergone thorough conformance and interoperability testing. This is the principal difference between the ETSI approach and the current understanding of interoperability. Note that both the QEs and the EUT may be very complex, comprising several devices, but conceptually they are always considered to be single entities.

Unlike conformance testing, the MOT does not require complex test equipment (note that this does not preclude the possibility that interoperability

Figure 6. A generic model for interoperability testing (based on TS 102 237-1)

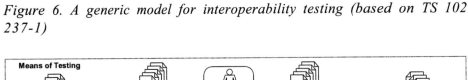
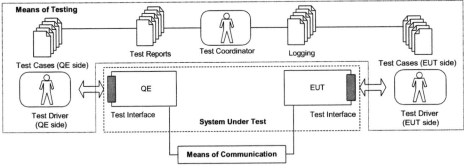

Figure 7: Illustration of interoperability testing

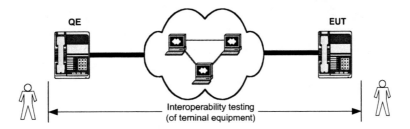

testing can be automated). The MOT includes the Test Operators as well as the interoperability test cases and mechanisms for logging and reporting, etc. The Means of Communication (MOC) between the QE and the EUT is considered to be neither part of the SUT, nor of the MOT. The following illustration shows interoperability testing of terminal equipment (EUT). The dotted lines in these figures indicate that the interfaces at which testing occurs are those interfaces offered to a normal user (e.g., the pushbuttons on a handset or dialog on a soft phone).

INTEROPERABILITY TESTING METHODOLOGY

On the face of it, the TIPHON™ interoperability methodology seems very similar to that of conformance testing. This is intentional. On the broad scale, many of the ISO/IEC 9646 concepts map well to interoperability testing. However, the difference is in the detail. This is summarised in Figure 8.

Figure 8. Illustration of the TIPHON™ interoperability testing methodology

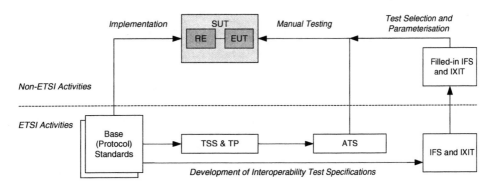

- The Test Suite Structure and Test Purposes (TSS&TP) are derived from the relevant base standards. They provide an informal, easy-to-read description of each test, concentrating on the meaning of the test, rather than detailing how it may be achieved. Each Test Purpose focuses on testing a specific functionality of the EUT that can be effected at the user interfaces offered by the SUT. Test Purposes are grouped into a logical Test Suite Structure according to suitable criteria (e.g., basic interconnection, functionality, test architectures, etc.).

- The Abstract Test Suite (ATS) is the entire collection of Test Cases. Each Test Case specifies the preconditions for setting up the test and the steps that must be taken by the test operators in order to perform the test. Note that if a suitable (usually proprietary) API is available, the test operator may be replaced by an automated script, written in some form of programming language, possibly TTCN.

- The Interoperable Functions Statement (IFS) is a checklist of the capabilities supported by the Equipment Under Test. In a way it is similar to a conformance testing ICS, but not so detailed. Some information in the IFS may be derived from the relevant ICS. The IFS can be used to select and parameterise test cases and as an indicator for basic interoperability between different products.

- The Implementation eXtra Information for Testing (IXIT) contains additional information (e.g., specific addresses, timer values, etc.) necessary for testing.

- The Executable Test Suite (ETS) is only necessary if scripts are defined in the ATS. These will be much simpler than conformance test cases, and their implementation will depend on the nature of the API. Again, it is possible that TTCN can be used.

TS 102 237-1 does not just define test specifications as described above. It also covers the specification of proformas for logging and reporting, and preparation for testing such as test plans and procedures, test execution, and certification procedures (if applicable).

Characteristics of Interoperability Testing

As described earlier, a complete system (product) is often based on a number of different standards, each covering only a part of the entire product. Interoperability testing tests the product as a whole, including the non-standardised components. Because the test operator only has control at user interfaces, interoperability testing tends to cover the normal behaviour of the product. Unlike conformance testing, coverage is broad but shallower.

Interoperability testing is relatively cheap in the sense that no specialised test equipment is needed. Interoperability test cases are also (for the most part) easier (and thus cheaper) to produce. However, the effort should not be underestimated and a useful interoperability testbed is not always trivial to set up.

Advantages and Disadvantages of Interoperability Testing

As discussed above, interoperability testing certainly has the advantage over conformance testing in that it tests the entire product and is probably cheaper to perform. It shows that functionality has been achieved, without demonstrating how. However, showing interoperability between two or more systems does not guarantee interoperability with other, non-tested implementations. Neither does it guarantee conformance. Two non-conforming implementations may still interoperate, but they will probably fail to interoperate with a correct (i.e., conforming product). Finally, and this is possibly the key point, interoperability testing usually covers only the normal sequences of product behaviour. Unlike conformance testing, interoperability testing cannot force testing of error behaviour and other unusual scenarios. For example, a typical DSA (Dynamic Service Addition) state machine in IEEE 802.16 has up to 46 transitions, of which about six would be exercised by interoperability testing. Conformance tests could be written that would exercise all the transitions.

TTCN-3 TEST SYSTEMS

The Tree and Tabular Combined Notation (TTCN) is a language designed specifically for the development of test specifications. It has been used in numerous technology applications including ISDN, GSM, DECT, and SIP.

As the name suggests, early versions of TTCN enforce a tabular structure on test specifications. Although very powerful for protocol testing, this approach lacks both readability and the flexibility required to extend the language into different application areas. Thus, ETSI has produced a third version (TTCN-3), which has been given the look and feel of a modern programming language (ES 201 873). It has even been given a new meaning for its TTCN abbreviation. It is now the Testing and Test Control Notation, which better reflects its objectives and purpose. In addition to protocols, TTCN-3 is an ideal language for specifying tests of services, APIs, and other software based systems. It is not restricted to conformance testing and can be used in many areas, including interoperability testing where some form of automated test driver may be required. TTCN-3 test suites are executed on specialised test systems, the principles of which are illustrated in Figure 5.

A TTCN-3 test system consists of three main components, the standardised parts of which are the TCI and TRI, defined in Parts 5 and 6 of ETSI ETS 201 873:

- The Test Manager is responsible for controlling the tests, handling distributed components, tracing and logging, external encoder/decoder plugins, test suite parameterisation and selection, etc. In a good tool the Test Manager will offer a GUI (Graphical User Interface) to the user.
- The TTCN-3 runtime system provides the execution environment for the compiled TTCN-3 test cases. Some of this support is generic to all test suites; some is specific to the particular protocol being tested, most notably the encoding and decoding of messages.
- The Adaptation Layer interfaces the test system to the MOC (usually an underlying protocol stack). Again, some of this is generic, but mostly the adaptation layer is specific to the particular testing environment.

CERTIFICATION AND BRANDING

Certification has always been valuable. This is still so, but times have changed. Certification is no longer solely a regulatory process entailing expensive bureaucracy. Today, everyone has recognized the needs for speed, flexibility, reduced costs in certification, and concurrently, the imperative to earn the buyer's confidence in a certification label.

Certification is now largely conducted by industry fora. Manufacturers can use the results in their product branding, and there is usually a logo for packaging or product labelling. Certification activities are increasing in both the IP and telecommunications domains, one reason being that manufacturers want to ensure a recognizable level of quality and interoperability to merit consumer confidence in the logo and brand. A manufacturer can self-certify its product. Self-certification often culminates the product's development testing, and is an economical alternative frequently chosen by users and operators.

ETSI does not actually conduct certification processes itself. However, ETSI test suites and the methodologies described in this paper can be useful components in the kind of schemes mentioned above. In some cases, we can be a neutral custodian of certification test suites or an issuer of certificates following successful certification testing by an authorized laboratory or organisation.

TESTING ACTIVITIES IN ETSI

Test specifications cannot be developed in isolation. There is a strong relationship between standards development, product development, and testing. While there are no hard-and-fast rules, the important thing is that test specifications (and test tools) are in place when needed, and that they have been produced before the standards are locked down. One of the most useful side effects of the development of test specifications is the feedback that work has into the base standards.

ETSI has a long tradition of writing test specifications. ETSI Technical Bodies, such as TC TISPAN, 3GPP, and EP BRAN, are currently writing test specifications for technologies that include UMTS, HiperLAN2, Hiperaccess, Hiperman, VoIP (SIP, H.323), SIGTRAN, and IPv6.

There are also several entities specifically related to testing activities:

- TC MTS (Methods for Testing and Specification) is the technical body responsible for providing all other ETSI TBs with the methodologies and techniques for the development of high-quality test specifications (http://portal.etsi.org/mts/mts_tor.asp).

- The ETSI PTCC (Protocol and Testing Competence Centre) provides support and services to ETSI technical bodies on the application of modern techniques for specifying protocols and test specifications. The PTCC is also responsible for the technical management of the ETSI Specialist Task Forces (STFs) that develop conformance and interoperability test specifications for ETSI standards (http://www.etsi.org/ptcc).

- The Plugtests™ Service organizes interoperability events for ETSI members and non-members alike. It provides the logistical and organisational support (often at ETSI premises) for such events. Recent events include SynchFest, IPv6 InterOp, IMTC SuperOp, SIPit, SIGTRAN Interop, and Bluetooth UnplugFest. The intention with plugtests is to validate (debug) both the standards and early products or prototypes of those standards as they are developed (www.etsi.org/plugtests).

CONCLUSION

There is some division in the industry on whether conformance testing or interoperability testing is the best way to test information and telecommunications technology (ICT) products. It is our conclusion that it is not a question of one or the other. The approaches are complementary, not competitive.

Conformance testing does indeed have many advantages, not least the fact that such testing can explicitly force exceptional or invalid behaviour and other error conditions. Conformance testing gives a good indication that a certain product fulfils the requirements defined in a base standard. A conformant product is more likely to interoperate. Indeed, conformance testing can be focused and applied in such a manner that, while proving conformance, it can also explore some interoperability aspects. Even so, conformance testing does not guarantee interoperability.

On the other hand, interoperability testing is an excellent technique for showing that different products really do work together. Interoperability testing exercises the complete product and covers aspects too complex (expensive) to fully test through conformance. Well-defined interoperability tests can also imply a measure of conformance. Even so, interoperability testing does not guarantee conformance.

In the experience of ETSI, while it is not absolutely necessary to undertake both types of testing, the combined application of both techniques gives a greatly increased confidence in the tested product and its chances of interoperating with other similar products. This statement is substantiated by our experience with testing over the last 15 years. This is not just the view of ETSI. Many organisations and fora supporting specific technologies now recommend that certification should involve both forms of testing.

Finally, ETSI will continue to produce both conformance and interoperability test specifications for combined use by its members. We will also continue to organise Plugtests™ events that, depending on the participants' wishes, will provide opportunities for both conformance and interoperability testing.

REFERENCES

ETSI ES 201 873 Methods for Testing and Specification; The Testing and Test Control Notation version 3; Parts 1-6.

ISO/IEC 9646 Information Technology—Open Systems Interconnection—Conformance Testing Methodology and Framework—Parts 1-7.

TS 102 237-1; Interoperability Test Methods & Approaches; Part 1: Generic Approach to Interoperability Testing.

This chapter was previously published in the Journal of IT Standards & Standardization Research, 2(2), 34-48, July-December, Copyright © 2004.

About the Authors

Kai Jakobs joined Aachen University's (RWTH) Computer Science Department as a member of the technical staff in 1985. Over the years, his research interests moved away from the technical "nuts and bolts" of communication systems to socio-economic aspects, with a focus on IT standardization. He is the co-author of a text book on data communications and, more recently, five books on standards and standardization in IT. He has been on the program committees of numerous international conferences, and has also served as an external expert on evaluation panels of various European R&D programs, on both technical and socio-economic issues. He holds a PhD in computer science from the University of Edinburgh.

*　　*　　*

Daniel Beimborn studied business administration at Frankfurt University. Since 2001, he has been a researcher and PhD student at the Institute of Information Systems, Frankfurt University, where he is with the PhD program "Enabling Technologies for Electronic Commerce."

Knut Blind studied economics, political science, and psychology at the University of Freiburg and at Brock University. He holds a BA with distinction and a diploma in economics (1990 and 1992, respectively). He holds a doctoral degree (1995) and was honored with the F. A. v. Hayek Award (1996) of the Economics Faculty, University of Freiburg. From 1993 to 1995, he was a research fellow at the Institute for Public Finance, University of Freiburg. Since 1996, he has been a senior researcher at the Fraunhofer Institute Systems and Innovation Research, and in 2001, he was appointed deputy head of the Department of Technology Analysis and Innovation Strategies. In April 2003, he completed his habilitation on driving forces and economic impacts of standardization to the Faculty of Economics, University of Kassel. Besides standardization, he extended his research focus also to regulatory issues having an impact on new markets. Since April 2004, he has managed an institute-wide task force on innovation and regulation.

Tineke M. Egyedi is a senior researcher on standardization at the Delft University of Technology, The Netherlands. Her previous employers include KPN Research (The Netherlands), Royal Institute of Technology (Sweden), and the University of Maastricht (The Netherlands). In the capacity of consultant, researcher, and coordinator, she did several policy studies for Dutch ministries (e.g., trends in standardization) and participated in several European projects. She was co-organizer of, for example, the EURAS2001 (European Academy for Standardization) workshop, and recently chaired the IEEE Conference on Standardization and Innovation in Information Technology (SIIT, http://www.SIIT2003.org). Her present tasks include that of researcher (e.g., for Sun Microsystems), associate editor of the *International Journal of IT Standards and Standardization Research*, editorial board member of *Computer Standards & Interfaces*, and vice president of the European Academy for Standardization (EURAS). She participates in several international networks on standardization. She has published widely (http://www.tbm.tudelft.nl/webstaf/tinekee) and was a guest editor of special journal issues.

Gunnar Ellingsen has worked for several years as an IT consultant at the University Hospital of Northern Norway. He completed his PhD in information science at the Norwegian University of Science and Technology. Currently, he works as an assistant professor at the Department of Telemedicine, University of Tromso, Norway. His research interests center on the design and use of electronic patient records in hospitals. His theoretical adherence is within the fields of computer-supported cooperative work (CSCW), and science and technology studies. He has published articles in *Information and Organization*, the *Journal of Computer Supported Cooperative Work*, and the *International Journal of IT Standards & Standardization Research*.

Mark Ginsburg (PhD, New York University) is assistant professor of management information systems (MIS) at the Eller College of Business, University of Arizona (USA). His research interests are in the areas of collaborative computing, the visualization of large-scale digital libraries, socio-technical issues of virtual communities, and practical problems pertaining to document management, knowledge management, and e-business strategies. His research articles have appeared in *Communications of the ACM, Journal of Computer Supported Cooperative Work, IEEE Computer, Journal Autonomous Agents and Multi-Agent Systems*, and *Electronic Markets Journal*. From 1994 to 1996, he was the lead developer on the flagship EDGAR on the Internet project, the first large-scale dissemination of government data (corporate disclosure statements) to the general public. He has extensive consulting experience at financial services firms. He is a member of IEEE, ACM, and SIGGROUP.

Wolfgang König holds a diploma in business administration and a diploma in business pedagogic (1975 and 1977, respectively) both from Frankfurt University. At the same university, he was awarded a PhD in 1980 (hardware-supported parallelization of information systems) and completed his habilitation thesis in 1985 (*Strategic Planning of the Information Systems Infrastructure*). In 1991, he became a professor of information systems at Frankfurt University in the economics and business-administration faculty, and since then, he has run the Institute of Information Systems. Since the end of 2002, he has chaired the E-Finance Laboratory Frankfurt am Main. He serves as editor-in-chief of the leading German IS journal *Wirtschaftsinformatik*. He is a member of the board of external directors of several IS companies. His research interests are in standardization, networking, and e-finance.

Ken Krechmer has participated in communications standards development since the mid-1970s. He actively participated in the development of the International Telecommunications Union (ITU) Recommendations T.30, V.8, V.8bis, V.32, V.32bis, V.34, V.90, and G.994.1. He was technical editor of *Communications Standards Review* and *Communications Standards Summary* from 1990 to 2002. In 1995 and 2000, he won first prize at the World Standards Day paper competition. He was program chair of the Standards and Innovation in Information Technology Conference (2001, 2003). He is a fellow at the University of Colorado at Boulder (USA) and the International Center for Standards Research, and is a senior member of the IEEE. His current activities are focused on research and teaching about standards.

Gary Lea is senior lecturer in business law at UNSW@ADFA, a university college of the University of New South Wales that provides civilian studies programs for officers and cadets of the Australian Defence Force. He was formerly a research fellow at the Queen Mary Intellectual Property Research Institute, University of London (1997-2004), and has long-time interests in legal aspects of standardization.

Hun Lee is an assistant professor of management in the School of Management at George Mason University (USA). He has a PhD in strategic management from the Smith School of Business, University of Maryland. His research and teaching interests include competitive dynamics, internationalization, and new-venture strategy.

Ji-Ren Lee is an associate professor in the Department of International Business, College of Management at National Taiwan University. He has a PhD in strategic management from the University of Illinois at Urbana-Champaign. His research and teaching interests are in the areas of growth strategy, competence-based strategy, and business strategies in emerging markets.

Scott Moseley is a consultant with many years of experience in general engineering, software engineering, and protocol testing. His current projects include managing the ETSI (European Telecommunications Standards Institute) contribution to the European Commission's project for integrating UMTS middleware (OPIUM), writing specifications for interoperability events, writing broadband radio-access networks test specifications, analyzing IPv6 functional areas for testing, and working on ETSI's new *Making Better Standards* document.

Donald E. O'Neal is an associate professor of management in the College of Business and Management, University of Illinois at Springfield (USA). He has a PhD in strategic management from the University of Illinois at Urbana-Champaign as well as substantial industry experience at the senior management level. He writes and teaches in the areas of strategic management, boards of directors, and leadership.

Sangin Park is an assistant professor of IT and industrial policy at the Graduate School of Public Administration, Seoul National University. Prior to that, he was an assistant professor of economics at the State University of New York at Stony Brook (1996-2003), and a visiting assistant professor of economics at Yale University (Fall 2002) where he worked on industrial organization, econometrics, and applied microeconomics. His current research interests include standardization and network externalities, strategies and policies in digital convergence, e-commerce, the network industry, and structural economic policy. Dr. Park holds a PhD in economics from Yale University.

Mark Pruett is an assistant professor of management in the School of Management, George Mason University. He has a PhD in strategic management from the University of Illinois at Urbana-Champaign and has prior work experience in international business consulting and in land development. His research and teaching interests are in the areas of competitive strategy, technological change, and organizational economics.

Steve Randall is an independent consultant with several years of experience in software engineering and telecommunications. He is a fellow of the British Computer Society and is currently vice chairman of ETSI's technical committee Methods for Testing and Specification (TC-MTS). In this role, he has been active in investigating more formal approaches to the interoperability testing of communications products implementing ETSI's standards.

Mostafa Hashem Sherif has been with AT&T in various capacities since 1983. He has a PhD from the University of California, Los Angeles, and both an MSc and BSc in electronics and communications from Cairo University, Egypt. He has an MS in management of technology from Stevens Institute of Technology and is a certified project manager from the Project Management Institute (PMI). His areas of expertise are in software reliability for telecommunication systems as well as electronic commerce. His standards experience includes participation in ANSI, the ITU, and the ATM Forum. He is the author of *Protocols for Secure Electronic Commerce* (second edition, CRC Press, 2003; the French version was published by Eyrolles under the title *La Monnaie Électronique* and is co-authored by A. Sehrouchni, Paris, 2000) and a co-author of *Managing IP Networks* (IEEE Press, 2003) and of the forthcoming book *Managing Projects in Telecommunication Services* (John Wiley & Sons). He is also the standards editor for the *IEEE Communications* magazine and an associate editor of the *International Journal of IT Standards & Standardization Research*.

Nikolaus Thumm works as economic counselor at the Swiss Federal Institute of Intellectual Property. He holds a PhD in economic and social sciences from Hamburg University. Previously, he worked for the European Commission and the Institute for Prospective Technological Studies in Spain, and did research at the European University Institute in Florence and the Universities of Hamburg and Stockholm.

Henk J. de Vries (1957) is an associate professor of standardization at Erasmus University, Rotterdam School of Management, Department of Management of Technology and Innovation. From 1984 until 2003, he worked at the Dutch National Standards Institute NEN in several jobs included performing as committee secretariat, training, consulting, setting up a training department, developing informative publications, and finally working on research and development. Since 1994, he has worked at the university (full time since 2004). de Vries earned a PhD in standardization (*A Business Approach to the Role of National Standardization Organizations*). His research and education con-

cern standardization from a business point of view. For more information, visit
http://web.eur.nl/fbk/dep/dep6/members/devries.

Alfred G. Warner (PhD, Ohio State University) is an assistant professor of
management at Pennsylvania State University – Erie, teaching strategy and
innovation management. His research interests are centered around organizational learning themes, particularly how firms manage the effects of technical
change, the effect of innovation on industry structure, and the governance
choices firms make as technologies evolve. He has significant industry experience with a variety of manufacturing firms and consults with local nonprofit
organizations.

Marc van Wegberg is an assistant professor at the Department of Organization
and Strategy, Faculty of Economics and Business Administration, Maastricht
University. He studied social economics at Tilburg University and did his PhD
at the Department of Economics, Maastricht University. His academic background is in industrial organization, applied game theory, and strategic management. He is active in teaching and research on various topics, notably on
alliances, standardization, Internet economics, and multimarket competition.

Jürgen Wehnert's entire professional career has been dedicated to the use of
information technology in transport. He gained his first standardization experience in the late '80s, at a time where electronic data interchange had to be
normalized. Ever since, he has been active in practical standardization in
Switzerland and Germany, in European CEN working groups, and in the global
ISO. Today, he is convening CEN TC224, Working Group 11, for transport
applications on smart cards. Jürgen is qualified in electronics and education and
works as an independent consultant in Hamburg, Germany.

Tim Weitzel received his PhD from Goethe University, Frankfurt, where he
works as a researcher at the Institute of Information Systems. His research
interests include standardization and network analysis. He has been part of a
variety of research and consulting projects in the areas of standardization, e-
business, extensible markup language (XML), and EDI. His recent research
projects include *IT Standards and Networks Effects* (http://www.it-
standards.de) funded by the German National Science Foundation, and *Network
Controlling* as part of the E-Finance Lab (http://www.efinancelab.de), an
industry-academic partnership between Frankfurt and Darmstadt Universities
and Accenture, Deutsche Bank, Microsoft, Siemens, and T-Systems. His
scientific and technical publications, and links to his research work, lectures, and
so forth, can be found at http://tim.weitzel.com.

Ruben van Wendel de Joode is an assistant professor with the Faculty of Technology, Policy, and Management, Delft University of Technology. He is part of the Dutch Institute of Government (NIG), the research school for public administration and political science. His research focuses on the organization of open-source communities. He received two grants from the Netherlands Organization for Scientific Research (NWO) for research related to open-source communities. The first grant was to study the interplay between intellectual property rights and open-source communities. The results are published in *Governing the Virtual Commons* (Cambridge University Press, 2003). He has written numerous articles on open source, which have appeared in a wide variety of journals. He is the guest editor of a special edition on open source for *Knowledge, Technology and Policy*. For more information, see http://www.tbm.tudelft.nl/webstaf/rubenw.

Anthony Wiles is the manager of the Protocol and Testing Competence Centre (PTCC) at the ETSI in Sophia Antipolis, France. The centre supports ETSI technical bodies on the use of modern specification and testing techniques in ETSI standards. He was heavily involved in the development of the widely used conformance-testing standard ISO/IEC 9646, which includes the testing language TTCN. His current activities include the development and application of interoperability-testing methodologies, the development of TTCN-3, and the testing of VoIP and IPv6.

Index

conformance testing 324
consensus 3, 35, 138
consortia 92, 100, 121, 128
consortium standards 11
consumer interdependencies 285
coordination mechanisms 77
copyright 176
CORBA (see common object request
    broker architecture)
costs 207
critical mass 268, 285

D

de facto 167, 315
de jure 167
de novo firms 309
decision making 117
decision procedures 99
decreasing- or constant-returns mar-
    kets 52
degree of obligation 18
delegation principle 141
design-based standards 7
Deutsches Institut für Normung (DIN)
    9
developmental alliances 60
DGs (see directorate generals)
diffusion model 283
diffusion of innovation theories 285
digital imaging and communication in
    medicine 197
digital libraries (DLs) 231
DIN (see Deutsches Institut für
    Normung)
direct network externalities 312
directorate generals (DGs) 145
discharge letters 212
divergence 73
diversity 209
DLs (see digital libraries)
due process 35
dynamic model 256

E

$E = mc^2$ 2
Eastman Kodak 153
economic growth 167
efficiency boosters 146
enabling (participatory) standards 190
end-to-end telecommunication 187
equipment under test (EUT) 330
etiquette 8
ETSI (see European Telecommunica-
    tions Standardization Institute)
European Telecommunications Stan-
    dardization Instite (ETSI) 31, 322
European Union and European Free
    Trade Association 9
EUT (see equipment under test)
excess inertia 284
extrinsic functions 14

F

Farrell and Saloner model 260
fear, uncertainty, and doubt (FUD) 150
flexibility standards 196
FIPA (see Foundation for Intelligent
    Physical Agents)
formal standards 57
formal standardization 131
Foundation for Intelligent Physical
    Agents (FIPA) 197
FUD (see fear, uncertainty, and doubt)
funding 138

G

governmental standards 11
grand coalition 112

H

high-technology start-up firms 307
horizontal standards 6

O

object management group (OMG) 197
obligation 18
obligatory point of passage 96
OCS (see open citation system)
OIF (see optical internetworking
    forum) 197
ongoing support 42
open change 38
open citation system (OCS) 231
open documents 39
open interface 40
open IPRs 37
open meetings 34
OMG (see object management group)
open standards 31
open world 36
open-source software (OSS) 71
operating system (OS) 28
operational coordination 76
operational standards 17
optical internetworking forum (OIF)
    197
OS (see operating system)
OSS (see open-source software)

P

participatory standards 190
patent 154, 167
patent litigation 150
patent portfolio 176
penetration pricing 53
penguin effect 291
performance-based standards 7
perspective taking 226
platform innovations 185
Polaroid 153
practitioner community 95
preferences 295
procedures 3
product preannouncements 53
product standards 50

profitability 186
proof-of-technology consortia 92
property rights 167
prospective standardization 12
protection strategies 169
public availability 4
public standards 17
public telecommunication services 184

Q

QE (see qualified equipment)
qualified equipment (QE) 330
quality 220
quality standards 7
quantitative-analysis model 264

R

R&D-intensive companies 176
radical innovations 185
Rambus 155
random topology 294
reference standards 195
regulatory coordination 75
regulatory standards 17
relational models 286
requirements 3
requiring standards 7
research consortia 92
responsive standards 190
retrospective standardization 12
revision 138
rules 3

S

SDO (see standard development
    organizations)
sectoral standards 11
selecting standardization 12
semilogistic 285
similarity standards 8, 195
simulation design 289
social actors 95

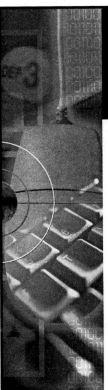